EARLY CHILDHOOD EDUCATION IN THE 21ST CENTURY

PROCEEDINGS OF THE 4TH INTERNATIONAL CONFERENCE ON EARLY CHILDHOOD EDUCATION (ICECE 2018), BANDUNG, INDONESIA, 7 NOVEMBER 2018

Early Childhood Education in the 21st Century

Edited by

Hani Yulindrasari, Vina Adriany & Yeni Rahmawati
Universitas Pendidikan Indonesia, Indonesia

Fonny Demeaty Hutagalung
University of Malaya, Malaysia

Sarita Gálvez
Monash University, Australia

Ade Gafar Abdullah
Universitas Pendidikan Indonesia, Indonesia

LONDON AND NEW YORK

Routledge is an imprint of the Taylor & Francis Group, an informa business

Typeset by Integra Software Services Pvt. Ltd., Pondicherry, India

Publisher's Note
The publisher has gone to great lengths to ensure the quality of this reprint but points out that some imperfections in the original copies may be apparent.

Published by: CRC Press/Balkema
Schipholweg 107C, 2316XC Leiden, The Netherlands
e-mail: Pub.NL@taylorandfrancis.com
www.crcpress.com – www.taylorandfrancis.com

First issued in paperback 2021

ISBN 13: 978-1-138-35203-2 (hbk)
ISBN 13; 978-1-03-223794-7 (pbk)

DOI: 10.1201/9780429434914

Early Childhood Education in the 21st Century – Yulindrasari et al. (eds)
© 2020 Taylor & Francis Group, London, ISBN 978-1-138-35203-2

Table of contents

Early Childhood Education in the 21st Century – Yulindrasari et al. (eds)
© 2020 Taylor & Francis Group, London, ISBN 978-1-138-35203-2

Acknowledgements

Prof. Dr. Asep Kadarohman, M.Si.
Prof. Dr. Didi Sukyadi, M.A.
Prof. Dr. H. Yaya Sukjaya Kusumah, M.Sc.
Prof. Dr. Hj. Anna Permanasari, M.Si.
Prof. Dr. Disman, M.S.
Assoc. Professor Vina Adriany, M.Ed., Ph.D
Professor Didi Sukyadi (Universitas Pendidikan Indonesia)
Professor Yaya Sukjaya Kusumah, M.Sc, Ph.D (Universitas Pendidikan Indonesia)
Professor Amita Gupta (The City College of New York, USA)
Professor Branislav Pupala (Tranava University, Slovak)
Assoc. Professor Marek Tesar (University of Auckland, New Zealand)
Dr. Jo Warin (Lancaster University, UK)
Dr. Simon Brownhill (University of Cambridge, UK)
Dr. Sonja Arndt (Waikato University, New Zealand)
Dr. M. Solehuddin, MA, M.Pd (Universitas Pendidikan Indonesia)
Dr. Ade Gaffar Abdullah (Universitas Pendidikan Indonesia)
Putu Rahayu Ujianti, M.Psi (Universitas Pendidikan Ganesha, Indonesia)
Hui-Hua, Chen, PhD (National Dong Hwa University, Taiwan)
Assoc Professor Dr. Dwi Hastuti (Institut Pertanian Bogor)
Pratiwi Retnaningdyah, PhD (Universitas Negeri Surabaya)
Fonny Demeaty Hutagalung (University of Malaya)
Sarita Gálvez (Monash University, Australia)
Assoc. Professor Tutin Aryanti, PhD (Universitas Pendidikan Indonesia)
Assoc. Professor Ahmad Bukhori Muslim, PhD (Universitas Pendidikan Indonesia)

Early Childhood Education in the 21st Century – Yulindrasari et al. (eds)
© 2020 Taylor & Francis Group, London, ISBN 978-1-138-35203-2

Committees

EDITORS

Hani Yulindrasari, Vina Adriany & Yeni Rahmawati
Universitas Pendidikan Indonesia, Indonesia

Fonny Demeaty Hutagalung
University of Malaya, Malaysia

Sarita Gálvez
Monash University, Australia

Ade Gafar Abdullah
Universitas Pendidikan Indonesia, Indonesia

SCIENTIFIC COMMITTEE/ADVISORY BOARD

Professor Didi Sukyadi (Universitas Pendidikan Indonesia)
Professor Yaya Sukjaya Kusumah, M.Sc, Ph.D (Universitas Pendidikan Indonesia)
Professor Amita Gupta (The City College of New York, USA)
Professor Branislav Pupala (Tranava University, Slovak)
Assoc. Professor Marek Tesar (University of Auckland, New Zealand)
Dr. Jo Warin (Lancaster University, UK)
Dr. Simon Brownhill (University of Cambridge, UK)
Dr. Sonja Arndt (Waikato University, New Zealand)
Dr. M. Solehuddin, MA, M.Pd (Universitas Pendidikan Indonesia)
Dr. Ade Gaffar Abdullah (Universitas Pendidikan Indonesia)
Putu Rahayu Ujianti, M.Psi (Universitas Pendidikan Ganesha, Indonesia)
Hui-Hua, Chen, PhD (National Dong Hwa University, Taiwan)
Assoc Professor Dr. Dwi Hastuti (Institut Pertanian Bogor)
Pratiwi Retnaningdyah, PhD (Universitas Negeri Surabaya)

STEERING COMMITTEE

Professor Dr. Asep Kadarohman, M.Si.
Professor Dr. Didi Sukyadi, M.A.
Professor Dr. H. Yaya Sukjaya Kusumah, M.Sc.
Professor Dr. Hj. Anna Permanasari, M.Si.
Professor Dr. Disman, M.S.
A/Professor Vina Adriany, M.Ed., Ph.D

ORGANIZING COMMITTEE

CHAIR
Hani Yulindrasari, PhD

TREASURER
Dr. Heny Djoehani, M.Pd
Rita Oktafil Marisa

SECRETARY BOARD
Yeni Rachmawati, Ph.D
Rizka Haristi
Marina Trie Ramadhani Gunawan
Suci Ramdhani

PROGRAMME
Dr. Euis Kurniati, M.Pd
Dr. phil. Leli Kurniawati, M.Mus
Sharina, M.Pd

PUBLICATIONS & DOCUMENTATIONS
Intani Prajaswari
Ayu Wulandari

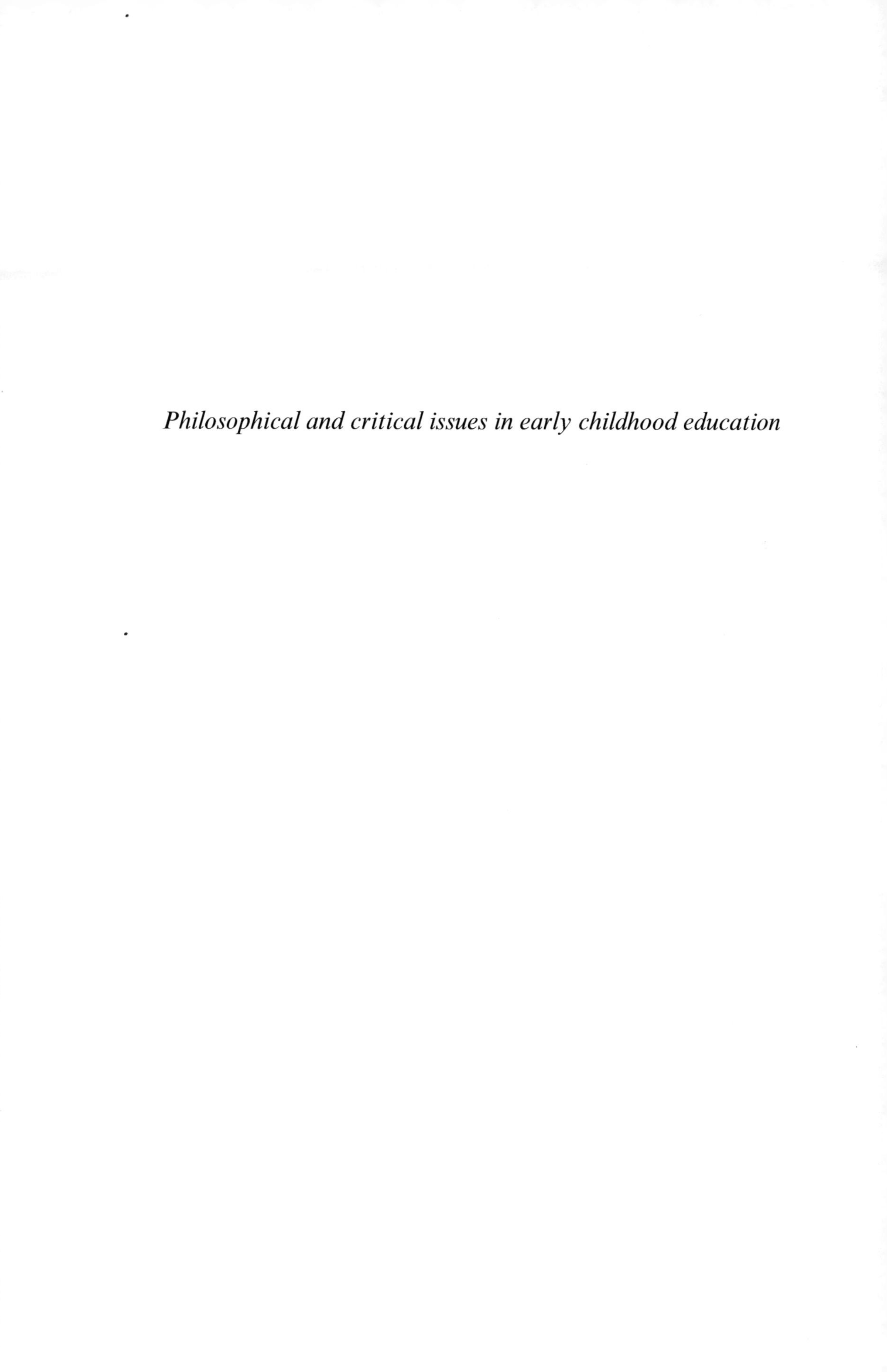

Philosophical and critical issues in early childhood education

Early Childhood Education in the 21st Century – Yulindrasari et al. (eds)
© 2020 Taylor & Francis Group, London, ISBN 978-1-138-35203-2

Pedagogical hybridity: Practicing within a postcolonial third space in early childhood classrooms as well as in teacher education classrooms

A. Gupta
The City College of New York, New York, USA

ABSTRACT: This paper is based on a qualitative study conducted with early childhood teachers within an Asian context that questions the use of Euro-Western assumptions to inform the preparation and practice of primary and pre-primary teachers in a non-western context. Conceptualized within the frameworks of postcolonial theory and globalization, the paper offers an alternate perspective on teacher education that will support the development of globally and culturally aware early childhood educators. Globalization works to bring global and local discourses together, with the former often being a western discourse which tends to dominate the latter. This leads to a colonized condition defined by Donaldo Macedo (1999) as being the imposition of an ideological yardstick against which members of weaker communities are measured and consequently fall short. Globalization also has a transactional nature to it and serves to bring diverse cultural elements together with regard to ideas, practices and policies. This transaction results in the creation of what Homi Bahba (1994) refers to as spaces of cultural hybridity. Rooted in the idea of cultural hybridity is the notion of pedagogical hybridity. Essentially this occurs within classroom spaces that are created when diverse pedagogical ideas and practices and policies are brought together. In this increasingly globalizing 21st century it is imperative that teachers are prepared in teacher education classrooms where culturally diverse notions on child development and educational philosophies are explored and acknowledged. In keeping with the theme of this conference of "Finding alternative approaches, theories, practices and frameworks of early childhood education in the 21st century" this talk will be focused on the globalized concepts of pedagogical hybridity and the pedagogy of third space by examining how that may apply to teaching practices in early childhood classrooms, and offering recommendations for teacher preparation programs.

1 INTRODUCTION

Primary and pre-primary education in Asia until recently was approached with an academically rigid curriculum and teacher-directed pedagogy. The 21st century, however, has witnessed a high level of global awareness and competition, and early childhood policy in many Asian countries has markedly shifted toward open classrooms and a learner-friendly pedagogy. Citizens of a globalized 21st century almost certainly need to be comfortable with diversity, flexibility and inclusivity in a manner that recognizes and respects the wider world, and school curricula now must include the diverse ideas and ways of thinking that represent the local and the global. Subsequently, it is imperative that teachers are prepared in teacher education classrooms where culturally diverse notions on child development and educational philosophies are explored and acknowledged.

This paper draws from a larger study of a series of inter-related qualitative inquiries that examine recent trends in EC policies and practices within the specific socio-cultural contexts and world-views of 5 Asian countries: India, Singapore, China, Sri Lanka and the Maldives.

I also draw upon the findings of my earlier study on early childhood teachers' preparation and classroom practice in some private urban schools in India.

In keeping with the theme of this conference of "*Finding alternative approaches, theories, practices and frameworks of early childhood education in the 21st century*" this paper will be focused on the globalized concepts of pedagogical hybridity and the pedagogy of third space by examining how that may apply to teaching practices in early childhood classrooms, and offering recommendations for teacher preparation programs.

2 GLOBAL AND NATIONAL INFLUENCES ON EARLY CHILDHOOD EDUCATION

The recent spotlight on the promotion of ECE as a developed field in Asia has been precipitated by global influences such as the United Nations initiatives of Convention on the Rights of the Child (UNCRC, 1990), Millennium Development Goals and Education For All (EFA 2015), Sustainable Development Goals 2030 (of which Goal 4 addresses access to quality ECE, increased supply of qualified ECE teachers), Developmentally Appropriate Practices (DAP) that was first published by NAEYC in 1987 and emphasized the concepts of learning through play, developmentally appropriate individualized learning, and developmentally appropriate child-centered classrooms.

Alongside this global push, nation governments of several Asian countries also promoted the development of a well-defined ECE field in the Global South by drafting national policies to prioritize ECE, with an emphasis on access, inclusion, and quality.

3 THEORETICAL FRAMEWORKS

The theoretical lenses utilized to frame this paper are *Globalization* and *Postcolonial* Theory. I view *Globalization* as the "Intensification of worldwide social relations which link localities in such a way that local happenings are shaped by events happening many miles away and vice versa" (Anthony Giddens as cited in Arnove, 2007). The most recent and current waves of globalization are *neoliberal* in nature, and marked by the "entry of foreign direct investment, and foreign corporations into national markets (Chatterjee, 2016). Additionally, the *globalization of education* refers to the impact that worldwide discussions, processes, and institutions have upon local educational practices and policies (Spring, 2009). Globalization can critically influence core educational decisions such as "what counts as responsive and effective education, what counts as appropriate teaching ... and who benefits from it throughout the world" (Apple, 2011, pp. 222–223).

The influence and assimilation of "foreign" ideas and practices into local contexts is certainly not a new phenomenon as seen from world histories of colonialism. Currently, a kind of neo-colonialism is being fostered through the neoliberal economic and political hegemonies of dominant powers that become established in nations generally perceived as weaker emerging economies. Thus globalization is closely connected to the notion of colonization which brings us to *Postcolonial Theory*.

Postcoloniality is rooted in the idea of the colonized condition viewed by Donaldo Macedo (1999) as the imposition of an ideological yardstick against which members of weaker communities are measured and fall short. Postcolonial theory addresses the two-way dialogues and transactions between seemingly binary ideas, opposing each other. When these binaries are viewed as cultures with fluid boundaries interacting with each other, then the exchange can appear as a form of cultural translation as ideas from one culture get modified and embedded into another culture (Bhabha, 1994). This process of transformation may lead to the creation of a grey area, a third space of *cultural hybridity*, which holds fresh possibilities (Bhabha, 1994, 2009). This approach further allows for a deeper understanding of the transaction and the knowledge production that occurs within the space of cultural hybridization (Tikly, 1999). In a similar process, when diverse educational elements of pedagogy and curriculum are brought together a space of *pedagogical hybridity* is created within which a *pedagogy of third space* is practiced. The notion of pedagogical hybridity is thus rooted in Bhabha's notion of cultural hybridity. Postcolonial theory enables an examination of the flow of diverse educational ideas and the curriculum that gets enacted within the space of pedagogical hybridity.

In Asia "the colonized condition can certainly be found in early childhood classrooms when schools in the "non-west" are evaluated by standards of pedagogy and curriculum that are based on an understanding of child development in the context of young children growing up in the "west" (Gupta, 2014, p.4). Early childhood teachers in urban India were observed to navigate between tradition and modernism... their voices in dialogue with the voice of the dominant discourse of the "west" (Gupta, 2006). Discourses not only refer to what is said and thought but also who has the authority to speak and when. Using a "western" discourse to describe educational philosophy and pedagogy undoubtedly provides credibility to schools in the developing world. Much more attention needs to be paid to the impact of globalization on the field of early childhood education and teacher education pedagogy and policy in postcolonial societies of the global South.

4 THE IMPACT OF THE GLOBAL NEO-LIBERAL AGENDA IN EDUCATION ACROSS ASIA

Neoliberal globalization has shaped early education in Asia in two primary ways: 1) by applying the market economy discourse to educational institutions and transforming them into commodities for consumers; and 2) by applying the discourse of the widely used American ECE guide, Developmentally Appropriate Practices (DAP). These "western" discourses seem to provide credibility to schools in emerging economies. Vavrus (2004) notes the legitimacy that is given to changes in the local systems through the borrowing of language and educational models from external countries. Unfortunately the projects that are selected for funding by world organizations end up having to subscribe to pedagogical practices that are shaped by a dominant Euro-America as "reforms from elsewhere are not necessarily borrowed for rational reasons but for political and economic ones" (Steiner-Khamsi, 2012, p. 4).

Several changes have been observed in the education sectors across Asia that fundamentally alter the definition of education. There has been an increase in the number of private early childhood and teacher preparation programs, along with an explosion so-called international preschools, & "global and world-class schools". Many of the private schools and franchises that have mushroomed under market economy policies are found to be unregulated and of sub-standard quality. Neoliberal forces have influenced the positioning of centers for children as businesses that are then aggressively advertised in market-based language. To keep competitive in the climate of globalization educators from Asian countries are sent on study tours to western English-speaking countries to learn and bring back innovations in education. There has been a movement away from the traditional, academically rigorous, teacher-directed approach toward a more learner-centered and inclusive approach, and significant prominence has been given to the discourse of DAP and a "play pedagogy".

Further, the globalized neo-liberal educational climate is creating additional inequities in educational and schooling experiences for children in Asia. Neoliberalism and market economy structures are shaped by capitalist values of a Euro-American "west" that are based on the ideas of individualism, consumerism, free choice, competition, and efficiency; a shift away from the concept of collectivism. This increased individualism has had a debilitating impact on public policy. Enrollment levels of children in pre-primary and primary schools has certainly risen but at the same time acute shortages in the availability of government-recognized schools and qualified teachers is resulting in overcrowding of classrooms. Teachers are inadequately prepared to handle the increasing socio-economic diversity in their classrooms. Most significantly, cultural and colonizing incursions are seen to occur when a "western" progressive early childhood discourse is viewed as the basis of "appropriate" pedagogy in all Asian classrooms.

Two examples of cultural and colonizing incursions are the problematic assessment of quality, and the application of the DAP-based play pedagogy in Asian classrooms. Both are discussed below.

4.1 *Problematic assessments of quality*

It is troubling that quality assessment of early childhood programs is measured against "western" standards in evaluations offered by external evaluators such as the following: "Preschool education in Sri Lanka has developed a style of its own that is uniquely out of step with the more widely accepted Early Childhood Education theories and practice valued in most developed countries" (King, 2010).

The "traditional teacher directed" EC classrooms observed were marked by tiny spaces, with large number of children who were seated for many of the activities. Despite the congested environment there was a high level of verbal and intellectual engagement marked by curiosity, bubbling eagerness to answer, eagerness to participate in songs and rhymes. This was very different from the more physical energy observed in many progressive Euro-American early childhood classrooms in the west.

This raises a critical question: is physical energy in a classroom more important than verbal and mental energy, and is that a context related decision? How important is the evidence of students' movements and physical energy in a classroom located within a country where physical and outdoor activities have lower priority as compared to a classroom in the United States where a very high value is placed on sports and outdoor activities? Conversely, are the students in an Indian classrooms, which is characterized by low physical energy but high mental energy, any less engaged, or experiencing a lower quality learning experience (Gupta, 2006/2013)?

4.2 *Applying the dominant discourses of DAP and play*

Inequities are created when dominant global discourses are applied to local contexts. The promotions of a developmentally appropriate early childhood discourse by world organizations of heft have placed a value on play in the early childhood education narrative in Asia and recent policy changes reflect this. This becomes problematic when the word PLAY is given different meanings in different cultural world views. For instance, in the Indian context the word for play in Hindi is "khel" which encompasses a range of activities: fun and frolic; games and sports; gambling; participating in fairs and celebrations; dramatization of stories; dance, music and rhythm; fierce competition of skills and abilities; preferred skills of cooperation, sharing, taking turns, following rules, but also survival skills such as harassment, deception, teasing and trickery which are inherent in successfully navigating the world of human relationships. Thus attempts to implement a play-based pedagogy, and assess an activity as acceptable play using the lens of DAP can be greatly challenged (Gupta, 2011).

Several challenges may be encountered in the implementation of a play-based learner-centered pedagogy in classrooms of the Global South: political contexts unsupportive of the democratic essence of learner-centered education; overcrowded classrooms with class sizes of 40-60 children; scarcity of basic supplies in schools such as furniture, running water, electricity and sanitation facilities; teachers untrained in the pedagogy of play and learner-centered approaches – none of these conditions support the individualized teacher-child interaction which is central to learner-centered pedagogy. Thus it might behoove educators to pause before trying to implement a western policy/pedagogy in a non-western context without adaptations.

4.3 *The concept of hybridity*

Alexander (2000) notes that "Life in schools and classrooms is an aspect of our wider society, not separate from it: a culture does not stop at the school gates. The character and dynamics of school life are shaped by the values that shape other aspects of... national life". Therefore, when national life begins to reflect globalization and a hybrid postcoloniality then classrooms will surely begin to reflect that hybridity as well. The findings of my earlier study revealed the existence of a postcolonial hybridity in the early childhood curriculum and classroom practices resulting from a transactional exchange of the three culturally diverse philosophical and pedagogical influences: an in-between area that I referred to as a third space of pedagogical hybridity (Gupta, 2006). The three influencing discourses were:

- the British colonial design and content of teacher education programs in India, a legacy which continued to linger long after India gained independence from the British rule;
- the dominant DAP influenced Euro-American discourse of early childhood education that teachers were being asked to implement in some private schools in India; and
- the underlying Indian cultural and spiritual values that deeply influenced the images of the child and teacher in Indian society.

The critical finding in this study seemed to be that teachers recognized the importance of working with the prescribed academic syllabus and helping students develop the skills to succeed in a competitive society. But they also recognized the importance of teaching the whole child and ensuring the development of social, emotional, and moral development. The teachers in this study seemed to have the freedom and flexibility to individually implement a more informal curriculum parallel to the rigid academic curriculum where issues in values, good attitudes, environmental protection, and diversity were being addressed. The teachers recognized that certain aspects of each approach had a place in their classrooms and in the overall success of the child's educational experience, and were comfortable teaching within the space of pedagogical hybridity.

5 CONCLUSION

Globalization is here to stay, and societies are becoming increasingly diverse where the global and local are juxtaposed in classrooms and outside. Thus we need to ensure that teacher education programs also reflect the same postcolonial hybridity that currently exists in schools and society. The goal is for teachers to be prepared with a deeper cultural awareness and sensitivity, and to be better able to address socio-cultural-economic diversities in their globalized classrooms. Teacher education programs can offer a more relevant experience to future teachers by preparing and educating teachers in a multicultural approach to education; by training them in qualitative and quantitative assessments; by deepening their understanding and knowledge of how to implement child-centered teaching within the local context; by exposing them to global discourses and research; by recruiting a more diverse teaching force, and preparing teachers to teach in diverse contexts; by offering teachers richer and more diverse clinical experiences; by holding teachers to higher teaching standards and accountability; and most importantly, by positioning teacher development as lifelong learning which does not stop after graduating with a degree from a teacher education college. This approach to teacher education is sure to result in a more culturally relevant and appropriate pedagogy and curriculum for teacher education and early childhood classrooms.

REFERENCES

Alexander, R. J. (2000). *Culture and pedagogy: International comparisons in primary education*. Oxford, UK: Blackwell

Apple, M. (2011). Global crises, social justice, and teacher education. *Journal of Teacher Education, 62* (2), 222–234.

Arnove, R.F. (2007) Introduction: reframing comparative education: the dialectic of the global and the local. In R.F. Arnove & C.A. Torres (Eds) *Comparative education: the dialectic of the global and the local*. Lanham: Rowman & Littlefield.

Bhabha, H. (1994). *The location of culture*. London, UK: Routledge.

Bhabha, H. (2009). An interview with Homi Bhabha: Cultural translation and interpretation: A dialogue on migration, identity, and ethical responsibility: An argument against fixed identities. *Sangsaeng*, (26), Winter 2009. Asia-Pacific Center of Education for International Understanding, UNESCO.
Chatterjee, I. (2016). Oxford University Press Blog accessed on 11/7/2016 from http://blog.oup.com/2016/05/globalization-in-india/
Giddens, A. (1990). *The consequences of modernity*. Stanford, CA: Stanford University Press.
Gupta, A. (2006). *Early childhood education, postcolonial theory, teaching practices in India: Balancing Vygotsky and the Veda* (1 Ed). New York, NY: Palgrave Macmillan.
Gupta, A. (2011). Play and pedagogy framed within India's historical, socio-cultural and pedagogical context. In S. Rogers (Ed.) *Rethinking play and pedagogy in early childhood education: Concepts, contexts and cultures* (pp. 86–99). Oxford, UK: Routledge.
Gupta, A. (2013). *Early childhood education, postcolonial theory, teaching practices and policies in India: Balancing Vygotsky and the Veda* (2nd Ed). New York, NY: Palgrave Macmillan.
Gupta, A. (2014). *Diverse early childhood education policies and practices: Voices and images from five countries in Asia*. New York, NY: Routledge.
King, D. (2010). *Failure of preschool education in Sri Lanka*. Retrieved from http://www.srilankaguardian.org/2010/01/failure-of-pre-school-education-in-sri.html
Macedo, D. (1999). Decolonizing indigenous knowledge. In L. Semali & J. L. Kincheloe (Eds). *What is indigenous knowledge? Voices from the academy*. New York, NY: Falmer Press.
Steiner-Khamsi, G., Silova, I., & Johnson, E. M. (2006). Neoliberalism liberally applied: Educational policy borrowing in Central Asia. In J. Ozga, T. Seddon, and T. Popkewitz (Eds.), *2006 World Yearbook of Education. Education Research and Policy: Steering the Knowledge-Based Economy* (pp. 217–245). New York/London: Routledge.
Steiner-Khamsi, G. (2012). Understanding policy borrowing and lending: Building comparative policy studiesIn G. Steiner-Khamsi & F. Waldow (Eds) *World yearbook of education 2012: Policy borrowing and lending in education* (pp. 3–18). London/New York: Routledge.
Spring, J. (2009). *Globalization and education*. New York: Routledge
Tikly L. (1999). Postcolonialism and comparative education. International Review of Education, 45(5/6), 603–621.
Vavrus, F. (2004). The referential web: Externalization beyond education in Tanzania. In G. Steiner-Khamsi (Ed), *The global politics of education borrowing and lending* (pp. 141–153). New York: Teachers College Press.

Early Childhood Education in the 21st Century – Yulindrasari et al. (eds)
© 2020 Taylor & Francis Group, London, ISBN 978-1-138-35203-2

Using philosophy to (re)value, (re)conceptualise and (re)vitalise ECE theory and practice

S. Arndt
University of Waikato, Hamilton, New Zealand

ABSTRACT: Pressures on early childhood education provision are immense. In Indonesia, as internationally, there is both a sense of crisis and of hope! Teachers and providers struggle, under pressure to perform, to achieve global 'quality' benchmarks, and meet universal standards of practice. Neoliberal policy and mindsets complicate already challenging prospects of professionalization, regulation, international aid, and conflicting theorisations of 'best' practice. This talk finds hope in challenging dominant theorisations. It challenges inequities arising for children in ECE, by reconceptualizing not the theories, but our thought. Grounded in the Kristevan notion that thought itself, perhaps, is a form of dissidence, the shifts I propose use philosophical thought to rethink the ways in which ECE theories, contexts and policy inform local ECE provision. I propose a model of thought that uses philosophy to (re)value local knowledges, (re)conceptualise dominant discourses, and (re)vitalise hope in ECE theory and practice.

1 INTRODUCTION

Rarangatia te kowhaiwhaitanga
O te tika, te pono, me te aroha
Hei oranga mo te iwi

Weave the tapestry
Of right, truth and love
As sustenance of your people
(Māori proverb)

Beginning this paper with this New Zealand indigenous Māori proverb grounds my paper in my intentions for rethinking approaches, theories and best practices in early childhood education. It reminds us never to lose sight of what is important, to remain engaged with what is important. Put another way, as Donna Haraway (2016) reminds us in her book title, it reminds us to 'stay with the trouble', to not become complacent, stagnant, or uninspired. From these underlying reminders, I embark on this paper from a position of wrestling with and thinking in diverse ways, unsettling the early childhood education status quo, to address, perhaps, through diverse thought, elements of that status quo that too often remain marginalised, forgotten or too difficult to think about (Arndt, 2017).

Affirming the 'thinking' part of the call of the ICECE conference in Bandung, 2018, I use this paper to make an argument for increasingly critical philosophical thought in early childhood theory and practice. This act echoes and attempts to keep alive recent calls for increased engagement, for example, with the "critical task for reformulating early childhood education" (Peters, 2007, p. 224), and for heightened philosophical thought in early childhood research and practice (Farquhar & White, 2013), as an ongoing task. It alerts us to thinking about thought, as

> not simply, or solely to evaluate means or decisions but to question ... consolidated criteria, practices and ideals. It is ... to bring hidden aspects to the fore, to accommodate

reflectively the new and the unknown ... To be critical means first and foremost to be imaginative of alternative realities and thoughtful about their possible value or non-value (Papastephanou & Angeli, 2007, p. 612)

This is not a task that we can solve and then forget about, but rather one that requires our continued engagement. Locally and globally, our societies are in flux, constantly evolving, affected and responding to national and international events and affected and responding to particular national and international orientations. Our early childhood education settings, both in Indonesia and in Aotearoa New Zealand, are impacted by the daily demands of keeping up, of developing programmes, maintaining some practices, striving for others, in the name of meeting internationally and locally prescribed outputs, demands and benchmarks (Adriany, 2018).

In arguing for a reconceptualization of thought – I follow the important work of the thinkers in RECE, the reconceptualist movement begun as a powerful response to the "disenchantment with the dominant discourses within the field of early childhood education" (Dahlberg, 2014). In this paper I suggest three points, not as a linear process, but as intertwined and ongoing engagements, in which we might focus at times more on one than the other, but at other times on all of them at once. These points are: to (re)conceptualise dominant discourses, to (re)value local knowledges, and to (re)vitalise hope in the field of early childhood education. I will begin with some thinking about reconceptualising dominant discourses, and throughout, I will elevate the importance of increasingly philosophical thought in early childhood education.

2 (RE)CONCEPTUALISING DOMINANT DISCOURSES

What does it mean, to increase philosophical thought? What does it look like, and how might we do it? First it is useful to think about what philosophical thinking might be. Harré (2001) sees philosophical thinking as an examination of life, and of how we live life. In terms of our thinking about young children's learning and our pedagogical practices, philosophy could then be seen as an ongoing engagement with how we teach, what we do – essentially, how we live life in our early childhood settings. It may then also involve thinking about how we live the theories, thought and realities that we experience in these settings, with the intention of being critical, as stated above by Papastephanou and Angeli (2007), so that particular ways of thinking are exposed and challenged and so that they help us to open our conceptualisations in different ways, to rethink what we are used to.

Such a way of thinking might, according to feminist, post-structural philosopher and psychoanalyst Julia Kristeva (1986), be seen as a form of dissidence. As she says, "true dissidence today is perhaps simply what it has always been: *thought*" (p. 299). And so perhaps our philosophical thinking about early childhood education involves being some kind of dissident, in this sense? Dissident thinking, Kristeva argues, involves a "ruthless and irreverent dismantling of the workings of discourse, thought, and existence" (p. 299). In other words, if we apply this idea to an examination of life in early childhood settings, this means that we must know about the 'discourses, thought and existence', in order to dismantle them.

Put another way, such thought depends on us knowing the structures and systems that govern our life. And it means we must know those structures, the rules, regulations and policies, by which we are governed *well*: by finding out what these systems, rules, regulations and policies mean to us; what they mean to our local early childhood settings, what they mean to teachers, to children, and to families. What do we know, about how policies impact on the people and places, and the teaching and learning that happens in these settings? What do we know about how they work in the wider system, for example, about who makes the rules, why, and what they expect to achieve? Furthermore, following Kristeva, it means asking if we are bold enough, and whether we think deeply enough about the ways that we can engage with them and rethink them, in ways that are meaningful, helpful and hopeful.

Adopting a 'heightened philosophical approach in early childhood research and practice' does not mean replacing one way of thinking with another. Rather, it means opening up to the complexities that 'examining life and the way we live it' reveals. This means that not only do we need to know our systems, the policies, guidelines, curriculum requirements, and so on, that govern us, very well, but also that we need to know the relevant *theories* well. That is, it means that we need to encompass the theories that guide our practices and orientations in early childhood education, into the 'life' that we are examining, including how they help us to articulate, to think through, and rethink what we know about the early childhood settings and the field overall. We need to know these theories well enough to apply them as required, to each situation, each context, each PAUD, Kindergarten, child, family, teacher and community, and rethink them in useful ways. Finally, we also need to know about ourselves, and our culture, language, and beliefs about teaching and learning. Our pedagogies arise from our individual and collective positioning, which, as Kristeva (1991) suggests in her theory on subject formation, are always in process. Given this, we will never be able to fully know ourselves, as our identities, beliefs and values are always in constant construction, in flux, and evolving.

Thinking philosophically can take diverse forms. It can be both how we think and what we think about, as both content and method (Standish, 2009), and it can lead to a blurring of the boundaries (Papastephanou, 2009) of those two. Thinking philosophically should not further entrench narrow, or surface level, hegemonising ways of thinking, if we want to reconceptualise early childhood education and how we live it. According to Deleuze and Guattari (1991), philosophy can be positioned as "the art of forming, inventing, and fabricating concepts". This opens up creative, conceptual and hopeful spaces for thought and new understandings. 'Doing' philosophy therefore involves a careful scrutinising of "the fine-grained complexities of social phenomena" (Davis, 2009, p. 371). This in turn means that it might lead to a more messy, open uncertainty, as it makes spaces for forming, inventing or fabricating concepts, asking questions, and being 'dissident', in a Kristevan sense.

Doing philosophy is not always easy, or comfortable. Indeed it can be seen as a kind of truth finding (Feinberg, 2014), where there can be multiple truths, and multiple ways to work with them. Opening up to the possibilities of multiple truths calls on us to let go of the expectation that certain knowledge or evidence will provide a magical solution that we can apply to fix all problems. In this sense, Ruitenberg (2009) says, thinking "only becomes philosophical when familiar words grow strange" (p. 426). In other words, how we see policies, how we see practices, or our own orientations, becomes unsettled, differently interpreted: it becomes a dialogue between language, culture and power. It depends then, on a "broad range of virtues ... open-mindedness" or on a particular "'epistemic humility'" (Vokey, 2009, p. 353). And further, thinking philosophically can be seen as a critical feminist practice that focuses on ethical interrogations, aimed at de-elevating marginalising or subjugating positioning, theories and practices. Theorizing subjectivities through language, for example, invites openness to radical change, aligning with post-structural perspectives on feminism, and demanding "attention to social, historical and cultural specificity" (Weedon, 1997, p. 132).

Philosophical thought, then, involves us in an ongoing questioning. It compels us to expose and interrogate power/knowledge relationships within the status quo, revealing those things that we may forget to think about, don't consider important enough to think about, or indeed are afraid to think or ask about. The 'critical task of reformulating early childhood education' that Peters (2007) earlier called us to take up, implicates us in an increasing sense of comfort with the discomfort of uncovering and negotiating multiple truths, as we find some answers at the same time as we open more questions. An important element within this thinking is our cultural groundedness.

3 (RE)VALUING LOCAL KNOWLEDGES

It has been argued that there is a contemporary lack of a comprehensive narrative in society today, resulting in a state where the complexity of society is flattened and histories and stories forgotten (Kristeva, 2000). In arguing for revaluing local knowledge I suggest a re-elevation of narratives of particular spaces, realities, orientations, insights, places and for openness to diverse, specific subjectivities. Such a re-elevation of local contexts and stories and knowledge calls together past, present and future, and it adds further depth to such concerns as those of teacher status and professionalism inherent in both Indonesia and internationally (Yulindrasari & Ujianti, 2018).

Recognising the depth of histories and lives of the land on which our early childhood settings are situated, for example, connects us to the place, and to knowledges and diverse ways of living and being that precede, but also live on, in these spaces. In the Aotearoa New Zealand context our disturbance of comfortable knowing, planning, is often inspired by indigenous Māori knowledges, as woven through the curriculum document, Te Whāriki (Ministry of Education, 2017). At the same time as those of us who re non-Māori acknowledge that we cannot necessarily access or represent Māori ways of knowing and being, they can motivate us in ways that push us further into a certain new comfort, with the discomfort of different ways of thinking and non-knowing. Local or indigenous thought can guide our navigation of the discomfort of multiple truths, for example, with the use of metaphor, to develop a picture of some of the complexities that a philosophical dismantling of discourses and the structures of early childhood education, as I have argued for, reveals.

Māori scholar Angus Macfarlane (2013), for example, likened his methodology to a metaphor of braided rivers, representing the diverse thoughts, policies, practices, demands and pressures that they represent. Braided rivers are particular geological formations, where many streams converge and separate in narrow streams, wide streams, deep and shallow streams, trickling or gushing, meandering, roughly in the same direction, each in their own ways (Arndt & Tesar, 2018). They metaphorically represent diverse cultural knowledges converging and detracting, to and away from one another. They traverse a metaphorical riverbed, which might be sandy, firm or soft, shifting and porous, that represents the local context, stories and narratives, and in a wider sense, our communities, histories and early childhood education cultural environment.

Imagined as a braided river, local contexts, knowledges and stories are integrally entwined in our early childhood educational whole. They may flow, gush, trickle, or disappear, and ground our rethinking of early childhood practices in the local people, places and things. Just as in Aotearoa New Zealand, in Indonesia local narratives can be in danger of disappearing in the porous, sandy riverbed. They may become pushed aside, overpowered, de-valued or lost, to make way for what we may see as re-colonising expectations to internationalise, economise, and 'make progress'.

In Aotearoa New Zealand, Te Whariki, promotes the aspiration for children to become "[c] ompetent and confident learners and communicators, healthy in mind, body and spirit, secure in their sense of belonging and in the knowledge that they make a valued contribution to society" (Ministry of Education, 2017, p. 5).

> Competent and confident learners and communicators, healthy in mind, body and spirit, secure in their sense of belonging and in the knowledge that they make a valued contribution to society.

Thinking philosophically creates spaces to reinsert and revalue local narratives, on cultures, rituals and practices that define what 'a valued contribution to society' is. And they define which dominant discourses and theories become destabilised as the values of the streams in the wider braided riverbed become elevated, recognised and removed from the margins. In this way, valuing local narratives raises the potential for revitalising hope.

4 (RE)VITALISING HOPE

Using philosophy as a way of thinking revitalizes hope in early childhood theory and practice. As Papastephanou (2009) says, it involves a sense of wonder, as it strengthens the importance of openness, of possibility, as aporia, not as a dead end, but rather as an opening to future potentialities. In reframing our orientations towards early childhood education theories and practices, and to the orientations that they represent, then also challenges us to become more open, for example, to our worldly interconnectedness and interdependence, in an ecological sense, to become alerted to how we affect and engage with the planet, earth and its environment (Latour, 2011; Ritchie, 2013). Multiple truths and perspectives on the greater forces and meanings affecting our very being, might elude our understanding, but awaken an awareness of perspectives, diverse life forces, and non-human influences, which we can recognise as important, but perhaps not 'know', in the early childhood life, that we are examining (Bennett, 2010).

By using philosophy as a way of thinking to challenge our existing thought, rather than fearing the new or unknown, or what is inarticulable, we may become hopeful. 'Weaving the tapestry' as the opening proverb suggests, affirms what indigenous thinkers (Mika, 2017; Andreotti, Akahenew, & Cooper, 2011) remind us of, as always already existing relationalities and greater forces, than those which we can know. It opens up a space, also, to similar conceptions of the wider 'tapestries' in the perspectives on relationalities that posthuman and new materialist thinkers (Bennett, 2010; Braidotti, 2013) have more recently impressed on Western philosophical and methodological thought.

5 CONCLUDING COMMENTS

How do we move forward, then, in hope and in a particular sense of awe? In the end, rethinking early childhood education to value, conceptualise and revitalise theory and practice differently opens up spaces for complex intersections of histories and realities. It involves each of us deeply considering underlying assumptions, ethical and moral orientations, towards children, childhoods, and towards what matters in their teaching and learning. It involves a reorientation towards early childhood theory and practice to retain what is important to be locally strong, relevant, and meaningful. Finally, using philosophy as a method to be critical, as noted in the earlier quote, by 'questioning the consolidated criteria', we must first establish: what is our consolidated criteria? So that by 'bringing hidden aspects to the fore' we can imagine alternative realities: What could our early childhood settings be? What is our ideal? What do we (dare to) hope for, in relation to our local cultural knowledges, values, teachers and children? What, indeed, might the possibilities be, inspired and pushed by our philosophical thought?

REFERENCES

Adriany, V. (2018). Neoliberalism and practices of early childhood education in Asia. *Policy Futures in Education, 16*(1), 3-10. doi:10.1177/1478210317739500.

Arndt, S. (2017). *Teacher Otherness in early childhood education: Rethinking uncertainty and difference through a Kristevan lens.* (Doctor of Philosophy), University of Waikato, Hamilton, New Zealand. Retrieved from http://hdl.handle.net/10289/11259.

Arndt, S., & Tesar, M. (2018). Narrative Methodologies: Challenging and Elevating Cross-Cultural Complexities. In S. M. Akpovo, M. J. Moran, & R. Brookshire (Eds.), *Collaborative Cross-Cultural Research Methodologies in Early Care and Education Contexts.* New York, NY: Routledge.

Andreotti, V., Ahenakew, C., & Cooper, G. (2011). Epistemological pluralism: Ethical and pedagogical challenges in higher education. *AlterNative: An international journal of indigenous scholarship, 7*(1), 40-50.

Bennett, J. (2010). *Vibrant matter: A political ecology of things.* Durham, NC: Duke University Press.

Braidotti, R. (2013). *The posthuman.* Cambridge, England: Polity Press.

Davis, A. (2009). Examples as method? My attempts to understand assessment and fairness (in the spirit of the later Wittgenstein). *Journal of Philosophy of Education, 43*(3), 371-389.

Deleuze, G., & Guattari, F. (1991). *What is philosophy?* (H. Tomlinson & G. Burchell, Trans.). New York, NY: Columbia University Press.

Farquhar, S., & White, E. J. (2013). Philosophy and pedagogy of early childhood. *Educational Philosophy and Theory*, 1-12. doi:http://dx.doi.org/10.1080/00131857.2013.783964.

Feinberg, J. (2014). *Doing Philosophy* (5th ed.). Boston, MA: Wadsworth, Cengage Learning.

Haraway, D. (2016). *Staying with the trouble: Making kin in the chthulucene*. Durham, UK: Duke University Press.
Harré, R. (2000). *One thousand years of philosophy: From Ramanuja to Wittgenstein*. Oxford, UK: Blackwell Publishers Ltd.
Kristeva, J. (1991). *Strangers to ourselves*. New York, NY: Columbia University Press.
Kristeva, J. (1977/1986). A new type of intellectual: The dissident. In T. Moi (Ed.), *The Kristeva reader* (pp. 292-300). Oxford, UK: Blackwell Publishers Ltd.
Macfarlane, A. (2013). *Me Whakawhiti - Crossing Cultural Borders Indigenising Research and Practice*. Paper presented at the Cutting edge conference, Rotorua, New Zealand. http://www.cmnzl.co.nz/assets/sm/8632/61/CuttingEdgekeynote2013Lu.pdf.
Mika, C. (2017). *Indigenous education and the metaphysics of presence: A worlded philosophy*. London, England: Routledge.
Papastephanou, M. (2009). Method, philosophy of education and the sphere of the practico-inert. *Journal of Philosophy of Education*, *43*(3), 451-470.
Papastephanou, M., & Angeli, C. (2007). Critical thinking beyond skill. *Educational Philosophy and Theory*, *39*(6), 604-621. doi:10.1111/j.1469-5812.2007.00311.x.
Peters, M. A. (2007). Editorial. Philosophy of early childhood education. *Educational Philosophy and Theory*, *39*(3), 223-224.
Ritchie, J., & Skerrett, M. (2014). *Early childhood education in Aotearoa New Zealand: History, pedagogy and liberation*. New York, NY: Palgrave Macmillan.
Ruitenberg, C. (2009). Introduction: The question of method in philosophy of education. *Journal of Philosophy of Education*, *43*(3), 315-323.
Standish, P. (2009). Preface. *Journal of Philosophy of Education*, *43*(3).
Vokey, D. (2009). 'Anything you can do I can do better': Dialectical argument in philosophy of education. *Journal of Philosophy of Education*, *43*(3), 339-355.
Weedon, C. (1997). *Feminist practice and poststructuralist theory*. Malden, MA: Blackwell Publishing.
Yulindrasari, H., & Ujianti, P. R. (2018). "Trapped in the reform": Kindergarten teachers' experiences of teacher professionalisation in Buleleng, Indonesia. *Policy Futures in Education*, *16*(1), 66-79. doi:10.1177/1478210317736206.

Early Childhood Education in the 21st Century – Yulindrasari et al. (eds)
© 2020 Taylor & Francis Group, London, ISBN 978-1-138-35203-2

From global policy to everyday preschool practice

B. Pupala
Trnava University, Trnava, Slovakia

ABSTRACT: Increasingly, the theoretical and empirical research on early childhood education, especially that concerned with critical pedagogy, has highlighted the challenges globalised policy creates for this sector. The main features of globalised policy are its economised view of education's mission and a narrowing of the content in favour of human and social capital; in short, everything that points to the neoliberal transformation of the entire education sector. The pressure emanating from supranational authorities is manifest and has a marked impact on national education policies, both in countries with a strong tradition of early childhood education and care, and in countries where this sector is gradually emerging and developing. However, these policies do not directly impact on everyday preschool practices. If we compare preschool practices in different countries and environments (on the European, Asian and African continents) then we see that practices do not directly reflect policy goals, but are counterbalanced by a variety of strategies and in different ways. What matters most in determining the existing and future forms of preschool education is not the political solutions and decisions, but the direct encounters between policy and practice, and the key role teachers play in that.

1 INTRODUCTION

The transfer of global education policy to local education systems is a key topic in international comparative research on education. Researchers are generally concerned with exploring the various ways in which global education policy impacts on local settings and how they respond to the more well-known global schemes, challenges and demands. Many of these studies belong to the more critical stream of education research. They often highlight a worrying tendency for local policies to be assimilated into uniform global schemes and that education, particularly in the Global South, is being subjected to a second wave of Westernization or even that we are seeing the emergence of neocolonial mechanisms within education governance (Gupta, 2015).

A particularly sensitive area, in this regard, is that of early childhood education and care. This should come as no surprise, as the very nature of ECEC means that strong links are maintained between it and its local contexts. Without going into a more indepth analysis, it is clear that the daily work that goes on in ECEC settings is more closely interlinked with local culture, customs and traditions, languages and geographies than is the case in other education institutions. There is also a deep connection to the local family and community culture. Equally, compared to primary school education, the national histories of ECEC tend to be more closely tied to the national traditions and policies within which the different types of ECEC have emerged and continue to evolve in the various corners and parts of the world.

Bearing in mind the importance of the local context for ECEC, on the one hand, and the way in which it intersects with globalizing ideas on the form it should take, on the other, there is clear immediate and future justification for understanding and describing how local and global factors interrelate within ECEC governance and practice. Needless to say, with the increasing influence of international education policy, ECEC is increasingly considered to play an important role in the success of an education system as well as to provide the initial springboard to education pathways and future life careers. It is also seen as an important investment that benefits the whole of society. All this is linked, above all, through the idea that ECEC is key to both social and economic progress, and capable of producing economic benefits at both the national and international levels. In recent decades, efforts have been made to strengthen, expand and reform ECEC precisely because it is a feature of international, and hence national, economic and social strategies and is considered to be an important area of social investment. International support for expanding ECEC, motivated by these concerns, is largely seen in increased participation levels in preschool education, but also in attempts to improve the institutional infrastructure, raise preschool teacher qualification levels and, in many countries – particularly in Europe – in the introduction of compulsory preschooling, especially in the year preceding the first year of primary education.

It is common knowledge that the main player in the global education policy field is the OECD, an organization which promotes the kinds of ECEC policies described above. These are bound up with economic strategies in which education's role is seen primarily as one of producing the human capital for dynamic labour markets. In the Global South, the World Bank is another key player, as the development and expansion of ECEC is tied to its investment strategies. In Europe, this same policy is promoted through the European Union and the European Commission, which adopt measures, communiqués and strategic documents, where the aim is to make preschool education more important in relation to Europe's economic expansion and competitiveness.

These supranational challenges and expectations are then incorporated into national ECEC policies and provide motivation for redrafting national ECEC programmes. Recently, attempts to bring about global governance and indirectly control ECEC across continents have been made by the OECD, through its International Early Learning and Child Well-being Study (IELS), known also as baby PISA. These have been received with great criticism, mainly by academics from around the world (Moss et al., 2016; Urban & Swadener, 2016), and are a very good example of this particular global vision of ECEC, which emphasizes success at subsequent education levels, and the acquisition of competencies considered crucial to today's economy and labour market.

2 LOCAL ENCOUNTERS

There is little doubt that, under the influence of powerful global players, and regardless of continent or locality, national and local ECEC policies are evolving and becoming more similar, or, in other words, global policies are being transferred to the local level. This is not a direct process, nor does it simply mean the unilateral adoption of new schemes within different environments. Although national ECEC policy documents may give the impression that the global policy demands are being rapidly implemented into national cultures, in fact the contexts, cultural backgrounds and existing traditions of preschool education act as a filter on the impact of global policies within local conditions, giving them both a local flavour and trajectory. When global policies are transferred to the local level, they take on different forms, depending on the many historical, political, economic and pedagogic factors. In this article, we shall highlight some of the transfer strategies we noted when conducting international research on ECEC in different national contexts.

The strategies we identify and describe are the result of our direct observations and of the research we conducted in Slovakia, Germany, Nepal, Indonesia and Kenya. Slovakia features on this list for the understandable reason that it is my home country. It is where I am actively engaged in preschool teacher education and in conducting research and policymaking on ECEC. As part of my work on the latter, I was the main author of the current national preschool education curriculum. Germany was selected both because it is a strong European country and because of my knowledge of ECEC in Germany through the collaboration of university colleagues and German academics on ECEC projects. In Kenya, as part of Slovak development aid, I and my colleagues were involved in educating local preschool teachers and worked with preschools in the marginalized county of Kwale. In the last three years our academic collaboration has led us to research ECEC in Indonesia as well as Nepal, where we have had the opportunity this year to carry out the first pilot investigations (2018).

Whilst it is true that this analysis touches upon very different environments, countries and contexts, in which ECEC has very different roots, this is because we deal with areas belonging to the Global North and also draw on our experiences of countries in the Global South.

Of course, Europe itself does not have a homogenous ECEC environment either. When covering Slovakia, we take note of the fact that its ECEC traditions were formed within the Communist environment of Central and Eastern Europe. When we turn to our partial exploration of Germany, though, we take greater account of its Western European traditions (those of former West Germany). Our descriptions of the strategies for transferring global ECEC policy to the local level are therefore quite varied. In some cases, they contain specific national features, while others could probably be applied in parallel within the national environment.

3 TRANSFER STRATEGY

We will describe five different transfer strategies that were revealed to us in the cases we observed. Two of these relate to Europe and differ from the remaining three which were identified in the Asian and African countries investigated. The first two strategies are a response to long-established ECEC traditions being confronted with new visions and demands, or new policy declarations, ideas and expectations. The way in which these confrontations are dealt with depends on the extent to which the new ECEC visions differ from the tradition in that country, and this leads either to (1) *a strategy of idea change* or to (2) *strategy convergence* (between the original and new one). A further three strategies can be found in areas where ECEC is still developing and national ECEC policies have no firm roots, and where the institutional network is still being developed. This is characteristic of countries in the Global South, and in our in-country observations of ECEC we generally identified the following strategies: (3) *the co-existence of parallel worlds*, (4) *mission seeking* and (5) *absorption*.

Strategy (3), *parallel world strategy*, is one in which there is an unquestioned discourse favouring the adoption of the global ECEC vision within the national context and in conjunction with (or alongside) the existing preschool status quo, based on local values and visions. Strategy (4), *mission seeking strategy*, is a response to a situation in which the global imperatives have yet to gain authority in an environment where there is still scope for its own tradition to develop, given the necessary support. By contrast, strategy (5), *absorption*, is one used when the only official authority is found in the globally promoted ideas that form the definitional pillar of transfer and the norms and imperatives essential to that.

As we have already noted, it is possible for all three strategies to operate in parallel within a single national setting; however, we came across them in the form of distinct strategies operating separately within their varied national settings. Further careful and more extensive research is required to ascertain that they do in fact co-exist in all similar settings. We also expect it is possible for additional strategies to be identified.

4 IDEA CHANGE AND CONVERGENCE

We shall look first at the two strategies observed in the European countries investigated. Although Germany and Slovakia are located on the same continent and are both geographically and culturally similar, they have quite different ECEC traditions. In Slovakia, ECEC developed and evolved under the Communist education policies found in Central and Eastern Europe, while, in Germany, ECEC is based on a decentralized system within the German social welfare sphere. As we have previously noted (Kaščák & Pupala, 2018), the emphasis in the German preschool sector was not on the educational function of the institutions, and it had a high degree of autonomy in relation to the general education system. In contrast, the Slovak system of ECEC developed within the Czechoslovak system (similar to that in other Communist countries at the time), and from the 1950s onwards it became part of the school system. It had a clear schooling function, was centrally managed and inspected, and had a detailed national curriculum. The German system, with its highly autonomous ECEC, and the Slovak system, with its centralized governance, began approaching the political challenges of ECEC global policy from quite different starting positions. As one might expect, the German response to the globalizing pressures to reform ECEC was to adopt a strategy of change, while it was enough for Slovakia to adopt the easier strategy of convergence.

The main, but not the only, defining characteristic of the globalization of ECEC is the rise of schoolification, accompanied by the standardization of education, the creation of rigorous national preschool education programmes and a focus on performance in the form of competencies, as noted earlier. This process of schoolification mainly occurs because global policy for early years education is currently focused on success during the later stages of schooling, with the aim of producing a qualified workforce for a well-performing economy. When looking at the effect preschool education has, special emphasis is placed on the link between good performance in the international PISA tests and subsequent pupil successes in these. The monitoring of pupil performance in the areas covered by the PISA tests leads to the educational component of ECEC being strengthened and linked closely to school education, and so ECEC, like school education, begins to focus more on specific competencies in literacy, mathematics and science and technology. This education content is considered to be crucial to building the human capital for the knowledge economy and is stressed by all the global players shaping international education policy.

At the beginning of the millennium, Germany began its own turn in ECEC as part of the wider European one (Fangmayer & Kaščák, 2016). It began moving away from its traditions and pursuing the schoolification of ECEC, strengthening the educational function as well as other components. The strong tradition of an autonomous preschool tradition in the federal states, or *Länder*, is weakening, while the preschool sector is becoming more homogenous through the adoption of a centralized ECEC policy which defines preschool institutions purely in terms of education institutions, where the goal is for children to acquire skills defined in terms of competencies. The curricula in the *Länder* are being harmonized and have undergone fundamental change. Much of the emphasis is on developing language, natural science and technical competencies. Teachers are also becoming interested in these areas, and parents expect preschool education to be more obviously linked to children's subsequent education pathways. As Mierendorf (2014) has stated, the consequences of this turn are a reduced family focus in early childhood and greater homogeneity.

In the last couple of decades, therefore, German ECEC has undergone a significant shift. The move towards homogeneity has supressed the autonomy within the ECEC sector, while its education goals have led to other traditional formative processes being foregrounded through the schoolification of ECEC via global education policy goals. Within this national ECEC policy and practice, Germany has opted for a strategy of change, in which its long tradition is replaced by neoliberal approaches. Nonetheless, it has retained its traditional social welfare mission and autonomy, since Germany is one of a number of European countries to have resisted the pressure to introduce compulsory preschooling.

Slovakia also has a long and rich tradition of preschool education. In its modern form, it was shaped within the confines of Communist ideology that informed education policy, as was the case in other Central and Eastern European countries. That meant that preschools were strongly promoted and incorporated into the education system, and, as part of the overall centralized organization of society, they were run by means of a central and detailed state preschool education programme (that included early years education institutions for the under threes). Although preschools were independent units within the education system, and the preschool state programme did not follow the logic of the school ones, the fact that they were part of the education system meant there was a natural link between preschool education and primary and secondary school education, and that there had long been a schoolification tendency. The arrival of the new discourse and economically motivated global education policy did not therefore require drastic change at either the national policy level or within the preschools themselves.

The neoliberal elements characterizing international ECEC policy have, of course, had a visible influence in Slovakia. The relative autonomy of the preschool education programme has now become a thing of the past. Since 2008, preschool teaching guidelines have been drafted around the same principles as those contained within the guidelines used in the rest of the education sector. When the new preschool curriculum was first drawn up, it explicitly detailed the competencies children must acquire in order to complete preschool (in the most recent revision, preschool education goals are no longer set out as competencies; instead, they are implicitly contained within the planned content). Since 2008, this national preschool education programme has been based on education standards, and these have been bolstered so as to correspond to the education areas favoured internationally. Although, the new central national curriculum has the overriding authority required in the neoliberal governance of ECEC, the ECEC institutions are responsible for creating their own teaching programmes that reflect the children's specific needs and opportunities. This is supposed to indicate greater local autonomy.

These global ECEC requirements and optics have been gradually and calmly absorbed into the preschool institutions, via Slovakia's education macropolicy. Generations of Slovak teachers have been taught to work within the demands of the central curriculum and have viewed their mission as contributing to the education of children. Thus, they consider the homogenizing ECEC sector to be a standard development and would be likely to resist any move towards diversification. They see no risk in preschool education becoming more similar to school education. Indeed, it has led to a strengthening of their professional teacher identity. In the same way, further emphasis on education areas such as language, mathematics and science and technology provides them with greater motivation to develop professionally. Even the bureaucracy associated with this, in the form of the work the teaching teams do in relation to creating and developing their own teaching programmes, has become core to their professional identity (Pupala, Kaščák, & Tesar, 2016).

This description shows that the pressures global policy has exerted on ECEC in Slovakia have not led to any significant upheaval, nor have they required any radical strategy change. In Slovakia, we have seen a strategy of convergence emerge, between tradition and the new demands, with the tradition already partly reflecting what is required and articulated by the new demands. It is also indicative of the fact that ECEC policy in former Communist countries is in some respects similar to the globally promoted outcomes of international ECEC macropolicy.

A similar situation to that in Slovakia (albeit with certain differences) can also be seen in neighbouring post-Communist countries with very similar ECEC traditions and roots. A new phenomenon associated with supranational ECEC policy in these countries is the shift towards a compulsory preschool year before entry into primary school. In Slovakia, preschool education is still not compulsory; however, it is probably merely a question of time.

5 ABSORPTION, CO-EXISTENCE, AND SEEKING A PATHWAY

Once we begin exploring the countries of the Global South, where the culture is different and the education systems and ECEC have followed different development trajectories, we find that the global transfer of preschool policy to the local setting has involved different pathways and strategies. The countries of the Global South are no more isolated or protected from these pressures than the countries of the Global North. Indeed, global ECEC policy is promoted here even more strongly, as the ideas and funding behind key ECEC development initiatives are realized through global players such as the World Bank. This influence is strongly felt in Indonesia and Kenya, while Nepal has retained more leeway, partly because it is of less interest to the West, having been more isolated historically and never colonized.

We shall deal with Kenya first. It has a longer tradition (relatively speaking) of ECEC provision than either Indonesia or Nepal. Almost immediately after gaining independence (in 1963), Kenya began promoting early childhood care, emphasizing the emancipatory ideal of making local settings more Kenyan and the important role of education in this process. Local initiatives to develop ECEC were also boosted by the national philosophy of *Harambee* (lit.

all pull together), which was about being able to rely on social solidarity and local community values and which had the potential to shape ECEC according to the renewed national and local principles. However, the dynamic development of ECEC was also strongly promoted by powerful international players, especially the World Bank. It strengthened the ECEC infrastructure, and Kenya became the first African country to gain a strong position in developing this sector. For example, with the help of the World Bank, it was able to build an institutional ECEC network through its National Centre for Early Childhood Education (NACECE) and regional District Centres for Early Childhood Education and Care (DICECE), and draft and implement the first national preschool education curriculum on the African continent. Kenyatta University in Nairobi launched the first doctoral programme in Early Childhood Studies with the help of the World Bank (Mbugua, 2004).

The funding and ideas from international organizations did, however, pave the way for Kenyan ECEC to open up to global visions of education, including preschooling. As we have shown in our research on preschooling in Kenya (Mbugua et al., 2018), education policy fully espouses a vision of education as a means of developing "human capital" for an economy with Western levels of affluence. This can be seen in recent education strategy, such as the *National Curriculum Policy* (2015), which closely follows the idea of harnessing education for economic goals, based on STEM subjects and developing competencies expressed as education standards.

Global visions for developing ECEC have not simply remained at the rhetorical level of national education macropolicy. Through the hierarchical governance of ECEC at the central and regional levels, these challenges find their way into ECEC practice in the regions and to preschool teachers. The homogeneity in ideas on ECEC provision in the very distinct regions of Kenya have led to discomfort among ECEC regional representatives and teachers (as we discovered when collaborating with regional ECEC administrators and preschool heads). Fulfilling these visions is a central imperative that preschools must follow. The regional aspects, values and variations of ECEC have weakened, and preschools have had to adapt. The means and strategy for transferring global ideas through national policies to local ECEC structures therefore tends to occur through a process of the local absorbing the global.

The macro level of national ECEC policy we described in relation to Kenya can also be found in Indonesia, although it has a much shorter tradition there. The two countries are similar in that both have had to deal with their colonial pasts, shape their postcolonial identities and have ECEC sectors supported through the World Bank. In their detailed examination of national policy documents in Indonesia, Formen and Nutall (2014) show that the "human capital" discourse, characteristic of the uniform global education imperative, has been the main discourse in Indonesian education policy since its national ECEC sector began developing at the beginning of the millennium. Indonesian education macropolicy is eloquently captured in the highly symbolic metaphor of the Golden Generation of 2045, which envisages Indonesia as having become an economically developed democratic country by the time it celebrates the centennial of its independence. Education at all levels is key to shaping and preparing this "golden generation", a generation of economically productive human capital to be shaped primarily through investment in human resources (Jokowi, 2018). The *Golden Generation* metaphor is a variation of the Golden Collars, the latest generation of elite workers of the knowledge economy, an idealistic vision that is the product of the neoliberal concept of contemporary global education policy (Kaščák & Pupala, 2012). The Indonesian national ECEC curriculum incorporates this *Golden Generation*, found in the emphasis on developing key competencies in pupils across the whole education system, and these global ideas are also felt in the focus on training future teachers. Even the traditional Indonesian pedagogical notion of *developing character* is being harnessed to the idea of training a new golden generation (Rokhman et al., 2014).

Although global education ideas have already become a firm part of contemporary Indonesian education macropolicy, those promoting the economizing of education tend to remain at the rhetorical level of macropolicy and are not transferred to local ECEC practice (Formen, 2017). We found evidence of this on our visits to preschools in Indonesia, which enabled us to identify the *parallel worlds strategy*. It reflects a situation where, on one hand, there is the declarative side of education policy with its own discourse of global ideas and theories and, on the other, ECEC practice continues to pursue its own life within the local forms of ECEC rooted in the sharing of culturally embedded values and ideas as to its mission and the nature of child care in preschool institutions. A nursery school banner setting out its vision and mission illustrates this (from the author's own photograph archive and translated into English from the original Indonesian):

Nursery school mission:

To ensure that the children have a balanced halal diet that provides stamina.
To develop the children's potential in relation to their interests.
To foster their faith and devotion to Allah SWT.
To proffer good examples based on Islamic principles of teaching.
To provide a stimulating environment for the growth and development of children in the family, at school and throughout.

There is nothing in this nursery school mission to suggest its activities are devoted to conveying and pursuing the macropolicy goals of ECEC, nor that there is any attempt to directly transfer them into its activities. These ideas (which of course are not necessarily a direct reflection of the reality of education practice in the school) reflect traditional elements of Indonesian culture that are largely found within Islam. Examples of the kind of support found in these principles also emerged during our discussions with preschool heads and teachers. They also emphasized that parental and local community expectations were in line with these nursery school missions, and we found at least one case where this was also true of the local authorities responsible for inspecting ECEC institutions. We also found that, with the parallel worlds strategy, actors operating under local conditions were able to actively filter the transfer of universal international ideas on the nature of ECEC despite macropolicy being saturated with them.

The remaining strategy is that of *mission seeking*, and it was in evidence in the last of the countries investigated: Nepal. Nepal is somewhat on the backburner when it comes to globalizing international pressures, partly because it has historically been isolated and has no colonial past and partly because it has a relatively short history of general public education (which began in the 1950s). Its ECEC has no long tradition behind it and faces great challenges. Foreign ECEC models are therefore seen more as potential models than as imperatives.

During our first encounter with ECEC in Nepal (in urban areas such as Kathmandu) we ascertained that its education market is informed by traditional liberal ideas rather than neoliberal ones, with its education and commercial branches being based on market principles. Demands regarding ECEC come from the "consumers", the parents, rather than the state. ECEC is a private service for the rich and is therefore only available to a limited extent. ECEC provision is very much in the public eye, with adverts for education services ever present on building facades and in the streets. Private investment in education represents an individual investment for the future and for the promise of a better life. However, the kind of private adverts we saw in Nepal's urban areas are not typical of the whole country; but, they are of a type that can be found all over the world.

To this, we can add that parental demands (from the rich) often correspond to the neoliberal (global) vision of education that is about training and educating quality "human capital". This is desired by both state and parents alike. However, for the parents it is more a question of "individual human capital" and the personal successes of their children, whose future will not necessarily be linked to the economic success of the country, especially given it is a country in the Global South. Indeed, their children's successes are more likely to occur outside the country and in an environment which already offers economic and personal prosperity (hence the frequent preference for English language preschools). The state is keen to produce "human capital"; however, it does not portray it as an individual value but as a universal outcome for all those entering the education system.

There is no obligation for private preschools organized around the supply and demand principle to be inspected by local specialists; indeed, they often consider the quality of these preschools to be unsatisfactory (Upreti, 2012). These preschools are truly autonomous and operate on a self-regulating basis, which can be advantageous in some cases, but risky in others.

In recent decades, Nepal has portrayed an increasing interest in ECEC and has adopted measures to expand this sector across the country (preschool classes in primary schools and ECEC community centres). These initiatives often lack funding and parental interest and have underqualified staff. The preschool education section of the Ministry of Education and Sport supports the expansion of ECEC as do the local District Child Development Boards (DCDB). Preschools are not part of the general education system and the framework curriculum has no real authority. The preschools or classes operate more on the basis of local teacher interventions and their ideas about what children should do in preschool. The schooling approach (schoolification) that is frequently very visible in preschool classrooms in Nepal has not been adopted as part of global ECEC demands, but is a result of the firmly embedded idea that institutional early learning is about helping children acquire the 3Rs (reading, writing and arithmetic). This is an idea that may be held by teachers and that is especially associated with parents (Upreti, 2013).

Despite Nepal's very short history of ECEC provision and the limits on expanding it nationwide, progress has clearly been made and it has set out on its pathway. There is no conflict in Nepal between traditional and globalizing pressures; instead the country is seeking its own pathway. It is one that is limited, on the one hand, by the financial resources available to it and, on the other, by the models that it will wish to pursue. It will be interesting to follow developments in Nepal in the coming years.

6 CONCLUSION

We have described a number of strategies used in various countries in different parts of the world to transfer global ECEC policy to local conditions. Transference is achieved in many different ways, and the kinds of relationships between national ECEC macropolicies and the macropolicies found on the ground differ greatly, resulting in the hybridization of global and local policies (Nguyen et al., 2009; Paananen et al., 2015). The strategies we have identified need not simply apply to the countries we found them in; parallel or additional strategies may be found in a single country or indeed in multiple countries. During the transfer process, local players, especially teachers, have an important role to play. Although the macropolicies found in countries across the world are clearly influenced by the global neoliberal discourse, local ECEC contexts and traditions are able to resist excessive and uniform pressures and to maintain a level of differentiation in ECEC that reflects the variety of the contexts within which it occurs.

ACKNOWLEDGEMENT

This work was supported by research projects VEGA 2/0134/18, VEGA 1/0258/18 and KEGA 009TTU-4/2018

REFERENCES

Fangmayer, A. & Kaščák, O. (2016). Transgressionen: Zum diskursive rekonfigurieren von institutionalisierten (frühen) kindenheiten. Ein slowakisch-deitscher vergleich von curricularen Dokumenten. In Nentwig-Gesemann, I., Fröhlich-Gildhoff, K, Betz, T., Viernickel, S. (eds.). *Frühpädagogik*, *9*, 141-172. Freiburg: FEL.

Formen, A. & Nuttal, J. (2014). Tensions between discourses of development, religion, and human capital in early childhood education policy texts: The case of Indonesia. *International Journal of Early Childhood*, *46*(1), 15-31.

Formen, A. (2017). In human-capital we trust, on developmentalism we act: The case of Indonesian early childhood education policy. In M. Li, J. Fox, S. Grieshaber (eds.) *Contemporary Issues and Challenge*

in Early Childhood Education in the Asia-Pacific Region. New Frontiers of Educational Research. Singapore: Springer.
Gupta, A. (2105). Pedagogy of third space: A multidimensional childhood curriculum. *Policy Future in Education, 13*(2), 260-272.
Jokowi. Improving Infrastructure for Indonesia Golden 2045 (2018). *Tempo.co*, Thursday, 24 May, 2018. Retrieved from http://en.tempo.co/read/news/2018/05/24/055918734/Jokowi-Improving-Infrastructure-for-Indonesia-Golden-2045.
Kaščák, O. & Pupala, B. (2012). *Škola zlatých golierov. Vzdelávanie v ére neoliberalizmu [Golden Collar School: Education in the neoliberal era]*. Praha: Sociologické nakladatelství (SLON), 208.
Kaščák, O. & Pupala, B. (2018). From South to North in the globalised world of ECEC: Varied national and local responses in selected countries. *Keynote paper presented at the 70th OMEP World Assembly and Conference*. Prague, 25–29 June 2018.
Mbugua, T. (2004). Early childhood care and education in Kenya. *Childhood Education, 80*(4), 191-197.
Mbugua, T., Pupala, B., Kascak, O., & Petrova, Z. (2018). Transforming public education in Kenya: Early childhood development and education in Kwale County. In C.S. Sunal, K. Mutual (eds.) *Transforming public education in Africa, the Caribbean, and the Middle East*. Information Age Publishing (in press).
Mierendorff, J. (2014). Annäherungen von kindergarten und schule. Wandel früher Kindheit? In Cloos, P., Hauenschild, K., Pieper, I., Baader, M. (eds.): *Elementar- und primarpädagogik. internationale diskurse im spannungsfeld von institutionen und ausbildungskonzepten* (23-37). Wiesbaden: Springer.
Moss, P., Dahlberg, G., Grieshaber, S., Mantovani, S., May, H., Pence, A., ... Vandenbroeck, M. (2016). The organisation for economic co-operation and development's international early learning study: Opening for debate and contestation. *Contemporary Issues in Early Childhood, 17*(3), 343-351.
National Curriculum Policy. (2015). Nairobi: Ministry of Education, Science and Technology.
Nguyen, P. M., Elliott, J. G., Terlouw, C., & Pilot, A. (2009). Neocolonialism in education: Cooperative learning in an Asian context. *Comparative Education, 45*(1), 109–130.
OECD. (2017). *Early learning matters*. OECD Publishing.
Paananen, M., Lipponen, L., Kampulainen, K. (2015). Hybridisation or ousterisation? The case of local accountability policy in Finnish early childhood education. *European Educational Research Journal, 14* (5), 395-417.
Pupala, B., Kaščák, O., Tesar, M. (2016). Learning how to do up buttons: Professionalism, teacher identity and bureaucratic subjectivities in early years settings. *Policy Futures in Education, 14*(6), 655–665.
Rokhman, F., Hum, M., Syaifudin, A., & Yuliati (2014). Character education for golden generation 2045 (National Character Building for Indonesian Golden Years). *Procedia – Social and Behavioral Sciences, 141*, 1161-1165.
Urban, M., Swadener, B. B. (2016). Democratic accountability and contextualised systemic evaluation. A comment on the OECD initiative to launch an International Early Learning Study (IELS). *International Critical Childhood Policy Studies, 5*(1), 6-18.
Upreti, N. (2012). *Preschool education in Nepal*. Child Research Net. Section: Asia, ECEC around the World, Reports from around the World. Issue Date: August 3, 2012. Retrieved from https://www.childresearch.net/projects/ecec/2012_03.html
Upreti, N. (2013). *Nepal – National plan and policies for Early Childhood Education and Care in Nepal*. Child Research Net. Section: Asia, ECEC around the World, Reports from around the World. Issue Date: March 8, 2013. Retrieved from https://www.childresearch.net/projects/ecec/2013_05.html

Early Childhood Education in the 21st Century – Yulindrasari et al. (eds)
 ISBN 978-1-138-35203-2

Beyond an economic model: Bringing social justice into early childhood education

Vina Adriany
Universitas Pendidikan Indonesia, Bandung, Indonesia

ABSTRACT: Practices of Early Childhood Education (ECE) in Indonesia is highly influenced by the economic model that perceives education as a form of investment. This model links with a human capital discourse that situates ECE as a vehicle to achieve a country's economic growth. The purpose of this chapter is to problematise the model by arguing that the model has overlooked more critical issues in ECE such as social justice issues. It is expected that this chapter would serve as an invitation to rethink the current model for ECE in Indonesia.

1 INTRODUCTION

For the past 20 years, the Indonesia government has paid more attention to Early Childhood Education (ECE) (Adriany & Saefullah, 2015). The awareness was driven to some extent by some foreign forces such as the intervention of the World Bank (Formen, 2017) and the global commitment to improve ECE such as depicted in the Sustainable Development Goals (SDGs) (Indonesia Ministry of National Development Planning (BAPPENAS) and the United Nations Children's Fund (UNICEF), 2017). This has led to several developments in the field of ECE such as the push for children's access to ECE, particularly for children in the rural area (Hasan, Hyson, & Chang, 2013).

Despite the progress, however, the government's model of ECE development is still very much confined to an economic model. The economic model of ECE is highly influenced by the human capital discourse which perceives ECE as a site to improve a country's economic development (Peach & Lightfoot, 2015). This discourse is based on theories from Heckman and Masterov (2007), who argue that every cent spent in ECE will bring a higher return in the future.

This chapter aims to problematise further the complexity and contradiction of the economic models. It seeks to see within that model, what discourse is privileged and what discourse is marginalised. The findings of this chapter will unpack dominant power relations in ECE in Indonesia. This chapter would also argue that by focusing on the economic models, other vital issues such as social justice issues are overlooked in ECE. It is expected that these findings will serve as an invitation to rethink and reconsider the current model of ECE in Indonesia.

2 THE ECONOMIC MODEL OF ECE

The economic model of ECE is very pervasive, not only in Indonesia but all over the globe (Lee, 2018; Penn, 2002). Beneath this model lies an assumption that sees education merely as a form of investment. This model is so powerful that using Foucault's concept of power and truth; it has become the regime of truth (Foucault, 1980). The word *truth* here does not equate with something which is true but, instead, illuminates something embodying more power. It becomes so powerful that one is rarely critically examining its practices.

The model is disseminated through another discourse. At the moment in Indonesia, there is one dominant discourse in ECE, that is, the golden generation discourse. The golden generation discourse is derived from an idea that by 2030, Indonesia will reach a demographic bonus. Indonesia's population is estimated to reach 297 population with 64% of it are a productive population, ages between 15 to 64 years old (BAPPENAS, 2017). A vast segment of the population is seen as having the potential to advance the country's economic model.

3 PROBLEMATISING THE ECONOMIC MODEL

This model as mentioned before is so highly prevalent that one rarely questions its problems. Informed by critical approaches in ECE (Burman, 2008; MacNaughton, 2005; Walkerdine, 1998), this chapter seeks to problematise the economic model of ECE.

The first problem associated with the model lies in the fact that within the model children are seen as a mean rather than an end. Because it is very central in the model that ECE links with the country's future economic development, children hence is perceived as a vehicle to achieve this goal. Children are constructed merely as an earner (Kaščák & Pupala, 2011). The government money is invested because economic benefits will be reaped in the future.

The second problem with the economic model is because the model is built on the neo-liberal and neo-colonial assumption of children and childhood. The neo-liberalism is an extension of the economic model that situates children as a rational individual (Dýrfjörð, 2012). An individual's success or failure is attributed to her internal factors rather than taking into account a more substantial sociological factor. This also means shifting the accountability from a government to an individual (Harvey, 2007). In Indonesia, this is evident as many high profiles ECE institutions are run by private organisations. While for middle-upper class parents, this means more opportunity to select the best education for their children, it leads at the same time to disadvantages for lower-class parents. The neo-colonialism postulation is also very strong within the economic model. For the neo-liberal agenda to be perpetuated, it is often disseminated through theories on children development (Burman, 2008). Many have criticised the approach as being culturally bias, as it is built on the Western and middle-class assumption of children and childhood (Penn, 2002, 2011).

Because of this neo-liberal and neo-colonial assumption, very often children, especially from the global North, are considered to be somehow different because they do not conform to the standard set by the theories of children development. In research conducted by the World Bank (2012), it was stated that Indonesian children in the villages have a slight delay in development. What is missing from the discussion is the fact that the World Bank has used an internationally validated instrument that probably is culturally insensitive toward some Indonesian children (Adriany & Saefullah, 2015). Hence, as Gupta (2006) and Tesar (2015) argue, children in the South continue to be thought of as different from children in the North.

The fourth problem with the economic model is that it may potentially overlook more important issues in ECE such as issues of social justice. As Lee (2018) argues the imaginary economic model of ECE is made to be visible because it makes social justice issues becomes invisible. Research conducted by Solehuddin and Adriany (2017) demonstrated how social justice issues are rarely discussed in an ECE setting in Indonesia.

Up to this point, this chapter has attempted to unpack problems associated with the economic model of ECE. In the following section, this chapter will elaborate on why we need to focus on social justice issues.

4 BRINGING SOCIAL JUSTICE TO ECE: AN ALTERNATIVE TO THE ECONOMIC MODEL

As mentioned earlier, the dialogue of social justice in ECE, particularly in Indonesia, remains scarce even though there are many problems related to social justice in ECE. Many people, including parents and teachers, often assume ECE as an innocent place where children would play and learn (Adriany, 2013). Usually, adults do this because they believe that young children cannot grasp complex issues and, hence, they want to protect children (Arnot, 2009). While, by doing these, they have silenced and depoliticised children (Martinez, 1998; Yelland, 1998).

There are various reasons why ECE should become a site to discuss social issues. In this chapter, I would like to mention only a few of them. The first reason is the fact that ECE has often become an institution that promotes racism, sexism, and intolerance (Arnot, 2009; Blaise, 2013; Formen & Nuttall, 2014). As Francis and Mills (2011) assert, school often becomes a vehicle that continues promoting social inequalities. This without any doubt contradicts the aim of education, which is to demolish all forms of racism, prejudice and hegemonic thought and behaviour (al-Hussein, 2000; Hollinsworth, 2006; Siraj-Blatchford, 2006; Calma, 2007; Lynn, 2007; Cannella, 1997; Sapon-Shevin, 2003).

The second reason why we need to talk about social justice in ECE is that up until now, ECE has prolonged the legacy of colonialisation. As elaborated earlier, neo-colonisation is apparent in the practices of ECE in Global South countries (Gupta, 2006). The legacy of neo-colonisation can be found in theories used in ECE as well as the fact that in a country like Indonesia, the development of ECE cannot be detached from the foreign intervention that comes with the neoliberal agenda (Gupta, 2018). This neo-colonialism often has led to a sense of alienation among ECE teachers because they feel like they have to teach something which is culturally irrelevant (Henry, 1996). Thus, this would disrupt learning as Gupta (2006) claims that ECCE teaching and practices which are culturally sensitive, would be likely to have a more significant impact.

The third reason to bring social justice into ECE is that almost everywhere, ECE is considered to be situated as one of the lowest ranks of the profession and, hence, the teachers often receive a small salary (Osgood, 2006). Because most of the ECE teachers are women, therefore it also becomes a gender issue (Newberry & Marpinjun, 2018; Yulindrasari, 2014).

5 CONCLUSION

This chapter intends to critically anylise the economic model within ECE as one of the most hegemonic models in Indonesia. This argues that the economic model has marginalised more critical issues such as social justice. Discussing social justice would allow ECE to become a fairer place not only for the children but also for the teachers.

For ECE to be able to convey social justice into the practice, teachers must first be equipped with an understanding of its importance (Christman, 2010). This process should be commenced since the teachers enrol in the teacher training institute. Hence, this chapter calls for a reconceptualising alternative model of ECE, but this is also an invitation to redefine the content of the curriculum in teacher training institutes.

REFERENCES

Adriany, V. (2013). *Gendered power relations within child-centred discourse: an ethnographic study in a Kindergarten in Bandung, Indonesia*. (PhD thesis), Lancaster University.

Adriany, V., & Saefullah, K. (2015). Deconstructing human capital discourse in early childhood education in Indonesia. In T. Lightfoot-Rueda, R. L. Peach, & N. Leask (Eds.), *Global Perspectives on Human Capital in Early Childhood Education: Reconceptualizing Theory, Policy, and Practice* (pp. 159–179). New York: Palgrave Macmillan US.

Arnot, M. (2009). *Educating the gendered citizen: sociological engagements with national and global political agendas*. London: Routledge.

Bappenas. (2017). Siaran pers: bonus demografi 2030–2040: strategi Indonesia terkait ketenagakerjaan dan pendidikan [Press conference: demographic bonus 2030–2040: Indonesian strategie relted to employment and education]. Retrieved from https://www.bappenas.go.id/files/9215/0397/6050/Siaran_Pers_Peer_Learning_and_Knowledge_Sharing_Workshop.pdf.

Blaise, M. (2013). Gender discourse and play. In E. Brooker, M. Blaise, & S. Edwards (Eds.), *SAGE Handbook of Play and Learning in Early Childhood* (pp. 115–127). London: SAGE.

Burman, E. (2008). *Deconstructing Developmental Psychology*. East Sussex: Routledge.
Cannella, G. S. (1997). *Deconstructing early childhood education: Social justice and revolution*. New York: Peter Lang.
Christman, D. E. (2010). Creating social justice in early childhood education: a case study in equity and context. *Journal of Research on Leadership Education, 5*(3.4), 107–137.
Dýrfjörð, K. (2012). *Infected with neo-liberalism: The new landscape of early childhood settings in Iceland.* Paper presented at the Creating Communities: Local, National and Global Selected papers from the fourteenth Conference of the Children's Identity and Citizenship in Europe Academic Network, London.
Formen, A. (2017). In human-capital we trust, on developmentalism we act: The case of indonesian early childhood education policy. In M. Li, J. Fox, & S. Grieshaber (Eds.), *Contemporary Issues and Challenge in Early Childhood Education in the Asia-Pacific Region* (pp. 125–142). Singapore: Springer Singapore.
Formen, A., & Nuttall, J. (2014). Tension between discourses of development, religion, and human capital in early childhood education policy text: the case of Indonesia. *International Journal of Early Childhood, 46*, 15–31.
Foucault, M. (1980). *Michael Foucault: power knowledge*. Hemel Hempstead, Hertfordshire: Harvester Wheatsheaf.
Francis, B., & Mills, M. (2011). Schools as damaging organisations—instigating a dialogue concerning alternative models of schooling. *Pedagogy, Culture & Society*.
Gupta, A. (2006). *Early childhood education, postcolonial theory, and teaching practices in India: balancing Vygotsky and the Veda*. New York: Palgrave Macmillan US.
Gupta, A. (2018). How neoliberal globalization is shaping early childhood education policies in India, China, Singapore, Sri Lanka and the Maldives. *Policy Futures in Education, 16*(1), 11–28.
Harvey, D. (2007). *A brief history of neoliberalism*. New York: Oxford University Press.
Hasan, A., Hyson, M., & Chang, M. C. (Eds.). (2013). *Early childhood education and development in poor villages of Indonesia: strong foundations, later success*. Washington, DC: International Bank for Reconstruction and Development/The World Bank.
Heckman, J. J., & Masterov, D. V. (2007). The productivity argument for investing in young children*. *Applied Economic Perspectives and Policy, 29*(3), 446–493. doi:10.1111/j.1467-9353.2007.00359.x
Henry, A. (1996). Five Black Women Teachers critique child-centered pedagogy- possibilities and limitations of oppositional standpoints. *Curriculum Inquiry, 26*(4), 363–384.
Indonesia Ministry of National Development Planning (BAPPENAS) and the United Nations Children's Fund (UNICEF). (2017). *SDG baseline report on children in Indonesia*. Retrieved from https://www.unicef.org/indonesia/SDG_Baseline_Report_on_Children_in_Indonesia(1).pdf.
Kaščák, O., & Pupala, B. (2011). Governmentality-neoliberalism-education: The risk perspective. *Journal of Pedagogy/Pedagogický casopis, 2*(2), 145–158.
Lee, I.-F. (2018). (Re) Landscaping early childhood education in East Asia: A neoliberal economic and political imaginary. *Policy Futures in Education, 16*(1), 53–65.
MacNaughton, G. (2005). *Doing Foucault in early childhood studies*. New York: Routledge.
Martinez, L. (1998). Gender equity policies and early childhood education. In N. Yelland (Ed.), *Gender in Early Childhood*. New York: Routledge.
Newberry, J., & Marpinjun, S. (2018). Payment in heaven: Can early childhood education policies help women too? *Policy Futures in Education, 16*(1), 29–42. doi:10.1177/1478210317739467.
Osgood, J. (2006). Deconstructing professionalism in early childhood education: resisting the regulatory gaze. *Contemporary Issues in Early Childhood, 7*(1), 5–14.
Peach, R., & Lightfoot, T. (2015). *Global perspectives on human capital in early childhood education: reconceptualizing theory, policy and practice*. New York: Palgrave Macmillan.
Penn, H. (2002). The world bank's view of early childhood. *Childhood, 9*(1), 118–132.
Penn, H. (2011). Travelling policies and global buzzwords: How international non-governmental organizatuon and charities spread the word about early childhood in the global south. *Childhood, 18*(1), 94–113.
Sapon-Shevin, M. (2003). Inclusion: A matter of social justice. *Educational Leadership, 61*(2), 25–29.
Solehuddin, M., & Adriany, V. (2017). Kindergarten teachers' understanding on social justice: stories from indonesia. *SAGE Open, 7*(4), 2158244017739340. doi:10.1177/2158244017739340.
Tesar, M. (2015). Te Whāriki in Aotearoa New Zealand: Witnessing and resisting neoliberal and neocolonial discourses in early childhood education. In V. Pacini-Ketchabaw & A. Taylor (Eds.), *Unsettling the Colonial Places and Spaces of Early Childhood Education* (pp. 98–113). New York: Routledge.
Walkerdine, V. (1998). Developmental psychology and the child-centred pedagogy: The insertion of Piaget into early education. In J. Henriques, W. Hollway, C. Urwin, C. Venn, & V. Walkerdine (Eds.), *Changing the subject: psychology, social regulation, and subjectivity* (pp. 153–202). London: Routledge.

World Bank. (2012). The Indonesia ECED project findings and policy recommendations. Washington, DC: World Bank. Retrieved from http://documents.worldbank.org/curated/en/126271468260084707/Proyek-Pendidikan-dan-Pengembangan-Anak-Usia-Dini-PPAUD-Indonesia-temuan-dan-rekomendasi-kebijakan.
Yelland, N. (1998). *Gender in early childhood.* London: Routledge.
Yulindrasari, H. (2014, 18–19 November). *Neoliberal early childhood education policy and women's volunteerism.* Paper presented at the Negotiating Practices of Early Childhood Education, Universitas Pendidikan Indonesia Bandung.

Early Childhood Education in the 21st Century – Yulindrasari et al. (eds)
© 2020 Taylor & Francis Group, London, ISBN 978-1-138-35203-2

Tracing the silence? An essay on the World Bank and discourses on women in the early childhood education and development projects

Y. Pangastuti
University of Auckland, Auckland, New Zealand

ABSTRACT: This chapter analyses a depiction of women as mothers and ECE workers based on two reports published by the World Bank's Early Childhood Education and Development (ECED) projects. Using Michel Foucault's discourse and Lynn Thiesmayer's silence, I argue that women are referenced frequently, but they are eclipsed behind discourses on children.

1 THE ECED PROJECTS

The massive expansion of Early Childhood Education (ECE) in Indonesia is partially facilitated by the works of many international development agencies, such as the World Bank (WB), UNICEF, UNESCO, Plan International, Save the Children, Child Funds, and some other bilaterally funded projects. This essay uses the WB's Early Childhood Education and Development (ECED) Projects as a sample. The decision is based on some considerations: First, ECED Projects are the largest externally funded early childhood education project, not only in Indonesia but also in the world. Through these projects, Indonesia becomes a significant 'comparable' case. Second, With Indonesia sits as one of its highest loan recipients (World Bank, 2017), the WB's capacity in steering policy agenda in the country should never be underestimated. The long connections between the WB's economists and Indonesia's of influential think tanks and policy elite circles enable circulation of ideas and thinking so great that it becomes embedded in the policy-making process (Hadiz & Robison, 2005). In education, Phillips Johnson and David Coleman (2005) dub the organisation as 'the strongest player in the world of multilateral education' (Jones & Coleman, 2005, p. 94).

The ECED projects are divided into two stages: the first stage is from 1997 to 2006 with the final funds of $14.85 million (IEG World Bank Group, 2007). This project was deployed to support the development of integrated early child framework in Indonesia (World Bank, 1998, 2003). It established new ECE centres, teacher recruitment and training on an integrated package of services. In order to reach poor communities, voluntary village health post (*posyandu*) and monthly health and nutrition service under the Family Planning's Parent-Child Program (or 'Bina Keluarga Balita', BKB) were integrated to the overall early learning scheme. Although the WB kept a low-profile claim about this project, it marked the warming up that lead to the fast-paced ECE expansion. A year after the project finished in 2005, the WB with the government of the Netherlands launched the most massive ECE project in the world. With total funds of $107.2 millions (IEG World Bank Group, 2014), the project used most of the money ($93 millions) to provide block grants to 5,990 poor villages with community-driven mechanism while the rest was used to finance project administration and evaluation, development of national standards and restructuring the education department at the national, provincial and regency level. The second stage project finished in 2013.

This chapter focuses only on two WB's publications related to the ECED Project: 'Early Childhood Education and Development in Indonesia: An Investment for A Better Life' (Chang et al., 2006), and 'Early Childhood Education and Development in Poor Villages of Indonesia: Strong Foundations, Later Success' (Hasan, Hyson & Chang, 2013). The first report was published in the last year of the first phase of the ECED project (1997–2006), in which 600 new ECED centres were constructed in Bali, Banten, West Java and South Sulawesi (Chang et al., 2006, p. 23). This report's position is to raise awareness among policymakers about the importance and the possible forms of 'feasible' ECED for Indonesia. The second report is more reflective and evaluative, giving answers to the questions that have shaped the ECED project since the beginning. With the significantly larger funding that affected its scope of 3,000 villages from 50 districts, the second report is almost three times longer than the first. It elaborates 'the pieces of evidence' to pursue the telling of 'the story of Indonesia's efforts to change the trajectory of development for poor children' (Hasan, Hyson and Chang, 2013, p. 32). The two reports not only summarise the ECED projects but also act as a source of information linking global and local evidence on ECED practices. Through these reports, Indonesia becomes a 'legible' case (Vetterlein, 2012); hence, comparisons are made to be possible, not only by using countries from the Northern hemisphere as a point of reference but also by positioning Indonesia as a global comparator against other countries (Hasan, Hyson & Chang, 2013). For practicality and due to space limitations, in this chapter the first document is referred to as 'ECED 2006' while the second is referred to as 'ECED 2013.'

2 CONSTRUCTING THE NORMS

In this chapter, I use Foucault's theory (1980, 1989) on discourse as a theoretical framework. In Foucault's theory, discourses cannot be separated from disciplinary mechanisms: in constructing boundaries of knowledge as well as in practices. Systematisation, regulations, and specifications are discourses' most important elements (Foucault, 1989; McHoul & Grace, 2015). Discourses also absorb, rearticulate, reproduce and resist power through a net-like network (Foucault, 1980). Through the interactions with power, discourses perform their functions in enabling or delimiting statements, producing knowledge and practices which are perceived to be more superior than others, known as 'truth' (Henriques et al., 2005). For Foucault, the truth is produced through discursive processes to constitute what is perceived to be good or bad, true or false, and right or wrong. Hence, objective truth never exists. Instead, the truth is always subjectively and relationally produced through a normalisation process (Paternek, 1987). By employing this discourse theory, I situate the WB as always 'involved' in a project and projection of 'truth'. The organisation produces, re-produces and operates within a framework based on what it assumes to be 'good' or 'bad' In this chapter, attention is specifically given to a question: 'how do ECED projects, as productions of truth, impact women as mothers and as ECE workers?' Due to the limitation of space, this chapter would only provide some key highlights, without looking into too many details.

The ECED projects set the tone by outlining what is considered as 'the development essentials' – a list of what a child should have in his/her life. These essentials include an environment that enables good health and nutrition; warm and safe, responsive relationships with caring adults; opportunities to play; to hear and use language, to explore books, stories, and other literacy materials; to practice self-regulation and function skills (Hasan, Hyson & Chang, 2013, pp. 23–24). These conditions are not part of many poor children's lives. On the contrary, children from low-income families are saddled with in multidimensional disadvantages. These children have a higher probability of dying before, during and right after their births, and, if they survive, they have higher chances to suffer from malnutrition, stunting, and preventable diseases. The sources for these deficiencies are interlinked with the lack of care practices in the household, in the community, and in accessing institutional services. The 'damage' from the 'wrong' practices is described to be 'irreversible' and too costly because it reduces children's possibility of having optimum brain development which is critical to their later productivity.

Regardless of the value of the projects, the WB surprisingly does not establish a framework that consistently connects poverty and childcare. Based on my field experience, low-income families' child caring behaviour is shaped by their situations of being poor, not the other way around. Women from poor families have to work double shifts: to handle the domestic tasks and to earn money to support their families; both tasks have to be done with minimal resources, skills, and opportunity. However, instead of working at the root causes of poverty, the ECED projects are offered mainly through cost-and-benefit analysis (Chang et al., 2006, p. 87). The statement, in my opinion, makes the WB's claim over the ECE program and poverty remain unresolved. From the reporting style, especially apparent in ECED 2013, families are assessed against a set of indicators that reflect certain lifestyles, such as habits of feeding, taking children's playing outdoors, or reading books without carefully look at the interweaving cultural, social and political contexts. Through deployments of various tests, deficiencies are not only illuminated but are also presented *as* the poverty itself, while the connections between those deficiencies *and* poverty as economic deprivations are obscured. As a consequence, low-income families and their communities are not given with incentives or support; on the contrary, they are judged for not performing the prescribed practices. Despite their difficult situations, the families are also asked to stretch their limited resources, to pay the school fees, volunteering, even lending or donating their space.

So where are the women? At a glance, the ECED projects' focus on young children seems to be removed from discourses on women. I argue that women as mothers and ECE workers are greatly affected by the rise of ECE. This phenomenon could be easily confirmed through statistics, political rhetoric, and observation in any ECE contexts; yet, women's presence is so normalised that they become 'invisible'. Their presence is swallowed into the policy language of gender neutrality, including in the reports of ECED projects. The invisibility of women, I argue, could be theorised through silencing discourse. Silencing is most productive when 'another discourse is used to designate and enforce the area of silenced material and eventually to fill it in' (Thiesmeyer, 2003, pp. 1–2). As in the case of many ECE projects, women are eclipsed behind discourses that put children as the main focus. Consequently, women are 'defocused', placed in the back to inhibit the 'blurry backgrounds.'

Nevertheless, women's roles in children's development, whether we like it or not, in reality, are vital to the practices of childrearing and child caring. In the context of Indonesia, various research across times have shown this direct relationship, from anthropological perspectives (Geertz, 1961; Broch, 1990), psychology (Farver & Wimbarti, 1995), policy (Newberry, 2012), to the current preschool practices (Adriany, 2013; Yulindrasari, 2017). Within the ECED projects, parenting and centre-based interventions are the targeted interventions. In both spaces, women are still the dominant caregiving faces as mothers at home, and as teachers in the centres.

To analyse, I start with this maternalism concept. As an ideology, maternalism necessitates and naturalises women's affections to children of their own or of others, like being female teachers in schools, as part of raising good citizens (Koven & Michel, 1990; Ladd-Taylor, 1993). My inspiration initially comes from Hollway's (2001) maternal subjectivity; Hollway states that motherhood is nothing but 'an empty category where the children's needs can be placed' (Hollway, 2001, p. 1). While Hollway's focus is more on 'mothers', Acker (1995) shows the connection of motherhood with ECE teaching through the persistent maternal analogues that shape teaching cultures. At ECE levels, teachers are frequently equalised as 'mother substitutes' (Goldstein, 1994; Moss, 2006) who willingly provide 'quasi-maternal nurturance to compensate for the depraved environments of the poor' (Walkerdine, 1986, p. 57). Therefore, any ECE's projection directed towards the children would certainly implicate women as mothers and teachers. However, it is important to note that, although the two roles possess linguistic genealogical similarities, especially in Indonesian context, my intention is not to conflate mothers and teachers under the same subjectivity of 'being women'. Instead, my idea is more to highlight how both mothers and teachers are drawn into maternalistic projects to be the instruments of the projects. The remainder of this chapter includes some of my observations.

3 THE STORY OF UNPAID CARERS: MOTHERS VS. TEACHERS?

Women and mothers are instrumental to any ECE projects. From the two reports, I find two types of statements that could point to this: (1) direct statements that link mothers' status and the likelihood of the children's 'proper' development. Promoting women's education to reduce under-five children mortality rate (Hasan, Hyson & Chang, 2013, p. 7) and description on breastfeeding and snacking habits (Hasan, Hyson & Chang, 2013, p. 6) are part of this type of statements; and (2) indirect statements which requires a process of inference that connect illustration of poor children's as an *implicit* result of 'improper' home-based care. These statements cover descriptions about the high rate of stunting incidence, poor brain development, and lack of verbal skills. Operating under the assumption of Indonesia's maternalistic culture, these results could indirectly be interpreted as mothers' caring incapability. At first, this argument may look near-sighted; however, the reading has to interact with reality. Many development projects on health and nutrition, parenting and child protection are known to primarily target mothers, to 'improve' their capacities on caring knowledge and practices. Fathers and other caregivers are classed as secondary. Hence, instead of looking at women as part of wider imbalance social structures, these projects establish a casuistic relationship between the ill-practices of motherhood to children's growth as something that is direct and close.

In order to address all sorts of deficiencies, ECED projects are offered to normalise the situations. The leads for this change are the educators who are recruited from the communities and trained to be ECED teachers. These teachers are meant to do 'real' works: to interact with children through play, develop learning materials, scaffold children's active learning, do some home visits provide the right balance between child-centred versus teacher centre, and other non-classroom activities (Hasan, Hyson & Chang, 2013). In addition to the teachers, Children Development Workers (CDW) are recruited to supplement parenting activities (Hasan, Hyson & Chang, 2013). Unlike the teachers, CDW, based on my knowledge, is not an official part of Indonesia's early childhood system. The reports, unfortunately, do not sufficiently elaborate on how these workers could help in addressing the multidimensional problems of poor children's development as illustrated in the background of the projects. Hence, there is a gap between what is considered as the 'problem' and how the 'answers' are laid out in details.

Like mothers, ECE workers are also directly and indirectly referenced through statements. The paragraph above reflects some examples of direct statements. Some type of 'teachers' remains invisible throughout the texts. As many research suggests, Indonesia's ECED is highly dependent to voluntary works performed mainly by women as unpaid workers (Newberry, 2010, 2018; Van Ravens, 2010; Yulindrasari & Ujianti, 2018). First, based on my visits to some ECED centres, I understand that ECED projects and the government allocated funds for wages. However, in some centres that I visited, I was told that most centres had four teachers, although only two were officially registered as ECED teachers. Supposedly, the government covered the additional teachers' expenses; yet, unfortunately, that was not the case. Instead, the wages had to be split into four to cover the unregistered teachers. Therefore, teachers received an amount of money less than the budgeted and reported. Second, the reports recognised the presence of 'volunteers' (sometimes also referred as 'cadres' or *Kader*); it still acknowledges the disadvantage of being 'volunteer teachers' (p. 111). However, despite the recognition, the ECED projects *chose to ignore* the systematic connection between women voluntarism and ECE services that produce large numbers of unpaid teachers and be more concerned with the extension of their training experience.

4 CONCLUSION

I have described how women as mothers and ECE workers are directly and indirectly referenced and left out of the discourses on children. Regardless of the narrative that positions them as opposition, both mothers and ECE workers are unpaid workers. The ECED reports capture practices on ECE that, based on Elaine Unterhalter's (2012) argument, tends to work primarily with ideas based on lines' (Unterhalter, 2012, p. 253) instead of capturing the intersectional gendered, classed, and raced reality. In ECED, despite their critical roles, mothers and ECE workers are pushed to the backstage where they become the 'Invisibles'. With this attitude, education becomes a process that denies, subjugates and marginalised women, detached education from its fuction to empowerment and social justice capacities (Unterhalter, 2016). As a conclusion, the ECED projects can be categorised as 'gender-blind' projects, because they lack interest in 'how the supply of labour is produced' (Kabeer, 2005, p. 32).

REFERENCES

Acker, S. (1995). 'Carry on Caring: the work of women teachers'. *British Journal of Sociology of Education, 16*(1), 21–36. doi: 10.1080/0142569950160102.

Adriany, V. (2013). *Gendered Power Relations within Child-Centred Discourse: An Ethnographic Study in a Kindergarten in Bandung, Indonesia.* Doctor in Philosophy thesis, Lancaster University.

Broch, H. B. (1990) *Growing up agreeably: Bonerate childhood observed.* Honolulu: University of Hawaii Press.

Chang, M. C., Dunkelberg, E., Iskandar, S., Naudeau, S., Chan, D., Cibulskis, R., Namasivayam, P., Josodipoero, R., Soekatri, M. & Warouw, E. (2006). *Early Childhood Education and Development in Indonesia: An Investment for A Better Life.* Washington, DC: The world Bank.

Farver, J. A. M., & Wimbarti, S. (1995). Indonesian children's play with their mothers and older siblings. *Child Development, 66*(5), 1493. doi: 10.2307/1131659.

Foucault, M. (1980). *Power/knowledge: selected interviews and other writings 1972–1977.* New York: Vintage.

Foucault, M. (1989). *Archaeology of knowledge.* London: Routledge Classics.

Geertz, H. (1961). *The Javanese family: A study of kinship and socialization.* New York: Free Press of Glencoe.

Goldstein, L. S. (1994). 'What's Love Got to Do With It? Feminist Theory and Early Childhood Education', in paper presented at the annual meeting of the American Education Research Association. New Orleans, LA, 1–25.

Hadiz, V. R., & Robison, R. (2005). Neo-Liberal Reforms and Illiberal Consolidations: The Indonesian Paradox. *Journal of Development Studies, 41*(2), 220–241. doi: 10.1080/0022038042000309223.

Hasan, A., Hyson, M. C., & Chang, M. C. (2013).]*Early Childhood Education and Development in Poor Villages of Indonesia: Strong Foundations, Later Success, Directions in Development: Human Development.* Washington DC:The World Bank. doi: 10.1596/978-0-8213-9836-4.

Henriques, J., Hollway, W., Urwin, C., Venn, C., & Walkerdine, V. (2005). *Selections from Changing the Subject: Psychology, Social Regulation and Subjectivity, Feminism & Psychology.* London, New York: Routledge. doi: 10.1177/0959353502012004003.

Hollway, W. (2001). From motherhood to maternal subjectivity. *International Journal of Critical Psychology, 2*, 13–38.

IEG World Bank Group. (2007). Report Number: ICRR12794 ICR Review Proj ID: P036049. Washington, D.C.

IEG World Bank Group. (2014) Report Number: ICRR 14534 ICR Review Project ID: P089479. Washington, D.C.

Jones, P., & Coleman, D. (2005). *The United Nations and education : multilateralism, development, and globalisation.* Edited by D. Coleman and I. Ebrary. London, New York: Routledge Falmer.

Kabeer, N. (2005). *Gender equality and women's empowerment: A critical analysis of the third millennium development goal 1. Gender & Development, 13*(1), 13–24. doi: 10.1080/13552070512331332273.

Koven, S., & Michel, S. (1990). Womanly duties: Maternalist politics and the origins of welfare states in France, Germany, Great Britain, and the United States, 1880–1920. *The American Historical Review, 95*(4), 1076. doi: 10.2307/2163479.

Ladd-Taylor, M. (1993). Toward defining maternalism in U.S. History. *Journal of Women's History, 5* (2), 110–113.

McHoul, A. W., & Grace, W. (2015). *A Foucault primer: Discourse, power, and the subject.* London: Routledge.

Moss, P. (2006). Structures, understandings and discourses: Possibilities for re-envisioning the early childhood worker. *Contemporary Issues in Early Childhood, 7*(1), 30–41. doi: 10.2304/ciec.2006.7.1.30.

Newberry, J. (2010). The global child and non-governmental governance of the family in post-Suharto Indonesia. *Economy and Society, 39*(3), 403–426. doi: 10.1080/03085147.2010.486217.

Newberry, J. (2012). Empowering children, disempowering women. *Ethics and Social Welfare, 6*(3), 247–259. doi: 10.1080/17496535.2012.704057.

Newberry, J. (2018). Payment in heaven: Can early childhood education policies help women too?, doi: 10.1177/1478210317739467.

Paternek, M. A. (1987). Norms and normalization: Michel Foucault's overextended panopfic machine. *Human Studies*

Van Ravens, J. (2010). *Holistic ECD for all in Indonesia: Supporting communities to close the gap*. Jakarta

Thiesmeyer, L. (2003). *Introduction: Silencing in discourse', in Thiesmeyer*, L. *(ed.) Discourse and silencing: Representation and the language of displacement*. Amsterdam,: John Benjamin Publishing Company, 1–33.

Unterhalter, E. (2012). Poverty, education, gender and the Millennium Development Goals: Reflections on boundaries and intersectionality. *Theory and Research in Education*, *10*(3), 253–274. doi: 10.1177/1477878512459394.

Unterhalter, E. (2016). Gender and education in the global polity'. *The Handbook of Global Education Policy*, 111–127. doi: 10.1002/9781118468005.

Vetterlein, A. (2012). Seeing like the world bank on poverty. *New Political Economy*, *17*(1), 35–58. doi: 10.1080/13563467.2011.569023.

Walkerdine, V. (1986). Progressive pedagogy and political struggle. *Screen*, *27*(5), 54–61. doi: 10.1093/screen/27.5.54.

World Bank. (1998). *Project information document: Indonesia early child development project* ID-PE-36049. Washington, DC.

World Bank. (2003). *A global directory for early child development project*. Washington, DC.

World Bank. (2017). *Annual Report 2011*: Lending Data. Washington, DC.

Yulindrasari, H. (2017). *Negotiating masculinities: The experience of male teachers in indonesian early childhood education*. Doctor in Philosophy thesis, University of Melbourne.

Yulindrasari, H., & Ujianti, P. R. (2018). "Trapped in the reform" : Kindergarten teachers' experiences of teacher professionalisation in Buleleng, Indonesia. *Policy Futures in Education*, *16*(1), 66–79. doi: 10.1177/1478210317736206.

Early Childhood Education in the 21st Century – Yulindrasari et al. (eds)
© 2020 Taylor & Francis Group, London, ISBN 978-1-138-35203-2

The legacy of colonialism within child-centeredness in Indonesia

A.P. Humara Baiin & V. Adriany
Indonesia University of Education, Bandung, Indonesia

ABSTRACT: The purpose of this chapter is to illuminate the extent to which the legacy of colonialism marks the practices of Early Childhood Education (ECE) in Indonesia. Using postcolonial approaches, this chapter will focus on understanding how the ECE curriculum in Indonesia perpetuate the legacy of colonialism. This chapter argues that the implementation of child-centeredness may situate Indonesian children as 'the other'.

1 INTRODUCTION

The high intensity with which the Indonesian government promotes Early Childhood Education (ECE) as a prominent education plan affects the increasing need for high-quality ECE as a requisite. Hence, child-centeredness is claimed to be the solution that can answer this requirement (The Directorate of Indonesia ECE, 2015). The main idea of child-centeredness is the delivery of ECE should be adjusted to the stage of child development. The maturation theory suggested by Gesell states that a child's developmental stage is determined by the maturity of his/her biological organs. Gesell further subjugates his research to classify children development patterns in the stages of development, which then becomes the foundation for research approach in child-centeredness (Lerner, 1983).

Child-centeredness is trusted as a research approach in ECE which then, in turn, can provide a qualified ECE NAEYC, (1997) supports this claim based on the objective value in stages of child development due to its naturalistic psychological aspect. However, critical theorists questioned the notion of objective. Burman (1991) challenges the objective values of child-centeredness by stating a crucial question: 'Why does child-centeredness always speak on behalf of all children in its theory?' The critical theory states that knowledge is not absolute and universal since it stems from value-filled discourse (Apple, 2012; Canella, G. S., Viruru, 2004; Lubis, 2015). The same is true for feminist perspectives that argue objective values of knowledge are biased and only benefit those in power (Alcoff, 1989; Harding, 1986, 1993, 1998; Reinharz, 1991). Postcolonial theory proves the domination of Western knowledge in child-centeredness approach only represents children from the North (Ghandi, 1998; Gupta, 2015; Jahng, 2013; Rajab & Wright, 2018; Viruru, 2005, 2001). The question then arises as to why does child-centeredness pervasively interfere with ECE in third world countries (Gupta, 2015) as seems to legitimate the child-centeredness as the most appropriate approach?

According to (Burman, 1991), child-centeredness is a discourse that provides relational construction of power and gender, in which power relation is born in a hybrid space as a discursive area that allows the less dominant group gains their subjectivity (Bhabha, 1994). Similar to Bhabha, Foucault argues the power in the modern era no longer exist in oppressive form, but it is hidden within the 'normal' label (Foucault, 1980). Thus, the child-centeredness intervention in less developed countries (Gupta, 2015) seen as an excessive force by putting the binary between children in the North and the South. As a result, the children that come from less developed countries are being labelled as 'the other".

Colonialization practice within child-centeredness is represented by the influence of western philosophy inside that approach in non-western ECE. Where there is a continuation of the dominance of Western culture inherited from the colonial era to the present, then there is a practice of colonialism (Ashcroft, B., Griffiths, G., Tiffin, 2002). Viruru (2005) argues that child-centeredness contains western philosophy that intervenes with non-Western culture. However, research aims to critically see the implementation of child-centeredness in Indonesia ECE using postcolonial lens are still limited (Adriany, 2018b; Adriany, Pirmasari, & Satiti, 2017; Adriany & Saefullah, 2015; Adriany & Warin, 2014). Therefore, by deconstructing power that is hidden in the hybrid space within child-centeredness discourses, this article is trying to demonstrate the legacy of colonialisation within ECE in Indonesia.

2 INDONESIAN CHILDREN AS THE OTHER

Child-centeredness see young children to undergo stages of development. The stages of development are influenced by Piaget theory which has been proven individualistic and asocial (Burman, 1991; Sharp & Green, 1975; Walkerdine, 2003). Indonesian children who live in the rural area are exposed by the unfair assessment within highly strict-imperative development discourse by World Bank's standard (Hasan, A. Hyson, M, Chang, 2013). As a result, Indonesian children are being labelled as the abnormal one or the other due to the lack of scores set by the standardised assessment (Adriany & Saefullah, 2015). The scores tend to overlook the socio-cultural background of Indonesian children.

Child-centredness claims that its theories and approaches are objective. However, the notion of objectivity has long been disputed (Lubis, 2015). Bhaba (1994) argues that objectivity must be understood as a constant negotiation between object and subject in the hybrid space. Foucault (1980) agrees with this notion by stating objectivity of knowledge is often used as a vehicle to normalise certain discourses. Looking at child-centredness as a discourse, it is used to normalise the development of Children in the North and used it as a standard against children all over the globe. In other words, by taking Foucault perspective, the excessive force which puts Indonesian children in an imperative strict development pattern is a hidden power of the modern era to be normal.

Foucault believes that the knowledge discourse produces the hidden power under the universality of knowledge objective. Viruru (2005) argues that western philosophy within child-centeredness approach claims the strict imperative development stages as a universal idea. This universal claim affects problems where it labels children who do not pass the stages of development becomes 'the other' (Walkerdine, 2003). One research that is conducted by Lee and Tseng (2008) in Taiwan ECE prove Walkerdine's statement of abnormal labelling. They prove that child-centeredness is domination of Global North knowledge which becomes the only way to generate qualifiable ECE. Thus, there is a binary opposition of the qualified ECE and the unqualified ECE, which can also be seen within the normal children and the abnormal children ('the other').

3 THE RESISTANCE AND NEGOTIATION PATTERN WITHIN CHILD-CENTEREDNESS DISCOURSE

The natural stage of development within child-centeredness discourse creates a liberal child construction that must be freed during studying (Walkerdine, 2003). This construction of liberal child generates a hybrid space like the research written by Subramanian (2015) in ECE in India. Subramanian highlighted the way Indian children play during free play time, how the Indian children love to imitate the way their teacher teaches them in the class. Learning-teaching process in ECE in India uses a teacher-centred approach so that their role in class interaction have a more significant portion. This approach is not recommended for child-centeredness, as the main issue from using teacher centeredness is that the teacher has a dominant role. In a child-centred perspective, the dominant roles of the teacher could harm the children inner creativity. But Subramanian pointed in the way Indian children imitated their teacher during playtime indicating that teacher centeredness is not erasing children's inner creativity for playing (Burman, 1991).

Huang (2013) proves the existence of hybrid space between teacher-centred learning that represents traditional Taiwanese education and the child-centeredness approach that serves modern education. A class that focuses on a teacher-centred approach articulates learning as transmitting the knowledge from the teacher to the children. Taiwan traditional education sees the creativity of children will grow after receiving the instruction from the teacher. This collides with the liberal child construction which puts teachers in Taiwan into the hybrid space. The teachers teach using Taiwan's traditional teaching value all the while being demanded to produce an output that corresponds to the foreign child-centeredness parameter.

In Indonesia ECE Adriany (2018a) proves child construction within the child-centeredness paradigm preserves traditional gender construction. With the reduced role of the teacher during free play, there is also less interaction between the teacher and the child who has the potential to remove the binary pattern. Burman (1990) agrees with Adriany who states that child-centred discourse restrains teacher's subjectivity due to its limitation of teaching. Walkerdine (2003) argues that this limitation is an impossible task: 'How do the teachers teach the children only by 'looking'?

By pretending to be the teacher, Indian children show an open relationship between the teacher and the child. Together with the teacher role, Taiwan ECE intends to preserve the values of local (traditional) wisdom. With the teacher's role during free play, the asymmetrical traditional gender construction among Indonesian children could be understood more, and hopefully stopped. Hence, the role of a teacher is no longer seen as an interfering force but rather as a figure of role model for the children. Role of teacher in traditional approach should be viewed as a form of teacher subjectivity to educate children. During teaching, teachers use their professional compatibility as the agent who is responsible for planning, organise, and evaluate the learning process.

4 CONCLUSION

Child-centeredness approach is an individualistic and asocial paradigm which abandon the socio-cultural background of non-Western children thus categorises them as the other. Therefore, the child-centeredness approach in the Global South countries needs to be balanced by different approach, such a socio-cultural approach. This approach originated from the interaction of children and teachers and from children who generates a tolerant learning mechanism (Stephen, 2010). From this definition, the socio-cultural approach is suitable to be applied in Indonesia ECE as a country with diverse ethnicities, languages, cultures and religions to balance the individualist child-centeredness approach.

REFERENCES

Adriany, V. (2018a). Being a princess: young children's negotiation of femininities in a Kindergarten classroom in Indonesia. *Gender and Education*, *0*(0), 1–18. https://doi.org/10.1080/09540253.2018.1496229.

Adriany, V. (2018b). The internationalisation of early childhood education: Case study from selected kindergartens in Bandung, Indonesia. *Policy Futures in Education*, *16*(1), 92–107. https://doi.org/10.1177/1478210317745399.

Adriany, V., Pirmasari, D. A., & Satiti, N. L. U. (2017). Being an Indonesian feminist in the North. *Tijdschrift Voor Genderstudies*, *20*(3), 287–297. https://doi.org/10.5117/TVGN2017.3.ADRI.

Adriany, V., & Saefullah, K. (2015). Deconstructing Human Capital Discourse in Early Childhood Education in Indonesia. *Global Perspective on Human Capital in Early Childhood Education*, (9), 159–179. https://doi.org/10.1057/9781137490865_9.

Adriany, V., & Warin, J. (2014). Preschool teachers' approaches to care and gender differences within a child-centred pedagogy: findings from an Indonesian kindergarten. *International Journal of Early Years Education*, *22*(3), 315–328. https://doi.org/10.1080/09669760.2014.951601.

Alcoff, L. (1989). Justifying feminist social science. In N.Tauna (Ed.), *Feminism And Science* (pp. 85–103). Bloomington: Indiana University Press.

Apple, M. (2012). *Education and Power*. New York and London: Routledge.

Ashcroft, B., Griffiths, G., Tiffin, H. (2002). *The Empire Writes Back*. New York: Routledge.

Bhabha, H. (1994). *The Location of Culture*. London: Routledge.

Burman, E. (1990). Differing with Deconstruction: A Feminist Critique. In I. P. and J. Shotter (Ed.), *Differing with Deconstruction: A Feminist Critique* (pp. 141–153). London: Routledge.

Burman, E. (1991). Power, Gender, and Developmental Psychology. *Feminism and Psychology*, *1*(1), 141–153. https://doi.org/https://doi.org/10.1177/0959353591011018.

Canella, G. S., Viruru, R. (2004). *Childhood and (post-colonialization): Power, education, and contemporary practice*. London and New York: Routledge.

Foucault, M. (1980). *Power/knowledge: Selected interviews and other writings 1972–1977. New York* (23). https://doi.org/citeulike-article-id:798470

Ghandi, L. (1998). *Postcolonial Theory: A Critical Introduction*. New York: Columbia University Press.

Gupta, A. (2015). Pedagogy of third space: A multidimensional early childhood curriculum. *Policy Futures in Education, 13*(2), 260–272. https://doi.org/10.1177/1478210315579540

Harding, S. (1986). *The Science Question in Feminism*. Ithaca, NY: Cornell University Press.

Harding, S. (1993). Rethinking standpoint epistemology: What is strong objectivity? In E. Alcoff, L., Potter (Ed.), *Feminist epistemologies* (pp. 49–82). New York: Routledge.

Harding, S. (1998). *Is Science Multicultural? Postcolonialisms,Feminisms,& Epistemologies*. Bloomington, IN: Indiana University Press.

Hasan, A. Hyson, M. Chang, C. (Ed.). (2013). Young Children in Indonesia's Low-income Rural Communities: How are they doing and what do they need? In *Early Childhood Education and Development in Poor Villages of Indonesia: Strong Foundations, Later Success* (pp. 37–96). Washington, DC: World Bank.

Huang, T. (2013). Cultural analyses in a Taiwanese kindergarten: A postcolonial reflection and study. *International Journal of Educational Development, 33*(1), 15–24. https://doi.org/10.1016/j.ijedudev.2012.01.010.

Jahng, K. E. (2013). Reconceptualizing kindergarten education in South Korea: A postcolonial approach. *Asia Pacific Journal of Education, 33*(1), 81–96. https://doi.org/10.1080/02188791.2012.751898.

Lee, I., & Tseng, C. (2008). Cultural conflicts of the child-centered approach to early childhood education in Taiwan. *Early Years, 28*(2), 183–196. https://doi.org/10.1080/09575140802163600.

Lerner, R. M. (1983). *Human Development: A Life Span Perspective*. United States of America: McGraw-Hill, Inc.

Lubis, A. (2015). *Pemikiran Kritis Kontemporer: Dari Teori Kritis, Cultural Studies, Feminisme, Poskolonial, hingga Multikulturalisme*. Jakarta: Rajawali Pers.

NAEYC. (1997). *Guidelines for preparation of early childhood professionals*. Washington DC: National Association for the Education of Young Children.

Pembinaan, D. P. A. U. D. (2015). *Kurikulum Pendidikan Anak Usia Dini Apa Mengapa dan Bagaimana*. Jakarta: Kementrian Pendidikan dan Kebudayaan.

Rajab, A., & Wright, N. (2018). The idea of autonomy and its interplay with culture in child-centered education: evidence from practitioners in preschools in Saudi Arabia. *Early Years, 5146*, 1–14. https://doi.org/10.1080/09575146.2018.1434134.

Reinharz, S. (1991). *Feminist Methods in Social Research*. Oxford: Oxford University.

Sharp, R., & Green, A. (1975). *Education and Social Control*. London: Routledge & Kegan Paul.

Stephen, C. (2010). Pedagogy: The silent partner in early years learning. *Early Years, 30*(1), 15–28. https://doi.org/10.1080/09575140903402881.

Subramanian, M. (2015). Rethinking play: A postcolonial feminist critique of international early childhood education policy. *International Journal of Educational Development, 45*, 161–168. https://doi.org/10.1016/j.ijedudev.2015.10.001.

Viruru, R. (2005). The impact of postcolonial theory on early childhood education Radhika Viruru. *Journal of Education, Volume 35*(1).

Viruru, R. (2001). Colonized through Language: The Case of Early Childhood Education. *Contemporary Issues in Early Childhood, 2*(1), 31–47. https://doi.org/10.2304/ciec.2001.2.1.7.

Walkerdine, V. (2003). Developmental psychology and the child-centred pedagogy: the insertion of Piaget into early education. In *Changing the subject* (pp. 166-216). London: Routledge.

Young, R. J. (2001). *Postcolonialism: A Historical Introduction*. Oxford: Blackwell Publishing.

Early Childhood Education in the 21st Century – Yulindrasari et al. (eds)
© 2020 Taylor & Francis Group, London, ISBN 978-1-138-35203-2

Optimization of moral education for early childhood according to Abdullah Nashih Ulwan

M.N. Fairuzillah
Universitas Pendidikan Indonesia, Bandung, Indonesia

ABSTRACT: This study devoted to describing the optimization of moral education for early childhood according to Abdullah Nashih Ulwan. Lately many negative behaviors have been found in children, such as lying, theft, abuse and insult. Parents and educators are responsible for their children's morality that can rectify their deviations and encourage good relation with others. In fact, morality is one of the results of strong faith and there is a close connection between faith and morals. Optimization of moral education becomes very important as basic ethical principle and attitudes in children. Nashih Ulwan had formulated the concept of moral education itself, and how parents and educators should be able to optimize moral education for early childhood.

1 INTRODUCTION

Education is very important in human life. Human growth is influenced by the education that the human experiences. Thus, education should be truly directed to expose human beings who are high quality and able to compete. Noble character and good morals beginning in early childhood have a big role to play in human life (Althof & Berkowitz, 2006).

Today, many negative behaviors have been found in children. The reality is that there are many cases of children who speak disrespectfully, exhibit physically and sexually violent behavior; even bullying cases have occurred in early childhood (Nasrulhaq, 2018).

Therefore, if adults cannot take responsibility and educate their children, and also do not know the factors that can cause a naughtiness in children and efforts to overcome them, so a generation will be born that suffers (Mitton & Harris, 1954).

Moral education can improve the souls of children, rectify their deviations, lift them from all humiliation and encourage good relations with others. However, all educators, especially parents, have great responsibility to educate their children with kindness and morality (Mustofa, 2012).

The purpose of this study are as follows: (1) to discover Abdullah Nashih Ulwan's point of view on the meaning of moral education; (2) to compare the ideas of Nashih Ulwan with other scholars about moral education; and (3) to identify the ideal method of optimizing moral education according to Nashih Ulwan's point of view.

2 METHODOLOGY

The research method used was descriptive data analysis (Afifuddin & Beni, 2010) which utilized a set of procedures to draw truthful conclusions from a book. The content analysis viewed data not as a collection of events, but viewed data as a symbolic phenomenon and is more familiar with meaning.

The method used in collecting the data was by documentation method, which was the method of collecting data by searching data on matters in the form or text, transcript, and so on (Sugiyono, 2010). The research data source was divided into two: (1) primary data: in the form of *Tarbiyatul Aulad fil Islam* book by Abdullah Nashih Ulwan, both translated into English (Ulwan, 2004) and Indonesian (Ulwan, 2004); and (2) secondary data: in the form of library materials that have studies regarding the topic of this research

3 RESEARCH AND DISCUSSION

Before the main discussion, the author discuss the biography of Abdullah Nashih Ulwan, who was born in Suriah in 1928. He was a prominent figure in Islamic education who was concerned about the field of child education in Islam, as he stated in his book *Tarbiyatul Aulad fil Islam (Child Education in Islam)*. Abdullah Nashih Ulwan came from the descendants of the ulama; his father, Said Ulwa, was a scholar and a highly respected physician in Halab, Suriah. He completed his Bachelor's and Master's degrees at Al-Azhar University, Cairo, in 1954, in the *ushuluddin* faculty with a specialism in education. Then his doctoral degree was completed in 1982 at the Al-Sand Pakistan University. Nashih Ulwan passed away at the age of 59 years, on August 29, 1987, in Jeddah. His ideas are still being developed by future generations (Wiyani, 2017).

According to Ulwan (1981), moral education is a basic ethical principle and the virtue of the attitudes that must be possessed and used by children to become a *mukallaf*, ready to navigate human's life accordingly to ethics. Good moral education should not only consist of moral knowledge but also moral feeling and moral action, that include belief in God. Therefore, moral education is related to habit, which is has to be practiced all the time.

Morality, attitude, and character are one of the results of strong faith (Istiadie & Subhan, 2013). The strength of Islamic moral depends on its source as it is derived from the Qur'an and Hadith. If children grow up and develop by resting on the foundation of faith, they will have the ability and knowledge in moralities, even being accustomed to be a noble characters (Che Noh, Razak, & Kasim, 2015). Therefore, the religious defense rooted in the children heart has separated the children from bad traits, habits of sin, and broken traditions.

The moral patience shown by Prophet Muhammad in the face of threats should be the example in shaping children, and also he taught the people various ways related to moral and character education (Ulwan, Tarbiyatul Aulad Fil Islam, 1981). For the Moslem, the prophet Muhammad is the best owner of morals in the world. As Allah stated on the Qoran, *"Indeed in the Messenger of Allah you have a good example to follow for him who hopes in Allah and the Last Day, and remembers Allah much"* (Al-Qur'an and Translation). Thus, the Prophet is a role model and he did not only act as Allah's Messenger, but also a teacher to his family and companions with practical principles and righteous ideas for training children on upright behavior and true morals.

When compared with the concept of moral education initiated by Kohlberg, an American psychologist known for his theory of moral development, there are some significant differences in assumptions (Jumiyati, 2016): (1) the moral source according to Nashih Ulwan is based on one's faith guided by the Qur'an and Hadith, whereas according to Kohlberg that moral values are based on human reasoning; (2) Nashih Ulwan did not reveal the stages of moral development in detail because morality was formed from the child born and moral perfection was seen from the piety of mankind to his God, so it was not described as Kohlberg revealed; and (3) the function of moral education, according to Nashih Ulwan is as a means to instill moral values to children, whereas according to Kohlberg is to help children find their moral values without the existence of moral rules and according to the stages they are going through.

From interview with Imam Zarkasyi, one of the founding fathers of Gontor Islamic Boarding School, Indonesian, it found that Imam Zarkasyi placed morality in the first level of highest obligated character of human, and categorized moral education into two; spiritual moral education and physical moral education. Yet, he did not believe in dualism, hence did not differentiate these two types of moral education because the physical moral education must be accompanied by spiritual moral education. In explaining morality to children or students, Imam Zarkasyi used six efficient methods: giving good examples, establishing conductive moral environment, educating by customs, educating by advice, educating by interest, and educating by punishment (Karnen, As-Shidqi, & Mariyat, 2014). Therefore, the moral education concept of Imam Zarkasyi is almost the same as Nashih Ulwan in terms of the involvement of spiritual aspects and moral education approaches.

Nashih Ulwan (2004) explained that there were four phenomena that can damage the moral of children and become the worst attitude, among them: (1) *lying*: this is one of the meanest attributes from a religious point of view. Educators should try their best to show the children the odious aspects of lying; (2) *theft*: this is no less dangerous than telling lies. It is a must for adults to implant in the children the sense of observance of religious behavior and fear of God. Otherwise, children will gradually commit fraud, theft, and betrayal. Even become a criminal who is feared and shunned by the society; (3) *abuses and insults*: both are misbehavior that occur due to two factors; a bad role model and broken relationship among children and parent. Therefore, parents and educators must set a good role model and prevent their children from being carelessly sociable; and (4) *indulgence and dissolution*, such behavior has become rampant among young generation currently.

Among the ways to keep children from deviation and mischief in optimizing the moral education for children according to Nashih Ulwan (2004) are: (1) *warning against blind mimicry*: let us choose what we can take from strangers, and leave what must be left behind. While, things that are forbidden are mimicry of customs, traditions, foreign cultures, and principles that can eliminate the characteristics of the companions, and even subvert to faith defenses; (2) *forbidding excessive enjoyment*: it means that is the people exaggerate in pleasure and luxury. This act will result in lazy to do their obligations and plunge them into deviation; (3) *dissolute music and singing*, needless to say that anyone with discernment would realize that listening to such prohibitions leads the young to the aberrations of sex, luxury, depravity and immorality; (4) *effeminacy*: using gold ornaments, wigs, or silk clothes for men is not permissible. Similarly, effeminacy, adopting virility, and being partly-dressed but partly-bare on the part of women, all of these are effeminate and dissolute manners, and lead to effacing virility and humiliating the human personality; (5) *unveiling, flaunting, intermixing and prohibited viewing*: if all levels of society optimize the basics of moral, especially applying it as early as possible, then it is certain that the community will achieve piety, virtue, and happiness.

Nashih Ulwan (2004) noticed that there were several fatal factors which result in immoral acts and juvenile delinquency, including: (1) parents who let their children hang out with their friends who have bad and evil traits without any direction and supervision; (2) parents who allow their children any facility that have obscene and pornographic elements; and (3) parents who neglect to close their family members' genitals and let their daughter travel, preen, and get along freely. Thus, from the description, it becomes imperative for parents to always educate and monitor their children in their activity and socializing.

In teaching and instilling morality to children, Nashih Ulwan in this regard gave his valuable ideas in optimizing the moral education and education in universal on his methods. Furthermore, he argued that moralities can be transformed into the children in five approaches (Darisman, 2017): (1) approach to exemplary; from this approach, educators must be a good example for their children; (2) approach to giving praise and advice; giving praise and advice to children will have a positive influence on themselves. It will move feelings and instincts, and also will make children happy and earnest to immediately improve attitudes and behavior; (3) approach to habituation; in optimizing children's moral education, educators can command their children to repeat what they have received in the form of practical lessons and without realizing they will get used to doing it frequently and perfectly; (4) approach to giving attention and supervision; If the parents see their children something good, then let them be encouraged to do it. Yet, if the parents see something bad or worried, it must be prevented, given a warning, and explained the consequences; and (5) approach to giving punishment; the punishment aimed to correct the offenders, take care of regulations and as warning. the punishment must be in accordance with physical and psychological development, especially for early childhood should be done with full tenderness (Atabik & Burhanuddin, 2015).

4 CONCLUSIONS AND RECOMMENDATIONS

Based on this study, the author tried to make several important conclusions and recommendations related to the topic, as follows:

First, according to Nashih Ulwan that moral education and its optimization are a set of basic ethical principle and characteristics that must be possessed and made habitual by children as early as possible to become a *mukallaf*. Moral education is not just to distinguish between the wrong and the right, but more than that, it becomes an understanding and is embedded in the child be noble. Therefore, parents/educators should educate and understand their children moralities beginning in early childhood.

Second, morality according to Nashih Ulwan is based on the faith which is guided by Qur'an and Hadith. In fact, Islam is very concerned about the education of children to the moral aspects and character building. Thus, parents/educators should build the faith to the children's soul and make the figure of the Prophet Muhammad the best role model.

Third, there are some significant differences in assumptions when compared with the concept of moral education initiated by Kohlberg. Yet, another case when compared with ideas of Imam Zarkasyi is almost the same as Nashih Ulwan in terms of the involvement of spiritual aspects and moral education approaches.

REFERENCES

Afifuddin, & Beni, A. (2010). *Metodologi penelitian kualitatif [Qualitative research methode]*. Yogyakarta: Pustaka Setia.

Al-Qur'an and Translation. (n.d.).

Althof, W., & Berkowitz, M. W. (2006). Moral education and character education: Their relationship and roles in citizenship education. *Journal of moral education*, 495–518.

Atabik, A., & Burhanuddin, A. (2015). Konsep Nashih Ulwan tentang pendidikan anak [Nashih Ulwan concept of children's education]. *Elementary*.

Che Noh, M. A., Razak, K. A., & Kasim, A. Y. (2015). Islamic education based on Quran and Sunna during Prophet Muhammad's era and its relationship with tenagers' moral formation. *Tinta Artikulasi Membina Ummah*.

Darisman, D. (2017). Konsep pendidikan anak menurut Abdullah Nashih Ulwan [The concept of children's education according to Abdullah Nashih Ulwan]. *Jurnal penelitian pendidikan Islam*.

Istiadie, J., & Subhan, F. (2013). Pendidikan Moral Perspektif Nashih Ulwan [Nashih Ulwan's moral education perspective]. *Journal of Islamic Education Studies*, *1*(1).

Jumiyati, S. (2016). Perbandingan pendidikan moral anak usia dini menurut Nashih Ulwan dan Kohlberg (tinjauan psikologis dan metodologis) [Comparison of early childhood moral education acording to Nashih Ulwan and Kohlberg (psychological and methodological review)]. *Thesis*. Yogyakarta: Universitas Muhammaiyah Yogyakarta.

Karnen, A., As-Shidqi, H., & Mariyat, A. (2014). The policy of moral education on KH Imam Zarkasyi's Thought at Gontor Modern Islamic Boarding School. *Jurnal Pendidikan Islam*.

Mitton, B. L., & Harris, D. B. (1954). The development of responsibility in children. *The Elementary School Journal*, 268–277.

Mustofa, Z. (2012). Factors affecting students' interest in learning islamic education. *Journal of education and practice*.

Nasrulhaq, A. (2018). *Pesan Bamsoet ke FKPPI: jaga anak bangsa dari kemerosotan moral [Bamsoet message to FKPPI: keep the nation's children from deterioration]*. Retrieved from news.detik.com: https://news.detik.com/berita/4176479/pesan-bamsoet-ke-fkppi-jaga-anak-bangsa-dari-kemerosotan-moral.

Sugiyono. (2010). *Metode penelitian kuantitatif, kualitatif, dan R & D [Kuantitative, Kualitative, and R & D research methode]*. Bandung: Alfabeta.

Ulwan, A. N. (1981). *Tarbiyatul Aulad Fil Islam*. Beirut: Dar as-Salam.

Ulwan, A. N. (2004). *Child Education in Islam*. Cairo: Dar Al-Salam.

Ulwan, A. N. (2004). *Pendidikan anak dalam Islam [Child education in Islam]*. Jakarta: Pustaka Amani.

Wiyani, N. A. (2017). Optimalisasi kecerdasan spiritual bagi anak usia dini menurut Abdullah Nashih Ulwan [Optimization od spiritual intellegenve for early childhood according to Abdullah Nashih Ulwan]. *Thufula: Jurnal Inovasi Pendidikan Guru Raudhatul Athfal*, 77–98.

Early Childhood Education in the 21st Century – Yulindrasari et al. (eds)
 ISBN 978-1-138-35203-2

The value of children

M. Amanah & E. Kurniati
Universitas Pendidikan Indonesia, Bandung, Indonesia

ABSTRACT: The purpose of this chapter is to explore the reasons behind the parenting style applied by mothers from a low economic group. The perspective of post-developmentalism is used to discuss the data. The data was gathered by using semi-structural interviews and analyzed by grounded theory. This study found that the value of children is embedded in parent's parenting style. How parents perceive their children and their selves are influenced by the religious and socio-cultural values. This chapter recommends parenting programs that are culturally appropriate.

1 INTRODUCTION

Parenting practice has received considerable attention in research for more than 60 years, especially its role on children's development (Roopnarine et al., 2014). Parenting has significant impact to children's development because the reciprocal interaction that happens between children and parents (Bronfenbrenner, 1994).

Parenting practice has received considerable attention in research for more than 60 years, especially its role on children's development (Roopnarine et al., 2014). Parenting has significant impact to children's development because the reciprocal interaction that happens between children and parents (Bronfenbrenner, 1994).

An issue that is still raised until now is the use of corporal punishment in parenting practices. Laws and policy to prohibit the use of physical punishment has been adopted by many countries such as Austria, Croatia, Siprus, Denmark, Finland, Germany, Israel, Italy, Latvia, Norway, and Sweden (Ghersoff, 2004). Even until now, there are 52 countries that ban any kind of corporal punishment. (*End of All Corporal Punishment of Corporal Punishment*, 2017). As well as in Indonesia, laws regarding corporal punishment to children has been regulated in National Constitution No. 35 2014 about Child Protection.

Parenting is indeed a dilemma when viewed from a socio-cultural perspective. Authoritarian parenting such as the use of physical punishment, threats, and verbal discipline is categorized as negative parenting. They believe that democratic parenting, focus on children, and demanding in children's maturity is the best parenting practice (Rubin & Chung, 2005). Meanwhile, some studies in Asian countries found that authoritarian parenting gives positive outcomes in academic performance in China (Chen & Wong, 2014) and Japan (Watabe & Hibbard, 2014). Moreover, Frey (2008) found that parenting practices in some ethnicities are contrary to the current laws and policy regarding child protection. He found that in certain ethnic groups and in certain religions the use of physical punishment against children is permissible.

Indonesia has a wide, complex, and diverse culture, ethnicity, language, economic condition and geography. This diversity makes Indonesia a culturally rich area. One of the challenges in understanding the Indonesian family is the parenting practices used. Because of the diversities of cultures, religions, and ethnicities in Indonesia, it is likely that Indonesian parents use a wide variety of parenting styles.

However, current research regarding parenting are conducted in Western countries and focused on the parenting styles and its impact on children's behavior and well-being (Smith, 2005; Ghershoff, 2014). Little has studied on the perception of parents itself regarding the parenting practice they apply. Moreover, as Chao and Tseng (2002) noted in their review of Asian parenting, Indonesia has rarely been included in studies of parenting. The perspective of what is a 'good parent' indirectly forms our mindset. As a result, it is often parents with certain parenting styles who are blamed for the occurrence of negative behaviors in children.

One of parent group from low socioeconomic status. Amanah (2018), in her study with more than 70% of Sundanese ethnic participants, found that parents from lower socio-economic groups tended to use verbal threats and disciplines such as criticizing and scolding children when children's behavior did not match their expectations. The author believes that there might be some other factors that become the 'truths' and those can explain why parents apply such kind of parenting style to their children. However, a deep explanation of why they doing it has not been explored in a previous study.

The post-developmental perspective is used to analyze the finding of this study in order to see the phenomenon not only from one point of view but also others factors, not generalize one theory for all problems at the same or different context (Morrow, 2006).

2 THEORITICAL FRAMEWORK

2.1 *Post-developmental perspective: A critic on 'normal' parenting*

Post-developmental perspective critics the concept that ideal norms and standards are formed based on "developed countries" and other countries should follow them (Ziai, 2007). By using this perspective, the possibility of different discourses and representations will be created which are not mediated by development constructs such as ideology, language, place and others. Therefore, it is necessary to multiply the source of information, especially from the perspective of the "object" of development itself (Ziai, 2007) in this study are parents.

Singer (1993) discussed any difficulties to handle cultural differences in the norms of parenting. This is related to the claims of experts who universalize it, when in fact it only refers to certain cultural norms.

> The style recommended by parent educators is the authoritative, democratic style, because it is through that children raised under that style will achieve, be dependable and responsible, and feel good about themselves.
>
> (Berger, 1995, p. 83)

Experts in the fields of education, health, and various parties such as the media also have a role in determining what is good for children and how children should be raised. This is due to the construction that children are vulnerable, incompetent, and need protection (James & Prout, 1990). In other words, there are many parties who give expectation as to what 'good parent' figure is. Therefore, parents with certain parenting styles are often blamed for the occurrence of negative behaviors in children.

Studies regarding parenting are conducted in western countries and focused on the parenting styles and its impact on children's development (Smith, 2005; Ghershoff, 2014). As a result, the understanding about parenting that has been applied in Indonesia is also more oriented towards Western culture. Adriany and Saefullah (2015) also revealed that parents in the village are often included in parenting training programs, but the parenting style promoted comes from Western cultures such as those from America or Europe.

3 METHODOLOGY

This study used a qualitative approach, that is, a case study, in order to get a detailed and complete understanding the case (Creswell, 2014) and to answer research questions about "how" or "why" (Yin, 2003). The participant is a mother who has three children and comes from low socio-economic status. Data were gathered by using semi-structured interviews with three meetings.

This study used a grounded theory approach based on Charmaz (2006). There are at least two main stages of coding in grounded theory, namely the initial stage which involves naming each word, sentence, or segment (open coding) and then a later stage of focus coding. Finally, themes are formed that reflect parenting.

4 RESULT AND DISCUSSION

Discussions of parenting cannot be separated from perspectives on the nature of childhood (Ambert, 1994). How parents perceive their children can impact directly on how they deal with children and help the children to develop and grow. In the result of this study it was found not only how parents perceive their children but also how parents perceive them as a parent.

Parents and children construction are influenced by values in their setting, such as religion value. Because Indonesia has the largest Muslim population in the World, Islamic religion has a big impact in Indonesian people:

> I think parent's role is really important. Really significant. Taking care of children, giving affections, and guidance too. Child needs to be guided and educated by parent. It is parent's responsibility.
>
> (Interview with Mrs. Yuni, 30 November 2018)

Mrs. Yuni's statement shows that there is Islamic religion concept in seeing parent's role. Religion is often being seen as cultural value system that has big influence on parenting in Indonesia (Riany, 2016). Islamic religion views that parents especially mother has significant role and responsibility in child rearing. It is stated on *Hadits* (a book of Islamic Role) that mother is the first person for children to learn. This concept is also showed when Mrs. Yuni was being asked about her opinion on spanking the children:

> Actually I feel sorry to my child when I spank her. I just want to teach her. Only to make her learn. After I angry on or spank her I always say sorry. Getting angry doesn't mean that I hate my child.
>
> (Interview with Mrs. Yuni, 28 October 2018)

> ... especially for our provision in 'akhirat'. A good child can save us in 'akhirat' (means afterlife) later.
>
> (Interview with Mrs. Yuni, 30 November 2018)

Islamic religion also views that children is a gift from God and parents are responsible for them. As a result, anything that parents give to the children when they are alive will be accountable in akhirat (the afterlife). Therefore, parents feel responsible to guide their children until they become adults. This was supported by Mrs. Yuni's statement:

> I will guide my child until they have their own family. Even after get married, I still want to guide them. After all, children need to be guided, until they become adult, until they can decide their own path.
>
> (Interview with Mrs. Yuni, 30 November 2018)

Socio-cultural is also become aspect constructed the value of children. In Asian countries, parents and older family relatives are considered have more authority than younger children (Javillonar, 1979). Parents are expected to have responsibility to taking care children during their lifetime (Yu & Liu, 1980). Furthermore, in Asian families, there is an interdependent relationship between parents and children. Mulder (1992) suggested that there is a hierarchy in parents and children relationship where parents are always as 'giver' while children are 'receiver'. As a result, there must be a reciprocal relationship between parents and children; even when children already have their own family, they need to fulfill their obligation to their parents. This is also mentioned by Mrs. Yuni:

Having a child is benefit for us, especially when we are already get old. I know, I have husband. But still, it is very helpful to have children.

(Interview with Mrs. Yuni, 30 November 2018)

5 CONCLUSION

The use of parenting styles is influenced by the value of children. Parents view themselves as people with responsibility to teach and take care of their children until their children become adults. A child is also viewed as a person that needs to be guided and as a parent's savior in *akhirat*. This perspective is fully influenced by the Islamic religion. Cultural values also play a role in constructing the value of children. It is characteristic of Asian families that parents are always seen as 'givers' while children are 'receivers'. Therefore, there must be a mutual relationship between parents and children. Parents expect that their children can help them later.

Those perspectives of parents and children value can be the reason why parents do such kind of parenting style to their children. They might think that they fully have responsibility to form their children to become good children. This research might be a new perspective for us in understanding parenting practice of Indonesian parents. Especially in planning for parenting program, it is better for us to consider the socio-cultural background of the family.

REFERENCES

Adriany, V., & Saefullah, K. (2015). Deconstructing human capital discourse in early childhood education in Indonesia. In T. Lightfoot, R. L. Peach & Leask, N. (eds). *Global Perspectives on Human Capital in Early Childhood Education*, 159–179.

Amanah, M. (2018). Parenting styles and socio-emotional development of young children among different groups of socio economic status: A study in city C, Indonesia. Master's thesis, Hiroshima University.

Ambert, A. M. (1994). An international perspective on parenting: social change and social constructs. *Journal of Marriage and the Family*, *53*(3), 529–543.

Berger, E. H., & Riojas-Cortez, M. (1995). *Parents as partners in education: Families and schools working together*. Englewood Cliffs, NJ: Merrill.

Bronfenbrenner, U. (1994). Ecological models of human development. *International Encyclopedia of Education*, *3*(2): 37–43.

Charmaz, K. (2006). *Constructing grounded theory: a practical guide through qualitative analysis*. New York: SAGE.

Chao, R., & Tseng, V. (2002). Parenting of Asians. *Handbook of Parenting*, *4*: 59–93.

Chen, W. W., & Wong, Y. L. (2014). What my parents make me believe in learning: the role of filial piety in Hong Kong Students' motivation and academic achievement. *International Journal of Psychology*, *49*(4), 249–256.

Creswell, J. W. (2014). *Penelitian kualitatif & desain riset, edisi ke 3 [Qualitative research & research of design, 3rd edition]*. Yogyakarta: Pustaka Pelajar.

End of All Corporal Punishment. (2017). Retrieved from http://www.endcorporalpunishment.org/progress/prohibiting-states/.

Frey, R. (2008). The practice of discipline: The child's right to a culture vs. the child's right to safety. In Babacan, H., Gopalkrishnan, N. (eds). In: *Proceedings of The Second International Conference on Racisms in The New World* Order. *The Complexities of Racism Proceeding*, pp. 16–26. Queensland: University of The Sunshine Coast.

Gershoff, E. T. (2002). Corporal punishment by parents and associated child behaviors and experiences: A meta-analytic and theoretical review. *Psychological Bulletin*, *128*(4): 539–579.

Javillonar, G. (1979). *Rural development, women's roles and fertility in developing countries: Review of the literature*. Chapel Hill, NC: Research Triangle Institute.

James, A., & Prout, A (Eds). (1990). *Constructing and reconstructing childhood: Contemporary issues in the social construction of childhood*. London: Routledge.

Mulder, N. (1989). *Individual and society in Java: A cultural analysis*. Yogyakarta: Gadjah Mada University Press.

Morrow, V. (2006). Understanding gender differences in context: Implications for young children's every-day lives. *Children & Society*, *20*(2), 92–104.
Rubin, K. H., & Chung, O. B. (2013) *Parenting beliefs, behaviors, and parent-child relations: A cross-cultural perspective*. New York: Psychology Press.
Roopnarine, J. L., Krishnakumar, A., Narine, L., Logie, C., & Lape, M. E. (2014). Relationships between parenting practices and preschoolers' social skills in African, Indo, and mixed-ethnic families in Trinidad and Tobago: The mediating role of ethnic socialization. *Journal of Cross-Cultural Psychology*, *45*(3): 362–380.
Singer, E. (1993). Shared care for children. *Theory & Psychology*, *3*(4): 429–449.
Smith, A. B., Gollop, M. M., Taylor, N. J., Marshall, K. (2004). *The discipline and guidance of children: A summary of research.* Retrieved from https://resources.skip.org.nz/assets/Resources/Documents/the-discipline-and-guidance-of-children.pdf.
Watabe, A., & Hibbard, D. R. (2014). The influence of authoritarian and authoritative parenting on children's academic achievement motivation: A comparison between the United States and Japan. *North American Journal of Psychology*, *16*(2): 359–382.
Yu, E., & Liu, W. T. (1980). *Fertility and kinship in the Philippines*. Berkeley: University of California.
Ziai, A (ed). (2007). *Exploring post-development: Theory and practice, problems and perspectives*. London: Routledge.

Child development

Early Childhood Education in the 21st Century – Yulindrasari et al. (eds)
© 2020 Taylor & Francis Group, London, ISBN 978-1-138-35203-2

The impact of using gadgets on early childhood

F.R. Intan & Y. Rachmawati
Early Childhood Education Study Program, Postgraduate School, Universitas Pendidikan Indonesia, Jawa Barat, Indonesia

ABSTRACT: In today's digital era, gadgets are a communication tool that emerge because of technological advances that initially have a major function to facilitate the lives of adults. However, many children have been introduced to gadgets by their parents for various purposes including playing games, taking photos, watching videos and many other related activities. Parents often neglect to control gadget use by their children. As a result, unintended and unavoidable negative impacts are emerging. One of the studies in Indonesia revealed that there were some negative impacts from the disruption to children's health. This chapter aims to find out the effects of using gadgets and the role of parents in the use of gadgets in early childhood. First, researchers will explain the various effects of gadgets on early childhood. Second, researchers will discuss about the use of gadgets in early childhood. Third, researchers will explain about the importance of the role of parents to supervise, control and pay attention to all early childhood activities in relation to gadget use.

1 INTRODUCTION

Today's communication technology is one of the big factors in human life. In every aspect of life today, technology use is rapidly advancing and its uses are increasingly sophisticated (De Lima & Castronuevo, 2016). Communication technology is defined as hardware equipment, and to exchange information with others (Strasburg & Donnerstein, 2010). According to Chusna (2017), current gadget users are not only adults. However, almost all groups, including early childhood ages and toddlers, use gadgets in activities that they do every day, including playing games, taking photos, and watching videos.

According to Jackson et al. (2012), almost everyone who uses gadgets spends a lot of their time in a day using gadgets. There are various studies that state the negative effects experienced by children who use gadgets excessively. Radesky and Zuckerman (2015) said that the use of gadgets in early childhood is increasingly intensive and will have an impact on development of children's behavior; in other words, the growth of children becomes disrupted in reducing and replacing the amount of face-to-face time between people around him, and changing the behavior of children to be indifferent. However, gadgets are needed and play an important role for humans. A survey conducted in several parts of the United States by Uswitch revealed that 3.5 million children under the age of 7 were addicted to gadgets. In the United Kingdom and the United States, in recent years there has been a five-fold increase in the use of gadgets by groups ranging in age from 0 to 8 years of age. In addition, two small research projects in Australia, with under 160 families of early childhood and their respective parents, show that two-thirds of these children (0–5 years) overuse gadgets (Neumann, 2014).

Sucipto and Nuril (2016) revealed that out of 36 children, 72% were introduced to or used gadgets. As many as 27% of two-year-olds have been introduced to gadgets, and 54% of parents allow children to use gadgets at the age of 3 to 5 years. In Indonesia the situation is not much different; according to one information technology expert from the Bandung Institute of Technology, ITB, Mahayana (2014) about 5–10% of gadget addicts are accustomed to touching their gadgets as much as 100–200 times a day. If the effective time of human activity is 16 hours or 960 minutes a day, therefore people who are addicted to gadgets will touch the device once every 4.8 minutes.

In Indonesia, this device fever has been going on since 2008, from the time that Facebook rose. The use of mobile phones has now exceeded 50%. Parental negligence is something that makes unwanted impacts appear. Research conducted by Chusna (2017) shows a significant difference ways of children using gadgest by children with parents working as opposed to parents at home. Parents who work give gadgets to their children to communicateeasier, while stay house parents give gadgets to their children so they not interfere parental activities at home. Instead of just diverting children, it turns out that gadgets are becoming addictive. From some of these strict unattended habits, parents are negligent of their real important role in introducing and accompanying children using gadgets. This means that there are problems with excessive use of gadgets and this has reduced productive time, either for playing or for learning.

2 THE IMPACT OF USING GADGETS

Excessive use of gadgets will adversely affect children. Children who spend time with gadgets for too long will have another negative impact, as can be seen in Strasburg et al. (2010), who states that gadgets can provide information about unsafe health practices. Ozgur and Seyhan (2010) as well as Plowman and McPake (2013) mention that in the relationship between learning and the integration of the use of gadgets, they still believe that gadgets can hinder children's social, emotional, physical and cognitive development. The attitude that is highlighted is usually in the form of playing or interacting with peers.

The researchers also believe that if children are increasingly dependent on gadgets, then their relationship with their parents will be deteriorated and can cause delays in the development of adaptation to new groups. According to Wijanarko (2017), children need to interact socially, starting with the family, as the child's first environment. Children who interact well in the family will interact well with friends and their surrounding environment (Ariani & Permana, 2018). Therefore, it is important for parents to always supervise and limit gadget use by their children.

Linda Blair, a clinical psychologist, said that staring at the screen of a gadget actually has a bad impact on children. The gadget screen lowers the level of melatonin, a natural substance produced by the body to rest or sleep. Not only that, the gadget screen is also allegedly able to increase the level of cortisol hormone which triggers stress and impacts so that it is difficult for children to concentrate properly.

According to Fahriantini (2016), children do not need a lot of time and effort to learn toread and write in a book or paper. But children will also be more eager to learn because the applications available on their gadgets often have interesting images; Even on the other hand Jackson et al. (2012) say that playing video games is a predictor of all steps of children's creativity. Regardless of gender or race, bigger videogames play associated with greater creativity.

2.1 *Use of gadgets in early childhood*

Children are reliable imitators. They are smarter than they seem, so don't underestimate children at that age (Hastuti, 2012, p. 95). If a child at an early age has been given a gadget as a toy, then it will affect the process of health development and more worrying is the disruption in the social development of early childhood emotions.

According to Ling and Yttri (2006), gadgets have changed the pattern of parenting. More than 50% of parents in the United States admit that they use gadgets to "care for" children when they are busy working. Parents are more likely to think that gadgets can also help to stimulate their children. In addition, there are also parents who provide gadgets to their children with the aim of keeping them quiet so that parents can do other activities. This leads to uncontrolled use of gadgets.

2.2 *The role of parents in using gadgets in early childhood*

Seeing the effects of introducing gadgets to children in the end really depends on the readiness of parents to introduce and supervise children when playing gadgets. In this case, Brooks (2011) revealed that some of the reasons parents give gadgets to children are simply to calm children when they are fussy or unable to sleep or eat as well as provide free time for parents to do household work or relax.

Parents need to apply a number of rules to their children in using gadgets. Parents can understand and explain the content on the gadget. Without assistance from parents, the use of gadgets will not focus on what parents teach. Usually it will deviate from what parents teach.

In that moment, parents must always invite discussion and even invite them to tell stories so that children can display or be creative with ideas that are on their minds (Venkatesan & Yashodharakumar, 2017).

3 CONCLUSION

Daily development of early childhood is currently more affected by technology, especially gadgets. Children more often interact with gadgets. Children can be estranged from the environment because of a lack of interaction. But technological advances can also help the creativity of children if their utilization is balanced with the surrounding environment. They know how to use technology to fulfill their desire to play. Therefore, parents need to develop time management to help control gadget use by their children. In other words, old people supervise and control when their children use gadgets so that they are not too dependent on gadgets and do not forget to socialize with their surroundings.

REFERENCES

Ariani, H. B., & Permana, G. (2018). Wisely using gadget for parents in family environment campaign design. *Bandung Creative Movement (BCM) Journal*, *4*(1).

Brooks, J. (2011). *The process of parenting*. Yogyakarta: Pustaka Pelajar.

Chusna, P. A. (2017). Pengaruh Media Gadget pada Perkembangan Karakter Anak. *Jurnal Dinamika Penelitian: Media Komunikasi Penelitian Sosial Keagamaan*, *17*(2), 315–330.

Fahriantini, E. (2016). Peranan Orangtua Dalam Pengawasan Anak Pada Penggunaan Blackberry Messenger Di Al Azhar Syifa Budi Samarinda. *Ejournal Ilmu Komunikasi*, *4*(4), 44–55.

Hastuti. (2012).*Psikolog Perkembangan Anak*. Yogyakarta: Tugu Publisher

Jackson, L. A., Witt, E. A., Games, A. I., Fitzgerald, H. E., von Eye, A., & Zhao, Y. (2012). Information technology use and creativity: Findings from the children and technology project. *Computers in human behavior*, *28*(2), 370–376.

Ling, R., & Yttri, B. (2006). *Control, Emancipation, and Status: Te Mobile Inteens' Parental and Peer Relationships*. New York: Oxford University Press.

Mahayana, D. (2014). *Pengaruh Gadget terhadap Perilaku Masyarakat Modern. Educational Survey*. Bandung: ITB.

Neumann, M. M. (2014). An Examination of Touch Screen Tablets and Emergent Literacy in Australian Pre-School Children. Australian Journal of Education, 58(2), 109–122.

Ozgur, E., Güler, G., & Seyhan, N. (2010). Mobile phone radiation-induced free radical damage in the liver is inhibited by the antioxidants n-acetyl cysteine and epigallocatechin- gallate. *International Journal of Radiation Biology*, *86*(11), 935–945. http://dx.doi.org/10.3109/09553002.2010.496029.

Plowman, L., & McPake, J. (2013). Seven myths about young children and technology. *Childhood Education*, *89*(1), 27–33.

Radesky, J. S., Schumacher, J., & Zuckerman, B. (2015). Mobile and interactive media use by young children: the good, the bad, and the unknown. *Pediatrics*, *135*(1), 1–3.

Strasburg, V. C., Jordan, A. B., & Donnerstein, E. (2010). Health effects of media on children and adolescents. *Pediatrics Journal*, *125*(4), 756–767.

Sucipto., & Nuril, H. (2016). Pola bermain anak usia dini di era gadget siswa PAUD Mutiara Bunda Sukodono Sidoarjo. *Jurnal Ilmiah Fenomena*, *3*(6), 274–347.

Uswitch. (2014). 3.5m children under 8 own a tablet. Survey lays bare kids' gadget http://uswitch.com /mobile/news/2014/01/3_5m_british_children_own_a_tablet_survey_lays_bare_kids_gadget_addiction/.

Venkatesan, S., & Yashodharakumar, G.Y. (2017). Parent opinions and attitudes on toys for children with or without developmental disabilities. *The International Journal of Indian Psychology*, *4*(4), 6–20.

Wijanarko, J. (2017). Pengaruh Pemakaian Gadged dan Perilaku Anak, terhadap kemampuan anak Taman Kanak-kanak Happy Holy Kids Jakarta. *Jurnal Institut Kristen Borneo*, *2*(1), 1–40.

Early Childhood Education in the 21st Century – Yulindrasari et al. (eds)
© 2020 Taylor & Francis Group, London, ISBN 978-1-138-35203-2

Sibling rivalry in early childhood

T. Rahmatika & Rudiyanto
Early Childhood Education Program, Universitas Pendidikan Indonesia, Jawa Barat, Indonesia

ABSTRACT: In a family in Indonesia, it is common for parents to have more than one child, so sibling competition occurs in a family. Sibling rivalry among brothers provides many opportunities for children to develop social and emotional aspects in their environment. One's relationship is seen as a unique relationship consisting of two characteristics, positive characteristics (love and affection) and negative characteristics (conflict and competition). Sibling rivalry among brothers arises from the fear of losing the love and attention of their parents. However, there are still many parents who consider competition as a normal thing that does not need to get special care. Based on this, the purpose of this chapter is to describe the competitiveness of concepts, causes, and influence of sibling competition in early childhood.

1 INTRODUCTION

The sibling relationship is one of the important components in the process of children's social emotional development. A good sibling relationship will have a good impact on the child's social emotional development. The arrival of a younger sibling can lead to jealousy and competition on the part of the first child. Jealousy and competition that occurs between siblings is called sibling rivalry. Sibling rivalry often occurs because of anxiety or fear of losing love from parents.

2 SIBLING RIVALRY'S CONCEPT

Sibling relationships are unique, but there can be conflicts and competition as well as love and warmth (Buist, Dekovic, & Prinzie, 2012). Sibling Rivalry is a feeling of hostility, jealousy, and anger between siblings, siblings or siblings not as sharing friends but as rivals who usually appear either before or after their siblings are born (Cholid, 2004; Chaplin, 2001; Musbikin, 2008). Competition from siblings (rival siblings) often occurs in families. Siblings rivalry usually appears at the age of 3–5 years and can reappear at the age of 8–12 years (Millman & Schaefer, 1989). In contrast to Dunn and Kendrick's opinion, sibling rivalrises are less common in children over the age of 8 because these children have a better understanding of their position at home (Dunn & Kendrick, 1982).

Cicirelli (1995) states that there are various types of siblings. Full siblings are those siblings with the same biological parents. Half sibling are two individuals with only have one biological parent in common. Step siblings do not have the same biological parents, but the relationship between them is united by the marriage ties of their parents. Adoptive siblings are those where at least one of the siblings has been legally adopted. Finally, fictive sibling are not family members but have been considered as family members based on certain desires. There are many studies that discuss sibling rivalry, but the discussion about young mothers' perception about sibling rivalry is still limited.

3 CAUSES OF SIBLING RIVALRY

Siblings rivalry is a form of competition between siblings, which occurs because someone is afraid of losing love and attention from parents. In addition, age differences among siblings can also trigger rivalry siblings (Millman & Schaefer in Setiawati & Zulkaida, 2007). The aspects of sibling rivalry include conflict, jealousy and resentment (Papalia, Olds, & Feldman, 2009). Novairi and Bayu (2012) state that there are two factors that influence sibling rivalry: external factors and internal factors. External factors include the parental attitudes such as comparisons and the existence of favoritism. Internal factors include character, children's attitudes, age differences, gender differences, and birth order. A study conducted by Stocker (1994) states that competition in sibling relationships is associated with higher levels of loneliness and depressive moods as well as lower levels of self-esteem. Research shows that conflict in the household is also one of the factors in the emergence of sibling rivalry because children witness the conflicts that occur between their parents (Radford, Corral, Bradley, & Fisher, 2013).

4 IMPACT OF SIBLING RIVALRY

The results of interviews conducted from two young mothers stated that the impact of sibling competition had a negative impact on children's behavior because the younger sibling imitated the rejection behavior carried out by his brother. It is estimated that one in three children aged 0–17 years may experience violence or assault from siblings while still children (Finkelhor, Turner, Shattuck, & Hamby 2015). Competition of siblings can have a negative impact on personal and social adjustment (Putri, 2013; Leung & Robson, 1991), and it also can result in a high level of anxiety (Wiehe, 1997; Graham-Bermann, Cutler, Litzenberger, & Schwartz, 1994), confidence and low self-esteem (Meyers, 2014). Other research shows that when children are victims of violence, they are predicted to experience the same things in their peers (Tucker, Finkelhor, Turner, & Shattuck, 2014).

5 CONCLUSION

Relationships with siblings are an important component in the process of children's emotional social development. But in a sibling relationship there are problems that arise. Sibling rivalry is one of the problems that often occur in the family environment. Sibling rivalry arises because of the fear of losing love from parents because of the presence of the younger brother. If sibling rivalry is not overcome early on, it will have a negative impact on the child, especially on the social emotional aspects in the future. This is certainly a challenge for every parent, especially mothers, to know what things should be done to solve this problem. There are two important points in the face of sibling rivalry. First, parents should understand the concept of fairness in treating children, not comparing them and as much as possible there is no favoritism. Second, communication between parents and children is very important so that closeness is established, fostering an attitude of sympathy, empathy and compassion within the family.

REFERENCES

Buist, K. L., Deković, M., & Prinzie, P. (2013). Sibling relationship quality and psychopathology of children and adolescents: A meta-analysis. *Clinical Psychology Review*, *33*(1), 97–106.

Chaplin, J. P. (2001). *Kamus lengkap psikologi*. Jakarta: PT. Raja Grafindo Persada.

Cholid N. S. (2004). *Mengenali stress anak & reaksinya*. Jakarta: Buku Populer Nirmala.

Cicirelli, V. G. (1995). *Sibling relationships across the life span*. New York: Plenum Press.

Dunn, J., & Kendrick, C. (1982). *Siblings: Love, envy, & understanding*. East Sussex: Harvard University Press.

Finkelhor, D., Shattuck, A., Turner, H., & Hamby, S. (2015). A revised inventory of adverse childhood experiences. *Child Abuse & Neglect*, *48*, 13–21.
Graham-Bermann, S. A., Cutler, S. E., Litzenberger, B. W., & Schwartz, W. E. (1994). Perceived conflict and violence in childhood sibling relationships and later emotional adjustment. *Journal of Family Psychology*, *8*(1), 85.
Leung, A. K., & Robson, W. L. M. (1991). Sibling rivalry. *Clinical Pediatrics*, *30*(5), 314–317.
Meyers, A. (2014). A call to child welfare: Protect children from sibling abuse. *Qualitative Social Work*, *13*(5), 654–670.
Musbikin, I. (2008). *Mengatasi anak-anak bermasalah*. Yogyakarta: Mitra Pustaka.
Novairi, A. & Bayu, A. (2012). *Bila kakak-adik saling berselisih*. Jogjakarta: PT Buku Kita.
Papalia, D. E., Olds, S. W., & Feldman, R. D. (2009). *Human development: Perkembangan manusia*. Jakarta: Salemba Humanika.
Putri, A. C. T. (2013). *Dampak sibling rivalry (persaingan saudara kandung) pada anak usia dini*. Doctoral dissertation, Universitas Negeri Semarang.
Radford, L., Corral, S., Bradley, C., & Fisher, H. L. (2013). The prevalence and impact of child maltreatment and other types of victimization in the UK: Findings from a population survey of caregivers, children and young people and young adults. *Child Abuse & Neglect*, *37*(10), 801–813.
Setiawati, I., & Zulkaida, A. (2007). Sibling rivalry pada anak sulung yang diasuh oleh single father. *Jurnal Fakultas Psikologi Universitas Gunadarma*.
Stocker, C. M. (1994). Children's perceptions of relationships with siblings, friends, and mothers: Compensatory processes and links with adjustment. *Journal of Child Psychology and Psychiatry*, *35*(8), 1447–1459.
Tucker, C. J., Finkelhor, D., Turner, H., & Shattuck, A. M. (2014). Sibling and peer victimization in childhood and adolescence. *Child abuse & neglect*, *38*(10), 1599–1606.
Wiehe, V. R. (1997). *Sibling abuse: Hidden physical, emotional, and sexual trauma*. California: Sage.

Early Childhood Education in the 21st Century– Yulindrasari et al. (eds)
© 2020 Taylor & Francis Group, London, ISBN 978-1-138-35203-2

Teachers' roles in ZPD-based sociodramatic play for children's prosocial behaviours development

E. Afiati
Department of Guidance and Counselling, University of Sultan Ageng Tirtayasa. Banten, Indonesia

U. Suherman & Nurhudaya
Department of Guidance and Counselling, Indonesia University of Education, West Java, Indonesia

ABSTRACT: Due to the concern of under-developed prosocial behaviour upon children, ZPD (Zone of Proximal Development) theory, which emphasizes the help from others to achieve a certain level of development, is utilized in this research to assist children to achieve the proper level of development through the implementation of sociodramatic play. Three kindergarten teachers were involved as the participants. Four weeks of recorded activities of a sociodramatic play were analyzed to identify teachers' roles in giving assistance to children. The result revealed that intervention or assistance was given when children showed low participation while it was hindered when they showed high participation in the play. Thus, when teachers' optimum role as assistance providers is fulfilled, ZPD-based sociodramatic play can serve as an alternative pedagogical strategy to develop children' prosocial behaviours.

1 BACKGROUND

Children were born without social character. In other words, children do not have abilities to connect to other people. To get the social mature, children need to learn about ways to adjust with others. This ability is acquired through many chances or experiences of connecting to others in their environment whether with adult or their peers (Yusuf, 2014). Children' behavior in their social life is really determined by how adults introduce social norms to live in society, support, and give model to their children on how to apply the norms on their daily lives (Yusuf, 2014). Children' social development is really influenced by their social environment, whether from their parents, relatives, other adult people or their peers. The social environment mediates the children social development (Yusuf, 2014).

It is stated that children's prosocial behaviour may be related to aggressive behaviour. When children in their childhood life have prosocial behavior, such as helping others, have good interaction and connection with their friends and adult people, there will be small chance to be antisocial when they grow up (Smith & Hart as cited in Alfiyah, 2014).

In learning context, the theory of ZPD (zone proximal development) by Vygotsky explains the importance of help from adult for children to achieve learning goal. Vygotsky (in Sujiono, 2009, p. 115) argues that ZPD is distance between the level of actual development which is realized through independent problem solving and the potential development level which is realized through problem solving with the assistance from adult or cooperation with other peers who have more abilities.

Vygotsky states that playing is social activity because children's social development depends on their interaction with people around them in various pleasant activities that serve as means of channeling ability to help children develop their view on environment. In addition, playing can also help children to improve their ability to self- manage their physical, social, and cognitive behaviours. Initially, children play individually. According to Piaget and Vygotsky (in Suyadi, 2016, p. 120) playing is an important aspect for children's learning.

According to Vygotsky, sociodramatic play is the only play which is good for kindergarten students (Bodrova et al., 2013). Vygotsky (1978) argues that children engage in sociodramatic play as they wish to imitate adults' roles and perform their activities that the child is too young to attempt in their real lives.

Smilansky (1990) defines sociodramatic as the cooperative activity of at least two children or more, in which each child plays and acts out a specific role. There are two main elements in sociodramatic play: imitating and make-believe. Imitation is the combination of action and oral, make-believe is highly dependent on verbalization. This play offers the good chance for children in developing their abstract thought (Piaget, 1962), defining their understanding on the world (McCain, Mustard & Shanker, 2007), solving problem in the safe context (Smilansky & Shefatya, 1990), having sensitivity to control their experience and action (Piaget, 1962), and learning how to connect with their peers in positive way (Saracho & Spodek, 2003).

Adults can help children in providing more themes and interest, enriching language and developing scenarios and facilitating exploration and literacy, but deciding when and how the intervention can be done is a challenge. Many teachers fail to support children in playing (Stantan-Chapman, 2014) so that children could not reach their highest potential (Bredekamp, 2013; Bredekamp & Rosegrant, 1992).

2 LITERATURE REVIEW

2.1 *Children's prosocial behaviours*

In social development, children are required to have abilities to meet the social requirements of the place where they live. In social development, there are prosocial and antisocial behaviours (Eisenberg & Mussen, 1989). Prosocial behaviours are activities related to others, whether with their peers, teachers, parents and siblings. Children build their personality through meaningful interaction with others.

Prosocial behaviours, according to Eisenberg and Mussen (1989), include: (1) sharing: individual can share with others whether in sadness or happiness. For example, visiting friends who are not in good condition, sharing ideas and experiences; (2) team work: the willingness to work together with others for a certain common goal. For example, completing projects together; (3) donating: willingness to donate including voluntarily giving their own goods to others who need help. For example, donating to natural disaster victims, poor people and to those who need help; (4) helping: willingness to help others in difficult situation and doing something to support others' activities. For example, helping others to carry their goods; (5) honesty: behaviour which shows the consistency between words and the existing situation. For example, avoid cheating and doing bad things that make other people loss; (6) empathy: showing care to others who found difficulties and telling others feeling during the conflict. For example, entertaining others who are sad by caressing their back or hands and saying, "please don't cry"; and (7) considering others' right and duty: giving ways to others to have access in their business. Showing that they care for others by giving attention and paying attention to others' people problems.

2.2 *Roles of teachers in learning process*

There are several roles of teachers in learning process, especially in developing children prosocial behaviour: (1) Optimizing children's opportunity to develop their proximal development zone or their potential through learning and development; (2) Doing learning process which is related more to potential development levels rather than their actual development; (3) Directing the learning to the use of strategy to develop the children inter-mental abilities rather that their intra-mental; (4) Giving optimum chance to the children to integrate their background declarative knowledge with procedural knowledge to do the tasks and solve problems; (5) Conducting teaching and learning process which is not only transferring but also co-construction (Vygotsky, 1978).

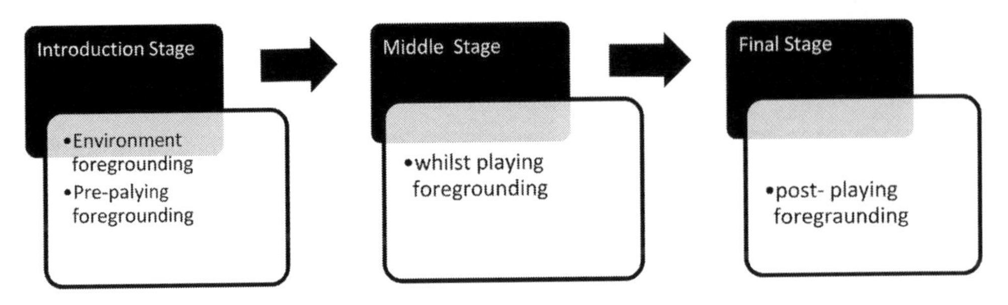

Figure 1. Stages of sociodramatic play adapted from workshop materials of childhood education (2012)

Vygotsky (1978) argues that children engage in sociodramatic play as they wish to imitate adults' roles and perform their activities that the child is too young to attempt in their real lives. Vygotsky meant only one kind of play, namely, the sociodramatic or a collaborative playtypical for preschoolers and children of primary school age (Bodrova et al., 2015). Vygotsky (1978) believed adult interaction can facilitate children's development within their zone of proximal development. A child "zone of proximal develop" is the range of tasks that the child can perform with guidance from others but cannot perform independently (Yellin et al., 2004). Bruner (1986) used the term *scaffold* as a metaphor to describe adult contributions to children's development and learning. Scaffolds are support mechanisms that teachers, parents, and others provide to help children successfully perform a task within their zone of proximal development (Rajapaksha, 2016).

2.3 *Stages of sociodramatic play activities*

Activities of sociodramatic play done by the teachers and students are designed in eight meetings in the classroom. Each meeting is divided into three stages of activities which had been designed in the form of Daily Learning Plan. Figure 1 shows the stages of sociodramatic play to be conducted.

2.4 *Research questions*

This research aims at finding out the teacher's roles in ZPD-based sociodramatic play in kindergarten. Research questions include:

1. How do teachers play their role as facilitators in ZPD-based sociodramatic play in kindergarten?
2. What are the challenges that the teachers face in implementing activities of sociodramatic play during the intervention implementation in the kindergarten classroom?

3 METHOD

3.1 *Research design*

This research employed both qualitative and quantitative approaches. The qualitative approach was conducted to validate the strategic action to develop prosocial behaviour of kindergarten students and the strategy to prepare teachers to conduct game-based learning to develop children prosocial behaviours through personal-social assistance to childhood education expert and written and oral counselling and guidance, interview, observational note, and teacher's reflection. In addition, we examined the effectiveness of the action using a pre-post test experimental design.

3.2 *Research site and participants*

This research was conducted in a kindergarten located in Cipocok Subdistrict. The research participants were three teachers of group B of TK Negeri Pembina Kecamatan. We used random sampling technique to select the participants for the experiment.

3.3 *Instruments*

To collect the data, the researcher used several methods, such as observation (this observation guide had passed interrater reliability test with Kappa coeffisien 0.6 meaning that this instrument is 'good' to be used), observational note, reflective journal, and structured interview to gather the data deeper. All observation activities were recorded and documented in the field notes. Structured interview was given to kindergarten teachers before and after training intervention and teaching intervention in the classroom.

3.4 *Intervention*

The intervention was in the form of training designed by the researcher. The training was intended to prepare the teachers' skill in developing children prosocial behaviours. The teachers are given materials, such as skills to assist the children, make sociodramatic lesson plan, make media for sociodramatic, and be the facilitator in sociodramatic implementation in the classroom.

After training, the teachers used sociodramatic play to develop children's prosocial behaviours.

3.5 *Data analysis*

The data was analyzed through several steps. Video recordings were played several times to observe the teachers' teaching practice after the intervention. It was carried out to identify the roles of teachers and kinds of prosocial behaviors such as sharing, cooperating, donating, helping, honesty, and empathy appeared in the classroom. Then, all teachers' activities were written carefully in the field notes. Finally, we analyzed the data descriptively.

4 RESULTS AND DISCUSSIONS

4.1 *Role of teachers*

The result revealed that the planning in the form of Daily Lesson Plan (DLP) designed by the teachers could be categorized as "good". In this case, the teachers could plan the learning material which arouse students' interest, the teachers used the theme "Universe" and change each meeting to make the activities varied. It was done to build the students motivation in learning. The teachers' lesson planning is shown in Table 1.

The table shows that the lesson planning done by the teachers are: (1) making DLP. Activities conducted in DLP making are determining core competence, basic competence, indicator, evaluation and learning steps; (2) choosing learning media. The activity is making media based on the sociodramatic scenario; (3) learning method. The method used by the teachers were more on the development through sociodramatic play by using several techniques, such as question and answer, discussion, reinforcement, task, and problem solving.

Lesson plans made by the teachers to develop students prosocial behaviour through sociodramatic play activities was based on the theme "Universe" and subtheme "Water, Fire, and

Table 1. Teachers' Lesson Planning

Measured Aspects	Pretest	Posttest	Improvement in Percentage
Making DLP	2.5	4	60
Making media for the play	2.25	4	77
Learning method	2.25	4	77
Learning evaluation	2.25	4	77
Total	9.25	16	73

Air" which were indicated by the play titles, like: "I Have Clean Teeth since I Brush my Teeth Regularly", "Laundry", "Selling Sausage", "Kinds of Wind", and "Selling Balloon".

The learning plan is categorized very good as the teacher could use it to solve the problem such as directing, controlling and assisting 12 children. The implementation of sociodramatic play to develop children prosocial behavior was done through several stages. First, they were doing environment foregrounding by setting the class and providing learning media based on the play title. Second, they prepared student to be ready for learning by a session of question and answer. Then, they explained the play scenario and rules, and discuss the roles that the students were going to play based on the scenario. In this stage, teachers read a story to inspire the children. Third, whilst playing foregrounding was done through directing the students to cooperate in the play, help each other, share and respect, be honest and empathy. At the same time, teachers guided children to follow a step-by-step of the determined drama scenario accordingly, encouraged children to practice new behaviours based on the scenario, and motivated them through affirmation, compliment, nod of approval, thumbs-up, smile and caress, as well as observed and documented the children' play and social development. The last or the fourth stage are the activities of doing post-playing foregrounding through closing activity, such as giving evaluation, reinforcement and follow up activity by giving the next day assignment. In this stage, teachers assisted children to identify the proper place and time to use their ability and the way to implement it in various situations. Such game can be repeated if the time is available by switching roles of the players.

This final activity was done through question and answer techniques, discussion and assignment. The learning implementation measured from pretest and posttest is shown in Table 2.

The teachers were able to conduct the learning by following the steps based on the rules of the play All aspects were designed thematically and aimed at improving children prosocial development. In this case, there was improvement on those who never used sociodramatic play to use this play by the researcher direction and assistance.

In sociodramatic play, the prosocial activities pattern worked well. The children showed their willingness to share with others, for example when Khanza who played a role as consumer would pay the sausage she bought but she did not have enough money, Gilang said, "*Here, use my money*". Then Gilang voluntarily gave her his fake money. Children's empathy will grow when facing difficult situation like condolence, children will feel their friends' happiness and sadness when they experience the same thing, the experience does not need to be real but can be made in sociodramatic play (Gunarti et al., 2008). Children had chance to help each other, such as when they finished playing, together they cleaned the room and packed up the toys and tools they had used..

While conducting sociodramatic play, the teachers gave some directions to children like "*Fely, these clothes are very* dirty, *could you wash it?*" then giving reinforcement with praise by saying, "Wow, Fely, now these clothes are clean and neat. Thank you for washing it". The dialogue shows that the teachers were directly involved in the play. The children also responded as if they were laundry staff by saying, "Yes, I will wash the clothes". It is identified that the teachers give guidance and direction to the children directly as well as indirectly by involving in the play (Meacham et al., 2014).

As a leader in the play, teachers develop several strategies to ask children to think what to do next by asking the questions in pretending dialogue. The teachers inauthoritatively asked

Table 2. Teachers' Learning Implementation

Aspects	Pretest	Posttest
Environment foregrounding	2	4
Pre-playing foregrounding	2	4
Whilst-playing foregrounding	2	4
Post-playing foregrounding	2	4
Total	8	16

them to think what to do next by asking them questions and statements in the pretending dialogue.

The teachers' involvement shows ZPD through giving modelling or demonstrating steps in the scenario, guiding children to do activities, giving questions, solving difficult problems through simple ways, and giving information. Bluiett (2009) posits, "When children gained knowledge and experience about how to perform the play, she gradually withdrew her support so that children could make the transition from external social interaction to internalized, independent functioning". The teachers in the theme play "Selling Sausage" seemed to give direction in steps, by giving questions and feedback. While playing, it seemed that the children did not really understand the theme because they never do sociodramatic play and never acted the roles in the theme. This dialogue shows how the teacher gives direction in several steps to the children:

> When Khanza, Agung, and Gilang stood and ordered sausage to the seller, the teacher said:
> *Teacher*: "Khanza, Agung and Gilang, please order the sausage to the shop keeper"
> When Arkarna and Dewa who took the roles as shopkeepers only sat and did nothing, the teacher said:
> *Teacher*: "Arkarna ... Dewa ... there are customers who had made an order, please grill the sausage."
> Then the teacher talked to Zahira who acted as the cashier:
> *Teacher*: "Zahira, have you counted how much money that Khanza should pay?"

It can be found that the teachers as adult people has main role to help children find the answers and know what to do (Vygotsky, 1978). The teachers helped the children to imitate the roles they play and helped them to act the roles properly by giving ideas on what to do based on each role.

4.2 *Challenges that the teachers encounter during intervention implementation*

It is identified that during observation and interview, teachers face many challenges in sociodramatic play intervention. The challenges are related to the implementation of sociodramatic play and media selection. In the first challenge, regarding the difficulty in implementing sociodramatic play, teachers found that they had difficulty to build children participation to act the roles, control large number of children (13 children), manage the time to set the room and make the necessary media. The teachers reflected that the challenges they face were because they never implemented sociodramatic play in their classroom before.

Following is an excerpt of an interview with the teacher about the challenges they met after the intervention in the classroom:

> During the implementation of sociodramatic play in the first meeting, I felt I could not do my role well, maybe I gave too many orders, could not control the children and I found it difficult to ask the children to act based on the scenario.
>
> (Interview, 27 April 2018)

The second difficulty is related to the room and media required for the next meeting. They made washing machine from used cardboards; hats from leaves, and toy-clothes from used paper.. The teachers also set the area by changing the position of table and chair based on the play scenario.

It was found out that the use of different media could increase the children's enthusiasm and participation to play. As one teacher said, "The children were very happy to play with different media, but a bit difficult to make media every day. However, I will give it a try."

It was stated that the challenges that the teacher faced during the sociodramatic play intervention were the lack of teachers and children experience in sociodramatic play in the class, and suitable media on every play theme.

5 CONCLUSION AND SUGGESTION

5.1 *Conclusion*

The children show more interest toward sociodramatic play compared to other activities which were usually done every day in the classroom. In the ZPD concept, teachers' role seemed natural when giving sociodramatic play intervention. Sujiono (2009) states that one of concepts which emphasizes learning for children with the help of adult is Zone of Proximal Development (ZPD), that is as the potential capacity of children learning that can be realized through the help of adult or more skillful people. The sociodramatic play can be a good pedagogical strategy in developing children prosocial development because of its communication and interaction inside the play. Moreover, the intervention can increase teachers' awareness and experiences, readiness in providing many kinds of media, and make the teachers skillful to act based on ZPD concept in sociodramatic play in the classroom. Therefore, assistance and emphasis need to be given to the teachers to raise their awareness based on ZPD concept in sociodramatic play.

5.2 *Suggestion*

Based on the research result, the recommendations offered to the teachers who will use sociodramatic play to develop children prosocial behaviour are: (1) teachers need to make lesson plan which can raise children interest and children development, (2) teachers need to enhance their creativities and imagination in providing and creating learning media, and (3) teachers need to have skills to assist and direct the children so that the play could be well organized even though with large number of children, (4) teachers need to choose a play theme close to the child, where the child has had direct experience with the theme.

REFERENCES

Alfiyah, S. (2014). Validasi modul bermain peran "Aku Sayang Kamu" untuk meningkatkan pengetahuan tentang perilaku prososial pada anak usia dini. Available online etd.repository.ugm.ac.id/downloadfile/74755/.../S2–2014–323534-chapter1.pdf. Accessed 1 February 2018.

Bluiett, T. E. (2009). Sociodramatic play and the potentials of early language development of preschool children. Available online http://acumen.lib.ua.edu/content/u0015/0000001/0000137/u0015_0000001_00001 37.pdf. Accessed 1 February 2018.

Bodrova, E., Germeroth, C., Leong, D. J. (2013). Play and Self-Regulation-Lessons from Vygotsky. *American Journal of play*, *6*(1), 111–123.

Bredekamp, S. (2013). *Effective practices in early childhood education: Building a foundation*.Upper Saddle River, NJ: Pearson.

Bredekamp, S., & Rosegrant, T. (1992). *Reaching potentials: Appropriate curriculum and assessment for young children* (Vol.1). Washington, DC: National Association for the Education of Young Children.

Bruner, J. (1986). Play thought and language. *PROSPECTS*: Quarterly Review of Education. Paris, France: UNESCO. Available online http://link.springer.com/article/10.1007/BF02197974#page-1. Accessed 1 February 2018.

Eisenberg, N., & Mussen, P. H. (1989). *The roots of prosocial behavior in children*. New York: Cambridge University.

Gunarti, W., & Muis, A. (2008). *Metode Pengembangan Perilaku dan Kemampuan Dasar Anak Usia Dini*. Jakarta: Open University.

Kontos, S. (1999). Preschool teachers talk, roles, and activity settings during free play. *Early Childhood Research Quarterly*, 14(3): 363–382. DOI: 10.1016/S0885–2006(99)0016–2.

McCain, M., Mustard, F., & Shanker, S. (2007). *Early years study 2: Putting science into action*. Toronto, ON: Council for Early Child Development.

Meacham, S., Vulkelich, C., Han, M., & Buell, M. (2014). Preschool teacher's questioning in sociodramatic play. *Early childhood research quarterly*, 29(4): 562–573. DOI: 10/1016/j.ecresq.2014.07.001.

Piaget, J. (1962). *The psychology of the child*. New York: Basic Books.

Rajapaksha, P. L. N. R. (2016). Scaffolding sociodramatic play in the preschool classroom: The Teacher's Role. *Medditeranian Journal of Social Science*, 7(4): 689–694. DOI: 10.5901/mjss.2016.v7n4p689.

Saracho, O., & Spodek, B. (ed.). (2003). *Contemporary perspectives on early childhood curriculum*. Greenwich. CT: Information Age Publishing.
Smilansky., S. (1990). Sociodramatic play: Its relevance to behavior and achievement in school. In E. Klugman & S. Smilansky (ed.) *Children's play and learning: Perspectives and policy implications*, 18–42. New York: Teachers College Press.
Smilansky, S., & Shefatya, L. (1990). *The Smilansky Scale for Evaluation of Dramatic and Sociodramatic Play. In S. Smilansky, Facilitating play*. Silver Spring, MD: Psychosocial and Educational Publications.
Stantan-Chapman, T. L. (2014). Promoting positive peer interactions in the preschool classroom: The role and the responsibility of the teacher in supporting children's sociodramatic play. *Early Education Journal. 43*(2), 99–107. DOI: 10.1007/s10643–014–0635–8.
Sujiono, Y. N. (2009). *Konsep dasar pendidikan anak usia dini*. Jakarta: PT. Indeks.
Suyadi, U. M. (2016). *Konsep Dasar PAUD*. Bandung: PT. Remaja Rosda Karya.
Vygotsky, L. S. (1978). *Mind in society: The development of higher psychological processes*. Cambridge, MA: Cambridge University Press.
Yellin, D., & Blake, M. E. (2004). *Integrating the language arts: a holistic approach*. Scottsdale, AZ: Holcomb Hathaway/York: Plenum.
Yusuf, S. L. N. (2014). *Perkembangan anak dan remaja*. Bandung: PT. Rosda Karya.

Early Childhood Education in the 21st Century – Yulindrasari et al. (eds)
© 2020 Taylor & Francis Group, London, ISBN 978-1-138-35203-2

Shaping healthy eating behavior in children

S.N. Fadhila, Rohita & Nurfadilah
Universitas Al Azhar Indonesia, Jakarta, Indonesia

ABSTRACT: The first thousand days in life is crucial for health, growth and development of children. Children's nutrition intake relates to healthy eating behavior. This study aims to find out how shaping healthy eating habits in children aged 2–3 years. This study uses descriptive qualitative research through observation, interview and documentation with data sources from two teachers of a 2- to 3-year-old group, one principal, and six children. The data analysis uses Miles and Huberman's model which involves data reduction, data presentation, and drawing conclusion. The result showed that shaping healthy eating behavior has been formed by teacher's role related eating habits, food choices, and meal times, which are (1) creating a pleasant environment; (2) maintaining an appropriate standards; (3) helping children eating food; (4) taking care children to stay on their meal; (5) choosing and serving food menus; (6) set up routine meals and snacks on daily activity schedules.

1 INTRODUCTION

Early childhood growth and development are important to be concern for parents and educators. Two hundred seventy days on pregnancy and 730 days after child born is a crucial time for child. During this period brain grows rapidly. The quality of child's growth and development is influenced by stimulations and nutritions. Parents need to pay attention to stimulation and nutrition given to children, especially nutrition during pregnancy. Iskandar says that good nutrients are obtained from healthy and variety food (Adriani, Merryana, & Bambang, 2014). Healthy eating behavior will help children fulfill their need of nutritions to grow and develop. Many parents prioritise more on the amount of food consumed by children with less variety of food. This phenomenon is supported by development food industry which is competing to offer fast food products. The habit of children in consuming excess fast food can cause several problems, including brain impaired, liver and heart disease, and obesity. According to Nutritional Status Monitoring data from Ministry of Health Republic Indonesia (2017) showed the number of infants with obesity problems increased from 4.3 percent to 4.6 percent. The increasing number of toddlers obesity problem is eating behavior.

Accordingly, healthy eating behavior becomes very important to be formed from an early age to help children grow up both physically and spiritually. In this study focused on children aged 2–3 years old. In addition, this is supported by the opinion of Gonzalez-Mena and Eyer (2011) that tastes and habits of children develop in the first three years of influence throughout their lifetimes. Furthermore, healthy eating behavior also is an important requirement in Indonesia accordance with Presidential Regulation No. 60 in 2013.

One of specific goals Holistic Integrative Early Childhood Development (HI ECD) contained in Presidential Regulation No.60 in 2013, namely fulfillment essential needs early childhood as a whole including health and nutrition; educational stimuli; moral-emotional; and parenting. The fulfillment of health and nutrition essential needs can be fulfilled through health and nutrition services by families, parents, government, and substitute caregivers. Substitute caregivers referred to Presidential Regulation No. 60 in 2013 is a person or institution that given the right or authority to childcare.

Research by Nurfadilah, Rohita, and Fitria (2017) on *Taman Anak Sejahtera* in Jakarta known that only one daycare has provided health and nutrition services, while other two daycare still need improvement in health and nutrition services in their care. This shows that health and nutrition services in child care at daycare have not been maximally provided. Other studies related to health and nutrition services by Rohita, Fitria, and Nurfadilah (2017), in this study was found that caregivers in daycare located in university, considered balanced nutrition for consumption by children. Accordingly background was mentioned earlier, the purpose this study to describe how the teacher's role in shaping healthy eating behavior in children aged 2–3 years. The teacher's role is related to eating habits, food choices, and meal times.

2 THEORETICAL FRAMEWORK

2.1 *Eating behavior*

There are several types behavior related to humans, such as learning behavior, consumer behavior, eating behavior, and so on. One of behaviors related with basic human needs is eating behavior. According to Sulistyoningsih in Nelvi (2015) eating behavior is behavior man or people in fulfilling need for food that includes attitude, trust, and food preference. Osorio in Hernandez, Bamwesigye and Horak (2016) defines eating behavior is a behavior related to eating habits, selecting foods that you eat; culinary preparations and quantities of ingestion. Grimm and Steinle (2011) defines eating behavior is a complex interplay of physiologic, psychological, social, and genetic factors that influence meal timing, quantity of food intake, and food choices.

Therefore, limitation of definition eating behavior is used combines three opinions by considering aspects that can be observed directly. When concluded, eating behavior is a human behavior in meeting food needs which include eating habits, food choices, and meal times.

2.2 *Healthy food*

Food becomes something that surely consumed by everyone. Food consumed will become recourses of energy for body. The body needs food intake from healthy and nutritious foods. As stated by Susilawati (2017) healthy food is food that containing all the nutrients needed by body to obtain energy.

Wijayati (2008) defines healthy food as food has nutritional balance. According to Adriani, Merryana, and Bambang (2014) balanced nutrition is a food consumed by toddler in a day and contains of energy substances, building substances, and regulatory substances according to needs of body. In general, an opinion that outlined above healthy foods are foods that contain nutrients. In spesifically, healthy food is that foods contain balanced nutrition include energy, building, and regulatory substances needed by body.

According to Adriani, Merryana, and Bambang (2014, p. 174) feeding balanced nutrition for children aged 2–3 years as follows: (a) give children a fresh cow's milk or formula milk twice a day; (b) give children a various family food according balanced nutrition three times; (c) nutritious snack are given 1–2 times a day; (d) give fresh fruits.

2.3 *Teacher's role in shaping healthy eating behavior*

Healthy eating behavior in children will be well formed, one of them with teachers at school. Pimento & Kernested (2010, p. 290) mentioned some role of teachers related to meal activities as follows: (a) choosing and serving nutritious food; (b) set up routine time for meals and snacks; (c) creating a pleasant environment; (d) maintaining appropriate standards of eating behavior at dining table; (e) helping children eating their food.

According to Robertson (2015), teacher can contribute setting up a positive eating environment. Teacher can create a positive eating environment. A positive eating environment can be formed with a pleasant atmosphere. The purpose to direct child interest and pay attention activities. There are various ways that teacher can do attract children's attention according Feurestein and Phina in Wilman, et. al (2009, p. 3) as follows: (a) make an eye contact between teacher and children; (b) saying a words with voice; (c) activate child's senses; (d) performs various functions or movement with aim helping children; (e) make attention object closer/far away.

Robertson (2015) also suggested opinion, if children not eat, adult should learn to relax, stay clam, and flexible. Teacher doesn't need force them when children doesn't want to consume their food. Encourage independent behavior in your child through offering choices in foods, utensils and respecting their "I am full" signal (Grim & Grosser, 2016). Ormrod (2011) says behavior could be set up as exected by way using extrinsict reinforcement. Extrinsict reinforcement such as motivation or rewards.

Teachers also need pay attention children's development in shaping healthy food eating behavior. Development eating skills of children aged 2–3 years that is drinks from a cup (no lid) without spilling, stabs food with fork, and uses spoon without spilling (Mielke, 2008). Teacher need realize development which appear on every child, in order to provide stimulation accordance with its development. Robertson (2015) also said that adults are also responsible for making sure children are at meal, keeping children on task, making sure they behave well, and set up time for meals and snack.

3 METHOD

This study is a descriptive qualitative research, to describe process of shaping healthy eating behavior in daycare. Sources of data were obtained from two teachers aged 2–3 years group, one principal, and six children through observation, interviews and documentation activities. The collected data was analyzed using data analysis of Miles and Huberman model through data reduction, data presentation, and conclusion drawing. So that it can answer problem that were present.

4 RESULT

4.1 *Eating habits*

Teachers created a pleasant atmosphere, teachers did it by eating together, make eye contact and spoke clearly with child, used movement to help children understand teacher's intentions such as pointing towards child's right hand, activatate child's senses, brought food choices closer and distract objects that disturb child. Children behavior that appears: (a) saw teacher's eyes when invited to speak; (b) showed behavior in accordance with teacher's instructions such as took cutlery, scoop out rice, ate and chew food; (c) chose food provided; (d) got a fruit; (e) paid attention when teachers brought objects/food choices closer; (e) continued pray when teacher kept glass played by child.

Teachers maintain standard of children behavior by remind them habituation before, during and after meals, children behavior that appears: (a) line up before washed hands; (b) washed hands, then go to dining room; (c) take cutlery e.g. glasses and spoons on rack independently; (d) sat neatly before pray; (e) raise hands and bow heads when teacher gives a gesture of prayer; (f) pray before/after meals and their meanings; (g) take one piece fruit/snack when teacher brings bowl closer; (h) holds spoon using right hand; (i) hold glass with two hands while drinking; (j) chew food; (k) stop eating when teacher reminds them that they have enough to eat; (l) consume water after eating; (m) arranged glass over bowl after eating; (n) returns cutlery on one place; (o) pushes chair.

Teachers help children consumed food such as get food, cut food into small pieces, peeling corn, scoop out food, give motivation, help feed their food, pour a vegetable sauce on a bowl, and stimulate their eating skills, Children behavior that appears: (a) ate by himself; (b) ate food that prepared on spoon by teacher; (c) consume and finish their food after teacher give motivation; (d) follow teacher scoop out food, pour water into glasses, chew food and drink using glass; (e) hold spoon by theirself.

Teacher maintain all child stay on their task by asking child to be patient following rules, child behavior towards role that teacher did: (a) participate eat accordingly schedule; (b) focus on eating their food while teacher remind them to clear their food.

4.2 *Food choices*

Healthy eating behavior can be established if intake obtained by children accordance with requirements of healthy food. The conditions healthy food reflected in food choices provided. Below this what teacher did: (a) make a monthly menu; (b) choose a menu according balanced nutrition, contains three elements of substances needed by body (energy, building, and regulatory substances); (c) menu consists of staple foods (rice), vegetable side dishes (tempe, tofu, corn) with different cooking methods, protein side dishes (fish, chicken, meat, anchovy, and eggs) with different cooking methods, vegetables with a variety (clear sauce, coconut milk sauce, and yellow sauce), and variety of fruit (in one week same type of fruit); (d) snack menu (sweet and savory snacks), as well as milk boxes brought from home.

4.3 *Meal times*

Healthy eating behavior related to regularity at meal times, teacher arranging routine meals and snacks on daily activity schedule of two snacks at 09.30 am and 03.30 pm, and lunch at 11:30 am. The behavior shown child following meal time according schedule.

5 DISCUSSION

Teachers have a very important role in providing support and facilitating children during meal activities to develop and optimize their potential including healthy eating behavior. Healthy eating behavior that formed is related to eating habits, food choices, and meal times. The teacher's role in child's eating behavior starts from their knowledge about health. Teacher have to give good nutrition to protect children by diseases. Good nutritional intake of children through food. Teacher not only has knowledge about health, but also has knowledge about nutrition to form food menu in a day. Not only that, teacher also needs skill in shaping children to consume healthy food. Help children eat their food by stimulating children's eating skills,e.g. help children holding a spoon using right hand, how scoop out rice using a spoon, and drink with glass using two hands. Not just practice and learning how to eat. Also forms a pattern eating habit on meal time activity.

6 CONCLUSION

Based on result, shaping healthy eating behavior in children aged 2–3 years can be formed by role teacher. The roles what teacher did at daycare to shaping healthy eating behavior in children aged 2–3 years includes eating habits, food choices, and meal times. The teacher's role is related to eating habits, that is (1) creating a pleasant environment; (2) maintaining appropriate standards of eating behavior; (3) helping children eating food; (4) take care of children to stay on their meal. The teacher's role related to food choices, which is choosing and serving a food menu by arranging monthly food that varies. Whereas, the teacher's role related to meal time is set up routine time for meals and snacks on daily activity schedules.

REFERENCES

Adriani, M., & Bambang, W. (2014). *Peranan gizi dalam siklus kehidupan [The role of nutrition in the life cycle]*. Jakarta: Prenada Media.

Gonzalez-Mena, J., & Eyer, D. W. (2011). *Infants, Toddlers, and Caregivers: A Curriculum of Respectful, Responsive, Relationship-Based Care and Education* (9th Edition) McGraw-Hill Education

Grim, E., & Grosser, J. (2016). *How to improve feeding skills in children*. Tallahassee Florida: Early Steps Children's Medical Service.

Grimm, E. R., & Steinle, N. I. (2011). Genetics of eating behavior: established and emerging concepts. *Nutrition Reviews*, *69*(1), 52–60.

Hernandez, J., Bamwesigye, D., & Horak, M. (2016). Eating behaviors of universitiy student. *Mendel Net*, 565–570.

Mielke, K. (2008). *Guidelines for the development of self-feeding skills*. Greenville USA: Super Duper Publication.

Ministry of Health Republic of Indonesia. (2017). *Buku saku pemantauan status gizi tahun 2017 [Handbook for nutrition status monitoring in 2017]*. Jakarta: Kementerian Kesehatan Republik Indonesia.

Nelvi, R. (2015). Hubungan antara dimensi kepribadian Big Five dengan perilaku makan pada mahasiswa UIN Suska Riau di Pekanbaru Riau [The relationship between the dimensions of the Big Five personality and eating behavior in UIN Suska Riau, Pekanbaru Riau]. *Thesis*. Fakultas Psikologi: UIN Sultan Syarif Kasim Riau.

Nurfadilah, Rohita, & Fitria, N. (2017). Pelaksanaan pengasuhan di Taman Anak Sejahtera [Implementation of care in Taman Anak Sejahtera]. *Jurnal Ilmiah Visi*, *12*(1), 18–23.

Ormrod, J. E. (2011). *Educational psychology*. Cambridge: Pearson.

Pimento, B. and Kernested, D. (2010). Healthy Foundations in Early Childhood Settings, 5th Edition. Toronto, Canada: Nelson Education

Robertson, C. (2015). *Safety, nutrition, and health in early childhood*. Independence Ohio State: Cengage Learning.

Rohita, Fitria, N., & Nurfadilah. (2017). Implementation of early childhood development integrative and holistic in Daycare. *Advances in Social Science, Education and Humanities Research (ASSEHR)*, *58*, 348–352.

Susilawati, F. (2017). *Makanan sehat [Healthy food]*. Jakarta: Kementerian Pendidikan dan Kebudayan.

Wijayati, E. S. (2008). *Mengenal makanan sehat [Knowing healthy food]*. Depok: Cerdas Interaktif.

Wilman, W., et al. (2009). *Modul pengembangan model teknik mediasi dalam meminimalkan permasalahan makan pada anak usia balita [Module development of the mediation technique model in minimizing feeding problems in toddler]*. Depok: Universitas Indonesia.

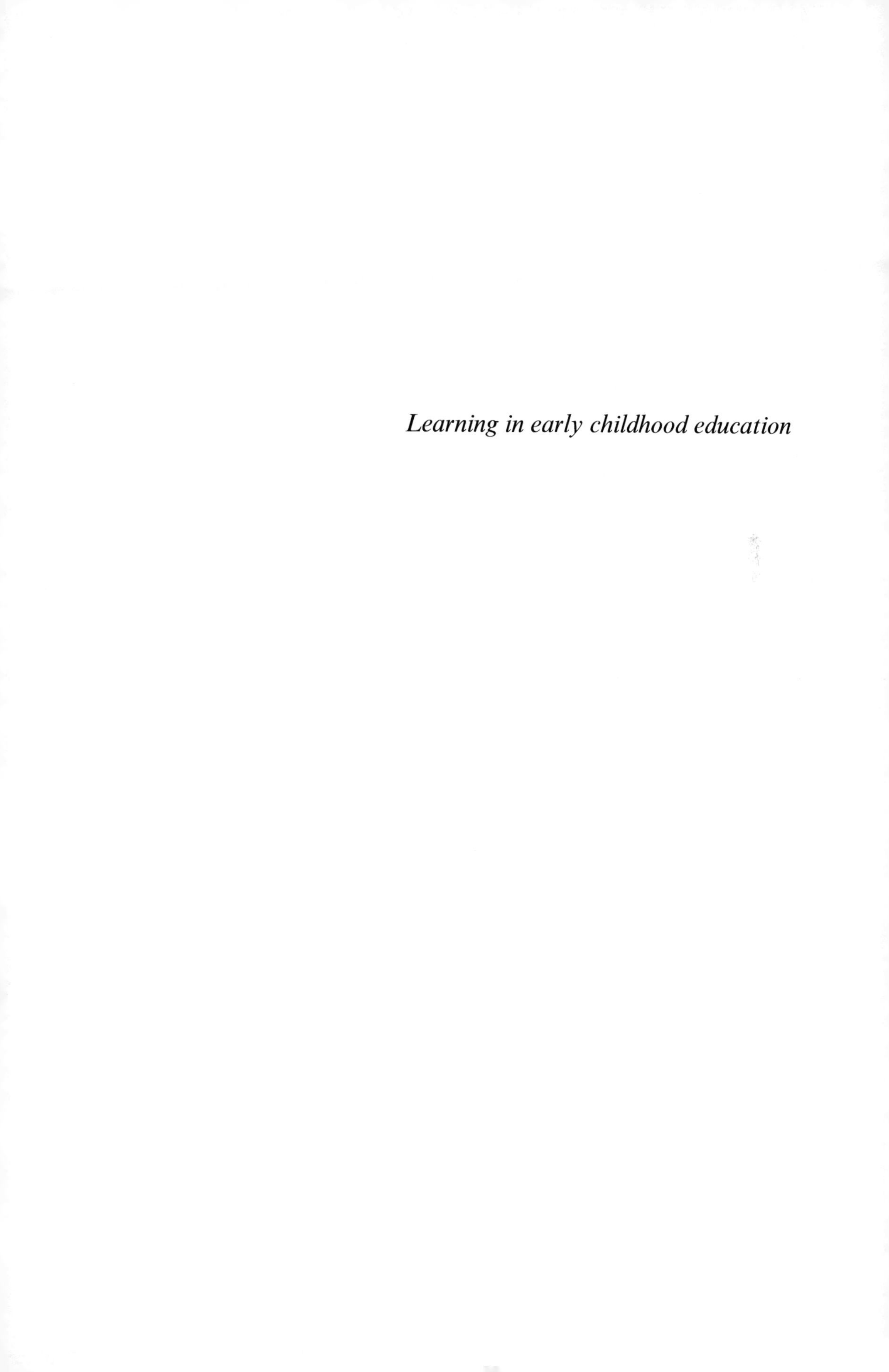

Learning in early childhood education

Early Childhood Education in the 21st Century – Yulindrasari et al. (eds)
© 2020 Taylor & Francis Group, London, ISBN 978-1-138-35203-2

Mathematics and children's construction in early childhood education

M. Noviyanti
Indonesia Open University, Banten, Indonesia

D. Suryadi, S. Fatimah & V. Adriany
Universitas Pendidikan Indonesia, Jawa Barat, Indonesia

ABSTRACT: To what extent mathematics can be taught to young children remains debatable in early childhood education (ECE). Many educators believe that young children are encouraged not to learn mathematics because they are still in the pre-operational stage where they find it difficult to understand complex and abstract concepts like mathematics. As a result, teaching *calistung* (how to read, write, and count) during ECE in Indonesia is prohibited. On the other hand, some also believe that mathematics needs to be introduced at early childhood in which children will get various advantages from learning math. This chapter aims to explore the debate on the pros and cons of teaching mathematics in ECE. Using literature review as the tool, this article argues that the polemic of teaching mathematics in ECE illuminates the different constructions about children and childhood in Indonesia.

1 INTRODUCTION

Mathematics is currently a central topic in debates circulating in early childhood education (ECE). The debate on whether mathematics should be taught to young children has raised concern, especially among parents, on whether their children should learn mathematics at kindergarten. Parents worry that being forced to learn mathematics will make children find it boring to learn in the next school stage. In addition to parents, teachers also feel that teaching mathematics in early childhood education is confusing.

Lee and Ginsburg (2009) stated that there are nine common misconceptions about early childhood education mathematics among teachers in the United States. One of those misconceptions is that teachers do not think that young children are ready for mathematics education. Whereas in Indonesia, the introduction of early childhood mathematics received a negative response (Semiawan, 1999). The Indonesian government has issued a Circular of the Directorate General of Primary and Secondary Education Number 1839/C.C2/TU/2009 concerning the Implementation of Kindergarten Education and Admission for Primary School Students. One of the points stated that was not permitted to teach mathematics directly in kindergarten.

On the other hand, some experts advocate that young children are ready for mathematics. Gardner (2011) stated that logical-mathematical intelligence is one of the nine intelligence potentials possessed by children. Kilpatrick (2001) added that based on several studies over the past 25 years, young children are actually familiar with mathematics. Even for young children, counting is not entirely a work of memorization, but rather it has been guided by their own mathematical understanding (NRC, 2001).

This chapter is an in-depth discussion on the pros and cons of teaching mathematics in ECE. Using literature review as the tool, this article analyses the polemics of early childhood education mathematics teaching by showing the differences in the construction of children and ECE in Indonesia.

2 CHILDREN'S CONSTRUCTION THEORY

Children's mathematical abilities are within the scope of cognitive development. Cognitive development is the growth of thinking from infancy to adulthood (Piaget 2002, Vygotsky 1978, Clements, Sarama, & DiBiase, 2003). Piaget and Vygotsky are two well-known figures in the theory of construction of children who are greatly influential in the field of children's cognitive development.

Jean Piaget explores the cognitive development of children to study genetic epistemology as his primary interest (Ojose, 2008; Piaget 2002). As for child constructivism, Piaget focuses on individual development (Piaget 2002; 2013; Ojose, 2008; DeVries, 2000) and does not recognize the important role of social factors in the construction of children (Ojose, 2008). Piaget (2013) stated that the development of a child occurs through the transformation of a continuous thought process and thought that cognitive development depends on chronological age (Hebe, 2017; Ojose, 2008). For this reason, he believes that cognitive development in human occurs gradually and goes through different stages (Hebe, 2017). Piaget divides the development into four stages, namely: (1) sensorimotor stage (0–18/24 months), (2) pre-operational stage (2–7 years), (3) concrete operational stage (7–11 years), and (4) formal operational stage (11–15/16 years) (Ojose, 2008; Piaget, 2013). From these various points of view, it can be concluded that Piaget's theory views children's construction as being built by children themselves without really depending on social factors. In addition, Piaget also focused on the stages of children development and suggested that the given intervention is adapted to the stages.

Piaget's developmental stage theory received criticism from several experts, among them is Lev S. Vygotsky. Vygotsky's theory tends to be centered on the argument that social relations with society and culture shape one's knowledge (Derry, 2013; Vygotsky, 1989). Some of the concepts of Vygotsky's theory relating to this research are social interaction that plays an important role in the process of cognitive development. Vygotsky views that social interaction or assistance from others can help increase understanding in children (Vygotsky, 1986; 2014, 1978). The second concept is the More Knowledgeable Other (MKO), referring to anyone who has a better understanding or higher level of ability in relation to a particular task, process, or concept (Vygotsky 1978; Derry, 2013). Next is the Zone of Proximal Development (ZPD). Vygotsky (2014) defines ZPD as the distance between the actual level of development as determined by independent problem solving and the level of potential development as determined through problem solving under the guidance of an adult or by working with a more capable partner. The last concept is Scaffolding, a term referring to the process that adults use to guide children through the ZPD (Derry, 2013). Vygotsky added that children in the early stages of learning need a lot of support or scaffolding in carrying out their tasks. Scaffolding facilitates learners' ability to build prior knowledge and internalize new information (Vygotsky, 2014; Derry, 2013). From his various points of view, Vygotsky stressed the importance of social factor towards the construction of children.

The difference between Piaget and Vygotsky's views has an impact on the debate in society, especially in the implementation of mathematics teaching for ECE. The debate also raised doubts among institutions and teachers of ECE to further explore mathematics.

3 DEBATE ON EARLY CHILDHOOD MATHEMATICS IN THE COMMUNITY

As revealed in the introduction, many favor the argument that it is not timely to give mathematics lesson to young children. Epstein (2003) argued that children are not ready for mathematics before entering elementary school. On the other hand, research conducted by Hembree (1990) suggested that children experience anxiety about mathematics when entering the period of junior high school to senior high school, and this also happens to young children. Teachers' strong emphasis on the academic results of mathematics will damage children's self-confidence, curiosity and intrinsic motivation to learn (Stipek, 2013). Kamii in (Clements,

Fuson, & Sarama 2017) claimed that children cannot connect each subsequent number with +1 operation until they reach the third grade of elementary school.

In Indonesia, calistung (how to read, write, and count) education in kindergarten has negative responses from early childhood education observers. Semiawan (1999), among others, revealed that "there has been a misunderstanding in some communities in the implementation of education for young children". Saniy (2014) stated that "the average learning outcomes of third grade elementary school students who received calistung in kindergarten are lower than those of students who did not get it." This is indicated by their average test scores.

In addition, the Indonesian government has issued a Circular of the Directorate General of Primary and Secondary Education Number 1839/C.C2/TU/2009 concerning the Implementation of Kindergarten Education and Admission for Primary School Students. There are three things highlighted in the circular: (1) Kindergarten education is not allowed to teach calistung material directly; (2) Kindergarten education is not allowed to give homework to students in any form; and (3) Every elementary school must accept students without any entrance tests. This circular certainly raises confusion among parents who still want to develop their children's intellectual potential without having to wait until their children reach 7 years of age.

The opinion of anyone who doubts the implementation of mathematics teaching for ECE is thought to be their interpretation of Piaget's theory (Clements, Fuson, & Sarama 2017; Lee & Ginsburg, 2009). As stated before, Piaget divided children's development stages into sensory motor, preoperational, concrete operational and formal operational. Out of the four stages, kindergarten children are in the pre-operational stage. According to Piaget, children at the pre-operational stage tend to see things from their own perspective and expect others to do the same (Hebe 2017). According to him, under the age of 7 years, children are not yet capable of operational-concrete thinking; consequently, they can neither understand abstract concepts nor have logical thinking needed in mathematics. In Piaget's view, pre-operational children fail to show not only the conservation of fluids but also the conservation of numbers, volumes, materials, lengths, and fields (Piaget, 2002, 2013; Hebe, 2017). Therefore, Piaget is worried if mathematics will burden children who have not been able to think structurally (Lee & Ginsburg, 2009; NRC, 2001; Piaget, 2002, 2013).

Piaget's theory of development has received criticisms from several experts. The theory has been criticized for its claim that all children in certain age groups reach the same level of cognitive development (Hebe, 2017). Siegler in (Clements, Fuson, & Sarama, 2017) criticized that children's cognitive development is a continuous learning process, not in the form of stages as Piaget put it. Children have achieved conservation at a younger age than what is claimed by Piaget (Devine, 2000). In relation to mathematics, Gelman (1972) believed that Piaget did not consider young children to have good competence in mathematics. Piaget's theory does not offer a complete description of cognitive development, which has an impact on the understanding that mathematics is not appropriate for young children (Ojose, 2008).

Furthermore, Piaget's argument on the categorization of age-based development is also in line with the child protection perspective. This perspective is more likely considered to be paternalistic because children's characteristics tend to be generalized without taking into account their diversity as individuals who live and grow up in diverse environments and parenting (Kirk, 1999). This view shows a romantic concept of childhood (James, Jenks, & Prout, 1998), detailing concerns over the loss of innocent 'childhood' (Burman, 2016, James, Jenks, & Prout 1998); emphasizing playing as the main business of childhood (Burman 2016) and the need to be protected from 'adult' ways of knowing things (Hebe, 2017). When children's rights are described as a form of protection rights, children tend to be seen as incompetent individuals that need protection from adults (Kirk, 1999). Piaget's developmental theory and child protection perspective clearly have raised people's concern about applying mathematics in early childhood education, thereby giving rise to the pros and cons of mathematics in early childhood education.

On the other hand, Vygotsky's theory is in harmony with the view advocating that mathematics can be applied to young children. The theory is also used by some groups of liberationists to interpret children's rights as a form of liberation, where a child's position as 'subject of rights' is questioned (Goldstein et al. 1996; Kirk, 1999). According to Sebba (2005), child liberation embodies the idea that children must have the same rights as adults by paying attention to their desires and freedom of choice. This view rejects age restrictions as the only measure of children's competence because in reality age does not always reflect the level of children's maturity (Kirk, 1999). Kirk added that there are several conditions that cause a child to be more mature and more competent than his age; hence, it is unfair if their competence and autonomy are solely measured by their age limits.

The role of adults is instrumental in guiding children's freedom in exploring knowledge. Vygotsky (1989) sees cognitive growth as a collaborative process of children developing more systematic, logical, and rational concepts as a result of conversations with friends and adults. In other words, Vygotsky's theory shows that other people and language play an important role in children's cognitive development.

In Indonesia, Vygotsky's theory is in harmony with the views of the Educational Assessment Centre (Puspendik) which analyzed the low performance of Indonesian students in 2016 regarding mathematical literacy, in the Programme for International Student Assessment (PISA) and Trend in International Mathematics and Science Study (TIMSS). An analysis by Puspendik revealed the factors that could potentially relate to low mathematics achievement scores, which are as follows: (1) Around 28% of students in Indonesia "did not" attend early childhood education, and this was the fifth highest number compared to other countries; (2) Only 27% of parents in Indonesia carry out activities that stimulate student numeracy and literacy skills, such as reading tales, singing alphabets, etc., while the international average is 44%; (3) The number of Indonesian teachers who find it difficult to follow curriculum changes is categorized as high (12.18%); and (4) Only 6% of students in Indonesia use computers in learning, while the international average is 37% (Puspendik, 2012).

From the results of the analysis, it can be concluded that developing mathematics in early childhood education is one of the important aspects for the development of children's abilities. Similarly, the role of teachers and parents in providing stimuli to children is important. Providing good stimuli will help children gain an advanced level of literacy skills. In Vygotsky's theory, teachers and parents are The More Knowledgeable Other (MKO), a model or a good role model for children (Vygotsky, 1978, 2014; DeVries 2000). The MKO is expected to provide stimuli to children to improve their numeracy and literacy skills. DeVries (2000) stated that Vygotsky had considered that stimuli is presented in conditioned and unconditioned responses that depend on environmental action (Vygotsky 1978, 1986, 1989; DeVries 2000); and based on a didactic teaching in education which focuses on guidance (DeVries, 2000; Kozulin, Boris, Vladimir, & Miller et al., 2003).

In addition, the MKO plays an important role to provide assistance to children in the form of scaffolding (Derry 2013; Kozulin et al. 2003; Van Der Stuyf, 2002). The scaffolding process in the development activities determines the success of the Zone of Proximal Development (ZPD). Vygotsky (1978) understands that ZPD is to describe the actual level and the next level of the children's development through the media of semiotic tools, environments, adults or peers.

With the right stimuli and scaffolding, it is expected that early childhood can learn mathematics concept well, considering that early childhood mathematics is very influential for the children's ability in the future. Several studies further suggest that young children's mathematical knowledge even predicts their achievement later in high school (Clements, Fuson, & Sarama, 2017; Duncan et al. 2012; Watts et al., 2014); and predicts future reading achievement and initial reading skills (Clements, Fuson, & Sarama, 2017; Duncan et al. 2012). The mathematical ability of young children also influences their abilities in the future, both mathematical abilities and other abilities (Duncan et al., 2012; Hardy & Hemmeter, 2014; Lino 2016; Watts et al., 2014). According to Lino (2016), children who get good developmental activities from 0 to 6 years perform better on academic tasks such as reading, writing, and mathematics. In addition, they become more capable of social interaction than other children and adults.

4 CONCLUSION AND RECOMMENDATION

Piaget's and Vygotsky's theories are the foundation for education practitioners to decide whether it is appropriate for young children to learn mathematics. Piaget's argument, especially regarding age categories that are associated with children's ability to complete mathematical tasks, has stirred some debates. In his view, mathematics is considered not feasible to be applied in early childhood education until children enter the pre-operational stage (7 years old). Based on this theory, children must wait for the proper age to build their knowledge. Piaget's opinion is in line with the child protection perspective that argues for children's protection rights but tends to view children as incompetent individuals who need protection from adults. In addition, Piaget's argument that children can construct their own knowledge without the influence of language and other people is deemed unreasonable.

Piaget's argument is refuted by Vygotsky, who stated that children can build their knowledge not only by themselves but also with the help of others. According to Vygotsky, people who have more knowledge than children (such as teachers) can act as the More Knowledgeable Other (MKO), who can help children understand a mathematical concept with good social interaction and scaffolding. If the process is done well, it will result in good ZPD. Vygotsky's theory is also in line with the perspective of child liberation which interprets children's rights as a form of liberation and rejects age restrictions as the only measure of children's competence.

Based on the literature review of theories and research results, it can be concluded that mathematics can be taught to young children. However, the introduction of mathematics should be done by the method of playing and making children happy. The influence of play on children's development is considerable (Clements, Sarama, & DiBiase, 2003; Derry 2013). If children are accustomed to memorable and enjoyable mathematics learning experiences at the pre-school level, they will tend to appreciate and get used to mathematics learning experiences when they enter elementary, middle school, and high school (Hardy & Hemmeter, 2014; Linder, Powers-Costello, & Stegelin, 2011). High quality learning and good classroom environment have been shown to encourage higher student achievement (Kilday & Kinzie 2008). During their golden ages, children must get the right stimuli so they can get to know mathematics. The key lies in the transfer of knowledge itself. The given intervention and how to teach mathematics to children must be really mature to avoid mathematical phobia that has occurred so far (Stipek, 2013). Types of interventions expected to produce the longest-term effects on children's mathematical achievements should also be explored (Clements, Fuson, & Sarama, 2017).

REFERENCES

Burman, E. (2016). *Deconstructing developmental psychology*. East Sussex: Routledge.

Clements, D. H., Fuson, K. C., & Sarama, J. (2017). The research-based balance in early childhood mathematics: A response to Common Core criticisms. *Early Childhood Research Quarterly*, *40*, 150–162.

Clements, D. H., Julie, S, & DiBiase, A. M. (2003). *Engaging young children in mathematics: Standards for early childhood mathematics education*. New York: Routledge.

Derry, J. (2013). *Vygotsky: Philosophy and education*.West Sussex: John Wiley & Sons.

Devine, D. (2000). Constructions of childhood in school: Power, policy and practice in Irish education. *International Studies in Sociology of Education*, *10*(1), 23–41.

DeVries, R. (2000). Vygotsky, Piaget, and education: A reciprocal assimilation of theories and educational practices. *New ideas in Psychology*, *18*(2–3), 187–213.

Duncan, G.J., Katherine, M., Ariel, K., & Ziol-Guest, L. (2012). The importance of early childhood poverty. *Social Indicators Research*, *108*(1), 87–98.

Epstein, R.M. (2003). Mindful practice in action (I): Technical competence, evidence-based medicine, and relationship-centered care. *Families, Systems, & Health*, *21*(1), 1.

Gardner, H. (2011). *Frames of mind: The theory of multiple intelligences*. New York: Hachette UK.

Gelman, R. (1972). Logical capacity of very young children: Number invariance rules. *Child Development*, 75–90.

Goldstein, J., Albert, J. S., Sonja, G., & Freud, A. (1996). *The best interests of the child: The least detrimental alternative*. New York: Simon and Schuster.

Hardy, J. K., & Hemmeter, M. L. (2014). *Systematic instruction of early math skills*. Doctoral dissertation, Vanderbilt University.

Hebe, H. N. (2017). Towards a theory–driven integration of environmental education: the application of Piaget and Vygotsky in Grade R. *International Journal of Environmental and Science Education.*
Hembree, R. (1990). The nature, effects, and relief of mathematics anxiety. *Journal for research in mathematics education*, 33–46.
James, A., Chris, J, & Alan, P. (1998). *Theorizing childhood.* Cambridge: Polity Press
Kilday, C.R., & Kinzie, M.B. (2008). An analysis of instruments that measure the quality of mathematics teaching in early childhood. *Early Childhood Education Journal, 36*(4), 365–372. doi: 10.1007/s10643–008–0286–8.
Kilpatrick, J. (2001). Understanding mathematical literacy: The contribution of research. *Educational studies in mathematics, 47*(1),101–116.
Kirk, S. (1999). The sexual abuse of adolescent girls: social workers' child protection practice. *Adolescence, 34*(136), 806.
Kozulin, A., Boris, G., Vladimir, S. A, & Miller, S.M. (2003). *Vygotsky's educational theory in cultural context.* New York: Cambridge University Press.
Lee, J.S, & Ginsburg, H. P. (2009). Early childhood teachers' misconceptions about mathematics education for young children in the United States. *Australasian Journal of Early Childhood, 34*(4), 37–46.
Linder, S. M, Powers-Costello, B., & Stegelin, D. A. (2011). Mathematics in early childhood: Research-based rationale and practical strategies. *Early Childhood Education Journal, 39* (1), 29–37.
Lino, D. (2016). Early childhood education: Key competences in teacher education. *Journal Plus Education, 14*(2), 7–15.
NRC, National Research Council.(2001). *Adding it up: Helping children learn mathematics.* New York: National Academies Press.
Ojose, B. (2008). Applying Piaget's theory of cognitive development to mathematics instruction." *The Mathematics Educator, 18*(1), 26–30.
Piaget, J. (2002). *Judgement and reasoning in the child.* London: Routledge.
Piaget, J. (2013). *The moral judgment of the child.* London: Routledge.
Puspendik, T. (2012). *Kemampuan matematika siswa SMP Indonesia menurut benchmark internasional TIMSS 2011.* Jakarta: Pusat Penilaian Pendidikan Badan Penelitian dan Pengembangan Kementrian Pendidikan dan Kebudayaan.
Saniy, M. M. A. (2014). Perbandingan Prestasi Belajar Matematika Siswa Sd Negeri Sampangan 02 Semarang Yang Mendapat Calistung Dan Tidak Mendapat Calistung Di Taman Kanak-Kanak. *Educational Psychology Journal, 3*(1), 14–18
Sebba, L. (2005). Child protection or child liberation? Reflections on the movement to ban physical punishment by parents and educators. *International Review of Victimology, 12* (2),159–187.
Semiawan, C.R. (1999). *Perkembangan dan belajar peserta didik.* Jakarta: Direktorat Jenderal Pendidikan Tinggi Proyek Pendidikan Guru Sekolah Dasar Departemen Pendidikan dan Kebudayaan.
Stipek, D. (2013). Mathematics in early childhood education: Revolution or evolution? *Early Education & Development, 24*(4), 431–435.
Van Der Stuyf, R. R. (2002). Scaffolding as a teaching strategy. *Adolescent learning and development, 52* (3), 5–18.
Vygotsky, L.S. (1978). *Mind in society*: M. Cole, V. John-Steiner, S. Scribner, & E. Souberman, Eds. London: Harvard University Press
Vygotsky, L. S. (1986). Thought and language. *Annals of Dyslexia* 14.1 (1964): 97–98
Vygotsky, L. S. (1989). Concrete human psychology. *Soviet Psychology, 27*(2), 53–77.
Vygotsky, L. S. (2014). Play and its role in the mental development of the child. *Soviet Psychology*, 5(3), 6–18. doi: 10.2753/rpo1061–040505036.
Watts, T. W, Greg, J. D., Robert, S. S, & Davis-Kean, P. E. (2014). What's past is prologue: Relations between early mathematics knowledge and high school achievement. *Educational Researcher, 43*(7), 352–360.

Early Childhood Education in the 21st Century – Yulindrasari et al. (eds)
© 2020 Taylor & Francis Group, London, ISBN 978-1-138-35203-2

The perception of kindergarten teacher: Socio-emotional skill as an important aspect of school readiness

S.R. Mashfufah, Rudiyanto & A. Listiana
Universitas Pendidikan Indonesia, Jawa Barat, Indonesia

ABSTRACT: The objective of this study is to determine the perception of kindergarten children about the importance of school readiness as one of socio-emotional aspects. This study used a qualitative method with phenomenological design; the participants are three kindergarten teachers of B Group. Data collection was done using interview techniques and field observation which then analyzed using grounded theory. The data collection process was carried out for three months, from June until September 2018. Based on the study it was revealed that kindergarten teacher(s) have a perception that it is important for children to possess school readiness as a socio emotional aspect, in which the children's ability to regulate themselves and independency are included.

1 INTRODUCTION

School readiness is the overall condition of students in respond to the process of learning at school (Mulyani, 2013). The school readiness here is more to children's readiness in entering elementary school, which includes five dimensions: physical health and wellness, social competence, emotional maturity, language and cognitive development and the communication and knowledge skills (Janus et al., 2007; Hair, Halle, Terry-humen, Lavelle, & Calkins, 2006). It is important for children to possess school readiness before they enter elementary school; especially in the dimension of socio-emotional skill, for it will provide the impact in their future, support them to succeed in the period of transition in a given learning formal environment (Fantuzzo, Bulotsky, McDermott, McWayne, & Frye, 2007).

The Indonesian Law of National Education System (2003) states Early Childhood Education (PAUD) should be able to facilitate the development of children's school readiness to enter elementary school. This study which shows that elementary school student who attended kindergartens have a higher level of readiness compared to those who do not (Halimah & Kawuryan, 2010).

In fact, however, there are still parents and school communities who have not shown a consistent attempt in preparing children to enter elementary school. It can be seen in the fact that Early Childhood Education focuses more on children's scholastic development than on social development. The previous studies suggest that children had higher score in academic knowledge than in socio-emotional maturity (Rahmawati, Tairas, Nawangsari, 2018; Nurhayati, 2017). Situngkir's study (20117) indicates that approximately 30 out of 60 parents of children aged 5–6 think that children need to independently acquire the skill of reading, writing and counting (RWC) on an everyday basis.

When children receive the load of learning beyond their capacity, they will experience academic stress (Suci, 2018). Stress is a distressing situation as a result of discrepancy between expectation and reality faced (Ibung, 2008). Boyce (2012) claims that stress on early childhood has such detrimental effect on brain architecture and other organ function that affect health, behavior (adaptive vs. maladaptive responses toward difficulties in the future), psychological (activated response toward chronic stress) and learning difficulties (linguistic, cognitive, and socio-emotional skills) in the future.

Teachers have a perception that RWC is identical with an academic-oriented learning and deprives the concept of playing for children, as if playing and learning are the different and separate concepts (Lutfatulatifah & Yulianto, 2017). Another study, however, suggested different things; pre-school teacher and first grade's teacher have similar perception that children's school readiness has various keys related to physical, social/emotional, cognitive and language, and children's self-help ability (Şahin, Sak, & Tuncer, 2013). Teachers' perception on the importance of socio-emotional skill in school readiness will affect the process and learning outcomes. The perception itself is an assessment process which involves cognitive, affection and conative, so as it affects behavior and attitude (Mahmud, 1989; Winardi, 2009; Robbins & Coulter, 2010). Therefore, this research find out how exactly kindergartens teachers perception toward the importance of socio-emotional ability as school readiness to attend kindergarten in Bandung, in 2018.

2 METHOD

This study used a qualitative method with phenomenological design. The participants were three kindergarten teachers in Bandung. Data collection was done using interview techniques and field observation which then analyzed using grounded theory. The data collection process begins with the arrangement of interview transcript and field notes, creating coding using open coding, axial coding and selective coding. This data collection was carried out for three months, from June to September 2018. The following are examples of the result of interview and field notes.

3 RESULT AND DISCUSSION

The result of interviews with three kindergarten teachers in Bandung suggested that reading, writing and counting skills are not the only important aspect of school readiness. Teachers believe that there are some other aspects of school readiness that are equally important: social competence aspect, emotional maturity, physical health and wellness, cognitive and language development, knowledge and communication, revealed in the following interview excerpt:

> Every each of them (developmental aspects) is important, for they are related to each other. Because it can't be individual. It cannot be, for instance, solely the social emotion, but it turns out that … uuumm … when they are given simple instructions, they cannot get it (don't understand), so, how can they go to elementary school then. For, it is such a unity and cannot be separated. That's the most important thing. For, it is easy to receive information while they have a healthy and strong body, they will be able to focus. They can directly digest (receive) all information, right? Not just part of it. Prepare them to that extent. What's the use of you mastered language well, while you still cannot wipe yourself cleanly, right?
>
> (Interview with Mrs. DT; Thursday, September 13, 2018)

It was found out that teachers emphasize several aspects of children's development as a requirement to entering elementary school, that is: the first, cognitive and language aspect, that is children are able to comprehend and follow instructions; the second, socio-emotional aspect, that is children possessed confidence and are able to be independent and exercise self-help practice; the third, physical motor, that is children are physically in the healthy condition. Thus, with all these aspects fulfilled children are assisted to fully and comfortably focus on learning by the time they entered elementary school. It is in line with the perception of kindergarten and elementary school teachers that children's school readiness holds key variations related to physical, socio-emotional, cognitive and language and self-help skills (Şahin, Sak, & Tuncer, 2013). Janus (2007) also argues that children aged 4–7 should meet the school readiness components, namely: physical health and wellness, social competence, emotional maturity, language and cognitive development, communication and knowledge skills.

However, this study revealed that kindergarten teachers are more concerned with socio-emotional skills that includes self-regulation and independency skills. The researcher explains these two matter as follow.

3.1 *Socio-emotional (self-regulation and independence)*

The findings of this study suggested that the three chosen kindergarten teachers possess the excellent insight on the importance of children's school readiness in the aspect of socio-emotional development. They explained that children are considered possess the readiness to enter elementary school when they have self-confidence, bravery, honesty, being able to express properly, regulate themselves, be responsible for their belongings and other's as well, queue to take turn, independent, socialize with peers, adapting to surrounding environment, respecting parents, solving their own problems, caring for friends, helping others without any encouragement, and completing their own tasks well. The social competence, according to Janus (2007) involves the entire social competence, responsibility and respect, learning approaches, and the readiness to explore new things. On the other hand, instead of showing the pro-social and helping behavior, children who have not yet fulfilled the aspect of emotional maturity will show anxiety and fear, appearance or expression of unhappiness or sadness, hyperactivity and lack of attention.

From this explanation, however, and the result of field observation, it was revealed that the three chosen teachers are emphasized more on the importance of children having self-regulation and independency. The researcher explains these two matters as shown in the following sections.

3.1.1 *Self-regulation*

Teachers' perception on the importance of socio-emotional aspect is obviously perceived from their elaborate explanation on children's self-regulation. They believe that when children are able to regulate themselves, it will easier for them to absorb information, as this interview excerpt indicates:

> ...he/she can tell what makes him/her sad ... uuumm ... I know what should be done to calmed him/her down. For kindergarten kids are not ready ... uumm ... like, emotionally. Throwing tantrums, or something ... how can they are able to take on information, right?
>
> (Interview with Mrs. DT: Monday, August 13, 2018)

When children are able to control themselves, they will begin to aware and susceptive upon existing matters in the surrounding environment; hence, it is allows children to receiving and maintaining information, assists them to designing solution and solving their own problems. Regulation is one of the forms of self-control in manage their emotion and cognitive, so that they are able to plan, organize and solve their problems, as well as behavior management (McClelland, Connor, Jewkes, Cameron, Farris, & Morrison, 2007). The other teacher believes that children's self-regulation skill will influence the problem-solving process in their everyday life, as this interview except indicates:

> Now he is able to overcome his fear when he met his third grade friends. He's able to cope with difficult situations. Though he continually asking me, "Is it right, Mommy? Is it right?" He keeps asking until I confirmed that it was right, though it is indeed right. The most important thing is that he's able to confront every problem without crying.
>
> (Interview with Mrs. E: Friday, June 22, 2018)

When children are able to manage their fear, they will aware to the surrounding environment, they begin to observe and apprehend information from their experiences; hence, their confidence in solving problem will gradually grow. It is in line with the result of another study reveals that teachers are believe that the temperament characteristic (shy and reserved) influence children's school readiness (McBryde, Ziviani, & Cuskelly, 2004).

The researcher believes as well that children who are able to know themselves will find that easier for them to learn to get to know and understand others; it may foster a good friendship and environmental relationship. This is in accordance with the result of other studies revealed that self-regulation may determine academics achievement and social relation (McClelland, Connor, Jewkes, Cameron Farris, & Morrison, 2007; Ursache, Blair, & Raver, 2011).

This importance of self-regulation is obvious when they are stimulating children's regulation skill through the question and answer method in story-telling activity, as these field notes indicate:

> When the story-telling activity is about to begin, the teacher asks children to take a seat in 5–7-8 formation and adjusts them based on the height. After the children are sitting orderly, the teacher starts to open the book of "KenapaAkuSedih?" (Why Am I sad?). The teacher asks, "Who has ever felt sad? I myself often feel sad, though I'm old." Some children raised their hands. The teacher then continues the story, "Look, this bear is sad. Just take a look at its face, how it looks like when he sad?" One of children answered, "He cries, there are tears," the teacher continues, "That's right. Look, the bear is crying, he feels sad. Do you know why he sad?" One of children answered, "He lost his Mommy, Ma'am." The other child raised her hand and said, "He's frost, Ma'am," the teacher continues, "That's right, the bear frost and lost his Mommy. Now I want to ask you, is there one of you feeling sad?" two children raised their hands, and the teacher allow them to explain the reason of their sadness. Another child answered, "I felt sad when I fell down and in pain." The teacher ask more, "Who else? How little of you who are ever feeling sad. I'm feeling sad when I run out of food. Is there other of you ever felt sad?" Some different children raised their hands; explain the reason of their sadness later on.
>
> (Interview with Mrs. DT: Monday, September 24, 2018)

The learning method utilization is the key for teacher to be able to make a classroom effective in order to achieve the learning objectives. By means of the question-and-answer method teacher provides opportunities for children to stimulate their curiosity on the feeling of sadness, therefore children are able to develop their knowledge more meaningfully. Children are stimulated to identify the characteristic of sadness and mimic the bear's sad face; seeking for the cause of and the solution for their feeling of sadness. From that identification process children learn to make a conclusion and obtain new insight on feeling of sadness. The question-and-answer method is itself a one of ways to convey or present subject material in a line of questions raised by teacher and must be answered by students, vice versa (Basrudin, Ratman, & Gagaramusu, 2015). This very method may also stimulate students' attention and participation in learning (Syamsiah, 2008).

It is in line with the teacher's statement that children's recognition of feeling is a requirement of self-regulation, as follows:

> Actually, some of our story-telling activities are indeed deviated, some don't. That was a really out of theme one, many of them child are still having difficulty to express themselves. They just crying, just crying. In fact, the teacher needs to know the reason, so that she will be able to help. Therefore, it all depends on the specific problem that occurs in each class. Yesterday we asked the children who among them often felt sad, or throw tantrum or something like that. Then we decided what stories will be delivered so that children will have a better understand about their sad feelings.
>
> (Interview with Mrs. DT: Monday, September 24, 2018)

This statement revealed that teacher is aware of children's need to regulate themselves. Teacher identifies the indicator of children's achievement that has not been achieved: children do not have an ability to recognize their sad feeling yet. Teacher realizes that the recognition of feeling is a children's process to find a strategy in overcome their feeling. Thereafter, teacher sets it as a learning objective; makes an effort by means of question-and-answer method through reading a story book about the sad bear. This method encourages children to develop knowledge on the matters they want to learn; through question and answer the acquired knowledge might be more meaningful. It is in accordance with the result of certain study suggested that question-and-answer method might improve children's learning outcomes (Basrudin, Ratman, & Gagaramusu, 2015; Mahdalena, Uliyanti, & Sabri, 2014).

3.1.2 *Independence*

Independence is better known as a situation in which children learn to work on a certain activity by themselves or not depend on others. The teacher believes that children's readiness in independency is just as important as preparing children to enter elementary school. This is in line with the teacher's good knowledge of independency, as follows:

> In my opinion, independency is self-help or taking care of themselves; choosing and wearing their own clothes, eating, taking bath by themselves, that's what we emphasized most to children. So they are aware of what equipment they must bring to school; what items do you have in your bag. Unfortunately, not all students here are having such awareness.
>
> (Interview with Mrs. DT: Monday, August 13, 2018)

Teachers can stimulate children's independence by habituating and motivating them, thus it might foster children's awareness in self-help. Habituation is a manifestation of teacher's trust in children; so as in the future, children will be able to carry out their own activities, such as eating using a spoon, changing clothes, wearing shoes and socks, and preparing the appliance that must be taken to school. Habitual self-help will in turn become a habit, so that children will find—when they grow up—that is easier for them to become independent.

The importance of children's readiness related to self-regulation and independency can be also seen from the achievement of students who have stimulated by the teacher. The researchers find that children— in general—are already be independent when performing self-help. Here are some results of field observations:

> At lunch time, children start washing their hands and waiting in line to orderly take plates, spoons and forks. Ms. EL invited and observed themtaking watermelon, rice, vegetables and chicken as needed. Children are sitting on chairs and eat on tables using spoons and forks. Ms. EL also instructed the child to take rice after finishing their fruit. The children were in an orderly manner waiting for their turn to get some rice. After the meal time was almost finished, I saw very little food crumbs above and below the children's table. Then some children reported to Mrs. EL that they had finished their lunch, even though there was still a little food left on their plates. Then Mrs. EL glances at the leftovers on the child's plate, then looked at the clock mounted on the wall and said "Here, let me help you," then Mrs. EL fed the children until there are no leftover anymore.
>
> (Interview with Mrs. EL: Thursday, September 6, 2018)

These observations suggested that the teacher understands that by carrying out handwashing and eating alone, children will get used to being independent. This is in accordance with other teachers' understanding that independence in carrying out daily routines is a manifestation of children's readiness to enter elementary school (Syarfina, Yeti, & Fridani, 2018). This independency will help children to get used to and attempt to do their own activities without the help of others and can foster confidence when they are able to carry out other tasks.

Independency, according to Rakhma (2017), is more about the ability to solve problems, such as initiative, solving everyday problems, diligently, and wanting to do something without the help of others. This is in line with the kindergarten teacher's knowledge that children are considered independent when they are able to solve their own problems in certain situations, as indicated in this interview excerpt:

> . . . Then, he can solve a simple problem by himself. For example, when he spilled drinking water, he should immediately say, like, "Ma'am,I need to get a cloth." Just like that.

They don't have to be told first. Something like, "Ma'am, I spilled my drinking water." "So, what should we do then?" Some just shook their heads ... some took the initiative, like; "get some cloths." It is obvious, right? It's simple things if it comes to kindergarten children.

(Interview with Mrs. EL: Thursday, July 26, 2018)

This statement revealed that children can be considered independent when they are able to account for their mistakes by solving their problems independently. Independent children are who are aware of their mistakes; when spilling drinking water, they will take the initiative to look for a cloth to dry the wet floor without previously asking what to do.

Teacher's perception on the importance of children's self-regulation and independency is based on knowledge of given conditions in the school environment where children are demanded to be able to complete basic academic tasks, adapt to more friends, and be more independent in practicing self-help. This can be seen from the teacher's statement that follows:

Later, in elementary school, they are already familiar with what the teacher instructed, already understand. Continue, for example ... about independence. They can cope with their parent's absent. Not depending. It is very important too. When he was in elementary school he was comfortable, he was able to ... if Mommy is not found around the, they could be independent, they had dared to face a problem. For example, with friends, it was very important to recognize their peers' characteristics. Who is not easy to cry, who is not easy for ...uumm ... what is it ... surrender, like: "it's better that I don't play, rather than get hit," or something like that. "We must face every obstacle, right? But there are more friends in elementary school. There might be just few in kindergarten. So, he knew his friend ... what their attitude was like, what their behavior was like; he already knew how to deal with them. It is pity if they cannot yet deal with this. Later in the elementary school there was a friend who was naughtier than his kindergarten friend. It is possible, right? Children are basically kind, but they are different. Some are easy to hit others, some are often throwing harsh words, and some are ignorant, right? Later, when they went to elementary school, they have self-confidence, be confident.

(Interview with Mrs. E: Friday, June 22, 2018)

It is undeniable that when children enter elementary school, they are inevitable have a new role as student and will automatically have new tasks. Moreover, children are also demanded to be able to develop relationships with more and diverse new people. The encounter one child with another is potentially to cause more friction or difference, for when children start new tasks, they will be pushed to interact with one another. According to Santrock (2002), children who are in the middle and late stages of comprehension development (elementary school students) will define themselves based on their characteristics psychologically. Even at the elementary school level children are already at the stage of social comparison, in which they begin to be able to compare themselves with others (Santrock, 2002). Hence, when children encounter friction with their friends, they will get used to solving their own problems, and when they successfully solved the problems, they can establish good friendships.

Creating a class climate, interaction conditions among students, and the substance of group functioning are symbols of the authority possessed by the teacher. Similarly, in preparing children to enter elementary school, teachers have an important role in organizing learning. The learning approach is one of the efforts of teachers to organize learning and bridge the demands that must be met by children when they entering elementary school. Kindergarten teachers believe that children will be stimulated to be able to have readiness in independence by means of the student center approach. The implementation of Project Base Learning (PBL) is a part of the student center approach. PBL is a learning strategy in which learning activity is centered on children in a long period of time, integrates with real issues, develops skills acquired from collaboration, and problem solving and learning can be accessed by all children (Chu, 2007). PBL learning is mentioned in these field notes:

When the children discuss the decoration for the assembly program, they are invited by the teacher to make the group and be given the task by the teacher to mapping the decoration requirements. Every child has a group and is assigned to describe what is on his mind in turn. Then Nita spoke "O, yeah. We need a piano; I'm drawing here, okay?" Then Nita drew a piano the on her group's map. After that, she spoke to her friend "Come on, it's your turn," while giving Beni a pencil "What else are we needed?" Said Nita, but Beni was remained silent as he thought for a moment, then he seems doubtful to draw on his mapping paper, and Nita reflexively said, "Come on, Beni. Here, I'll help you, this how do I do this" while holding Beni's hand. Then Mrs. DT reminded "Nita, just let Beni try by himself, Beni can draw on his own, "Beni then began to draw a chair in the mapping picture. Thereafter, they learn to share their map results in front of other friends.

(Interview with Mrs. DT: Monday, September 24, 2018)

The statement above suggested that Nita can see that Beni is facing difficulties in drawing, so Nita takes the initiative to help him. But Ms. DT gave more opportunity to Beni to try to complete her own assignment; so that Beniwasindirectly trained to be independent and fostered his self-confidence. PBL can also support children to develop knowledge and thinking skills, problem solving, and social interaction (Jonasen & Land, 2009). The notes of previous observations are clarified by the teacher's understanding as can be seen in the following statement:

Actually, in this assembly there is readiness in managing the self-management in preparing for one of the activities they will undertake.Forexample, preparing the appliances to be brought to their school. This assembly is a big event, but they are able to prepare everything themselves; song selection, dance selection, storyline, it starts from there ... so you can see the readiness in independency.

(Interview with Mrs. DT: Monday, September 24, 2018)

The teacher believes that by means of the assembly, children will be stimulated to perform self-control and seeking for solutions in order that the assembly can be well-planned and well-executed.

4 CONCLUSION AND RECOMMENDATION

Based on the findings and discussion above, it can be concluded that the teacher has a perception about the importance of children having social emotional abilities that include self-regulation and independence in entering elementary school. This can be seen from the teacher's explanation of the effects of self-regulation, such as children can receive information, have memory, planning problem-solving. While independence can affect the trust worthiness of children in carrying out self-help activities, completing academic tasks, and resolving social relations conflicts when in elementary school. The method of question and answer, story-telling, and PBL is also used by the teacher as an effort to fulfill the need for the importance of emotional social skills in children's self-regulation and independence. The next research recommendation is to investigate the perceptions of parents regarding children's school readiness.

REFERENCES

Basrudin., Ratman., & Gagaramusu, Y. (2015). Penerapan Metode Tanya Jawab untuk Meningkatkan Hasil Belajar Siswa Pada Pokok Bahasan Sumber Daya Alam di Kelas IV SDN Fatufia Kecamatan Bahodopi. *Jurnal Kreatif Tadulako*, *1*(1), 214–227.

Chu, S. K. W., dkk (2017). The effectiveness of wikis for project-base learning in different disciplines in higher education. *The Internet and higher education*, (33), 49–50. Doi: 10.1016/j.iheduc. 2017.01.005

Fantuzzo, J., Bulotsky, R. S., Mcermott, P., McWayne, C., & Frye, D. (2007). Investigation of Dimensions of Social-Emotional Classroom Behavior and School Readiness for Low-Income Urban Preschool Children. Retrieved from http://repository.upenn.edu/gse_pubs/124.

Hair, E., Halle, T., Terry-humen, E., Lavelle, B., & Calkins, J. (2006). Children's school readiness in the ECLS-K : Predictions to academic, health, and social outcomes in first grade ☾, *21*, 431–454. https://doi.org/10.1016/j.ecresq.2006.09.005
Halimah, N & Kawuryan, F. (2010). Kesiapan memasuki sekolah dasar pada anak yang mengikuti pendidikan tk di kabupaten kudus. *Jurnal Psikologi, 1*(1), 1–8.
Ibung, D. (2008). *Stress pada Anak (usia 6–12 tahun): Panduan bagi Orang Tua dalam Memahami dan Membimbing Anak*. Jakarta:Gramedia.
Janus, M., Brinkman, S., Duku, E., Hertzman, C., Santos, R., Sayers, M., & Walsh, C. et al. (2007). The Early Development Instrument: A Population-Based Measure for Communities. *Guide*. Retrieved from http://www.elcmiamidademonroe.net/KnowledgeCenter/reports/2007_12_FINAL.EDI.HANDBOOK.pdf
Jonassen, D. H & Land, M. S. (2009). *Theoretical Foundation of Learning Environments*. London: Routledge
Lutfatulatifah & Yulianto, S. W. (2017). Persepsi Guru tentang Membaca, Menulis, dan Berhitung pada Anak Usia Dini. *Jurnal Pendidikan Anak Usia Dini, 1*(1), 77–82. ejournal.unisba.ac.id/index.php/golden_age/article/view/2766/1954
McBryde, C., Ziviani, J., Cuskelly, M. (2004). School Readiness and Factors that influence decision making. *Occupational Therapy International*. *11*(4),193–208.
McClelland, M. M., Connor, C. M., Jewkes, A. M., Cameron, C. E., Farris, C. L., & Morrison, F. J. (2007). Link between behavioral regulation and preschoolers' literacy, vocabulary, and math skills. *42* (4), 947–959. Doi: 10,1037/0012–1649.43.4.947
Mulyani, D. (2013). Hubungan kesiapan belajar siswa dengan prestasi belajar. *Konselor*, *2*(1), 27–31.
Rahmawati, A., Tairas, M., & Nawangsari, N. A. (2018). Profil Kesiapan Sekolah Anak Memasuki Sekolah Dasar. *Jurnal Pendidikan Anak Usia Dini*, *12*(2), 201–210. https://doi.org/10.21009/JPUD.122.01.
Rakhma, E. (2017). *Menumbuhkan kemandirian anak*. Jogjakarta:Stiletto Book.
Robbins, S. P., & Coulter, M. (2010). *Manajemen, edisi sepuluh jilid 2*. Jakarta:Erlangga
Şahin, I. T., Sak, R., & Tuncer, N. (2013). A comparison of preschool and first grade teachers' views about school readiness. *Kuram ve Uygulamada Egitim Bilimleri*, *13*(3), 1708–1713. https://doi.org/10.12738/estp.2013.3.1665
Santrock, J. W. (2002). Life span development: perkembangan masa hidup edisi ke lima. Jakarta: Erlangga.
Situngkir, H. (2017). Studi Presepsi Orang Tua tentang Konsep RWC pada Anak Usia 5–6 Tahun di PAUD Gloria Desa Paropo Kecamatan Silahi sabung Kabupaten Dairi T. A 2016/2017. (Thesis, Universitas Negeri Medan, 2017). http://digilib.unimed.ac.id/27590/
Suci, G. W. (2018). Pandangan guru mengenai stress akademik pada anak usia dini. (Thesis Universitas Pendidikan Indonesia 2018)
Syamsiyah. (2008). *Penggunaan metode diskusi dan tanya jawab dalam meningkatkan efektifitas pembelajaran al-quran hadist kelas 2 b mts surya buana malang* (Skripsi, Universitas Islam Negeri Malang, 2008)
Syarfina, S., Yeti, E., & Fridani, L. (2018). Pemahaman Guru Prasekolah Raudhatul Athfal Tentang Kesiapan Bersekolah Anak. *JPUD-Jurnal Pendidikan Usia Dini*, *12*(1)
Boyce, W. T. (2014). The lifelong effects of early childhood adversity and toxic stress. *Pediatric dentistry*, *36*(2), 102–108.
UNICEF. (2012). *School readiness: A conceptual framework*. New York:United Nations Children's Fund.
Ursache, A., Blair, C., & Raver, C. C. (2011). The promotion of self-regulation as a means of enchancing school readiness and early achievement in children at risk for school failure. *Child Development Perspectives*, *2*(6), 122–128. Doi: 10.1111/j.1750–8606.2011.000209.x
Winardi. (2009). *Manajemen perilaku organisasi*. Jakarta:Kencana.
Peraturan Menteri Pendidikan dan Kebudayaan, pasal 1, 2014, PendirianStatuan Pendidikan Anak Usia Dini.
Peraturan Menteri Pendidikan dan Kebudayaan, Nomor 14, 2018, Sistem Pendidikan Nasional.

Early Childhood Education in the 21st Century– Yulindrasari et al. (eds)
© 2020 Taylor & Francis Group, London, ISBN 978-1-138-35203-2

Logical mathematical intelligence for preschool children

Atiasih & H. Djoehaeni
Universitas Pendidikan Indonesia, Bandung, Indonesia

ABSTRACT: In everyday life, humans always interact with mathematics. Every activity be related to a number, logic, and creativity in solving problems including in early childhood education. Then, there are some problems arising caused by some factors, firstly, mathematics is one of the subjects in the national exam. Secondly, the logical math learning media used in learning are still less relevant. This chapter employs a study of literature that analyzes research on early childhood logical mathematical intelligence. This chapter aims to make parents and teachers realize that learning mathematics is not just limited to numbers and memorization. It will be better if we conduct teaching and learning process of logical math through plays. Logical math intelligence, which is the ability to use numbers well and also reason properly, means intelligent in dealing with numbers and smart in the art of logical thinking

1 INTRODUCTION

Nowadays, controversy about mathematics for early childhood is increasingly emerging in the world of education. In a mainstream local newspaper, Pikiran Rakyat (2007), Mr. Ace Suryadi, the Director General of Non-Formal Education of the Department of National Education, states that learning to read, write, and count (calistung) in early childhood is one of the biggest mistakes in the Indonesian national education system. At an early age, calistung teaching will limit students interaction with their environment. However, if the willingness to learn calistung come from the children the meselves, then it is no problem.

If parents force their children to learn something without considering the development stage of the preschool children, it means they are unaware of the preciousness of the childhood. In addition, the controversy emerged in the last few years is still about the same topic in the current year. The controversy is that teaching mathematics is limited to teaching arithmetic, addition, and subtraction. In the teaching process, the cognitive development was not considered and it is also far from the thinking activities and developing the logical intelligence of the preschool children.

As educators, it is a must to realize that early childhood education is a stimulus effort to develop moral and religious aspects, cognitive, language, social emotional, and motor skills.The National Council of Teachers of Mathematics (Sriningsih, 2009) suggests principles and standards of mathematics for early childhood that includes (1) number and number operations, (2) algebra, (3) geometry, (4) measurement, (5) data analysis and probability (6) problem solving, (7) reasoning and verification, (8) communication, (9) connection and representation. Based on the nine standards of content and the standards of mathematics learning process, teachers should be able to plan carefully on how the standards of content are included in various aspects of children development. Presumably, the development of mathematical logic will support the preschool children mathematics learning by referring to the standards of content and learning process of mathematics and also by considering the stages of children development written in the national curriculum 2013.

Developing mathematical intelligence through logical math is a good way to give a new variation to preschool children. Mathematics is not only about numbers or calculating numbers but more than that, Musfiroh (2010) stated that mathematics emphasize on (1) Pattern Finding (2) Logical Relations Finding, (3) Reinforcement of Understanding Numbers, (4) Size Understanding, (5) Construction Skills, (6) Experimental Hypothesis Skills, (7) Problem-Solving Skills, (8) Clarification and Serial Skills.

2 LITERATURE REVIEW

This chapter is a literature review study which contains theories, findings and research materials obtained from various sources about logical math for preschool children. In writing this chapter, the first step is the selection of topics related to current issues and interests to be discussed more deeply.

2.1 *Logical math for preschool children*

Mathematics is an exact science that over time will experience progress as it goes towards modernization. Logical mathematical intelligence, which means the ability to use numbers well and reason properly, means intelligent numerically and intelligent in the art of logical thinking (Musfiroh, 2010). Logical mathematical intelligence is the ability to handle numbers and calculations, logical and scientific patterns of thinking. The relationship between mathematics and logic which both equally follows the basic law.

There are many factors that make mathematics difficult but we should realize that there are many things or even people that make it easy. What makes it easy when connected with the daily activities and inevitable of a child's or an adult's life (Pound, 2008). In everyday life, we will always interact with mathematics. Each activity must be related to a number, logic, and creativity in solving problems. Therefore, in order to know the meaning of mathematics itself, let us digest some statements of some experts as follows.

The mathematical intelligence emerges at an early age and will explode in adolescence and early adulthood and will decrease after the age of 40 years. Mathematics is all about pattern recognition. It involves recognition of a pattern of thinking, and a pattern of organizing logical proof. Math is also about symbol but not sound. Children with mathematical intelligence will build knowledge from an organized structure which the traits or theories are composed deductively based on the defined elements.

Research conducted by (Yanti, 2014) reveals that children will have higher levels of creativity and mathematical concept understanding if they are stimulated by using a constructive game. Ginsburg, Lee, Boyd (2008) stated that for the past 25 years, they have found evidence that children from birth to age 5 develop mathematics in their daily lives, such as developing ideas, shapes, sizes, locations, patterns, and positions.

The role of instructional media for the mathematical intelligence of preschool children has its own contribution which is expected to stimulate preschool children cognitive development, especially logical mathematical intelligence. In teaching materials of a program named PPG (Professional Teacher Education), which is created by Zaman and Eliyawati, it is clear that some researches suggest that the use of instructional media or learning media showed a significant increase in the learning outcomes of children. It can be concluded that instructional media is strongly recommended to improve the quality of learning and the number of outcomes of the children (Epoch & Eliyawati, 2010).

According to Jean Piaget (cited in Santrock, 2004), at the age of 0–7 years, children's development belongs to the sensorimotor and pre-operative period. At this time, the children will get resulta from their various senses. Thus, the children will understand the various concepts through concrete objects around him. There will be an association and assimilation in children mixed with the knowledge that children have possessed previously.

Logical mathematical intelligence plays a very important role in the children's life because mathematics is not limited to natural science alone but also social science, sports, arts, religion, handicrafts, and civics can be used as the ways to learn math. The definition of logical mathematical intelligence according to Said and Budimanjaya (2015) is the ability to calculate, measure and consider propositions and hypotheses, and complete numerical operations.

Let us take a look at other countries such as America, according to Chard et al. (2005) the decrease of students' achievements was caused by the unreadiness of teachers to teach mathematics. This is similar to the condition in Indonesia in which educators are not ready to receive new information and also there is an underestimation of early childhood mathematics. Whereas, in learning mathematics, preschool children learn very basic concepts which should be taught to preschool children.

2.2 *Developing logical math intelligence*

The development of logical mathematics can not be separated from the stimuli provided by parents and also the environment such as the teacher. In other words, this mathematical ability is indeed given by God as a grace that should be maintained and cherished but it must be improved as the children grow and develop.

Chard et al. (2005) stated that a skill that should be emphasized in the development of children's skills is conceptualizing mathematical understanding. After a concept is taught properly and gradually then they are given authentic tasks. Likewise, the logical mathematics for preschool children is not emphasized on the tasks in the form of paper test but how to build a mindset that can build their mathematical concept using logic.

Some components in logical mathematical intelligence are such as sensitivity to logical patterns and relationships, questions containing words such as "if-then" and causation, logical functions, and other abstractions (Musfiroh, 2010). Among the three components, if the logical math develops, then it will go through a process related to several things that are as follows:

1. Categorization or grouping, i.e. preparation by category; grouping based on certain criteria
2. Classification, that is the grouping based on certain rules or standards
3. Conclusion
4. Generalization, i.e. general inference of an event, thing, or data
5. Calculation, i.e. numerical activities, such as calculations; and
6. Hypothesis testing, which is the activity of children checking, observing and experimenting something so that they are able to know the truth or estimate what will happen.

3 DISCUSSION

One should be done by teachers to stimulate the logical intelligence of early childhood with the strategy of teaching. That way the teacher immediately know how to stimulate the logical intelligence of early childhood mathematics. It is undeniable that there are still many teachers who are unfamiliar with the number of children with the right strategy to use. For example, the number of children's ratios that many are not comparable with teachers so that giving the material to the child sometimes difficult to be accepted by the child due to lack of focus. Early childhood learning strategies should look first at the nature of early childhood education and characteristics of child development.

According to Masitoh, (2008), there are variables of learning strategy in early childhood that covers characteristic objectives, materials or theme of learning, activities, children, media and learning resources and teachers. The most important thing that must be considered for this logical mathematical intelligence to emerge is the teacher should be able to increase the sense of curiosity of children so that children's imagination to thrive.

3.1 *Implementation of logical math learning for preschool children*

The game of numbers concept in kindergarten aims to know the basics of counting learning so that in the future the children will be better prepared to follow the counting learning in the next level which is more complex. In particular, the game of numbers concept in kindergarten aims to allow children to think logically and systematically since early age through observation of concrete objects, images, or figures around them, to be able to adapt and engage in social life which requires arithmetic skills, possess an accuracy, concentration, abstraction and high appreciation, have an understanding of the concept of space and time and can estimate a possibility of sequence of events that occur around them and have a creativity and imagination in creating something spontaneously (Ministry of National Education, Directorate of Kindergarten and Primary School Development, 2007).

A concept is an initial step in understanding everything that children will do. Similarly, an early childhood mathematical concept and "the principles and standards for school mathematics" developed by educational groups of the National Council of Teachers of Mathematics (NCTM, 2000), namely: (1) Numbers, (2) Algebra, (3) Classification, (4) Comparison, (5) Arrangement, (6) Patterns, (7) Geometry, (8) Measurement, (9) Data analysis and probability, (10) Problem solving.

The implementation of logical math teaching for preschool children should be conceptualized first. Similarly, in the teaching process, teachers should prepare a daily activity plan. Prior to learning the teacher must determine what competence must be achieved by the children which could be measured by using some indicators. There are ten logical mathematical intelligence indicators for preschool children according to Musfiroh (2010, 11). First, the child has a sensitivity to numbers, likes to see numbers, quickly calculates possessed objects, quickly capables of using numerical symbols, correctly identifies numbers on money, and spells out numbers quickly. Second, children are interested in computers or calculators or even use a calculator. Fourth, children love games that require logic, strategy, and thoughts such as maze and chess. Fifth, children often ask something related to cause and effect and ask questions with identical ones by digging up information like "why", "how". Sixh, children can tell easy matters logically such as why they feel fear, why the stomach becomes full. Seventh, children can predict or make a simple hypothesis to estimate a result and prove the conjecture. Eighth, children will spend time choosing to play construction such as building blocks, putting numbers, and posting pictures. Ninth, children love the activities of something in serial, categorized and hierarchial. Tenth, children very easily understand the simple explanation of cause and effect and quickly capture information regarding the phenomena associated with logic. Eleventh, children love to see books containing images of nature, technology, or transportation.

Math can be said as a second language in which, as we know, there is a language of consideration and comparison. In addition, language and mathematics have a quality where the material is more than just instructions (Wakefield, 2000). Several things that can hone math skills through conversation and play are the operation, counting, sorting, organizing, distributing, sharing, and judging (Wakefield, 2000). A synergistic relationship between mathematics as a language and also logical intelligence lies in the everyday language spoken. It should contain ratios, numbers, and divisions so that it will feel easy and accustomed in applying to the daily life of children.

Research conducted by Greenes, Ginsburg, & Balfanz (2004) suggests that early mathematical skills of preschool children have been underestimated throughout history. Similarly, in Indonesia, the preschool children's mathematics skills are underestimated. Whereas if it is studied more deeply, the mathematical intelligence is not only about numbers but more than that, logical math intelligence is a very important thing for the continuity of development and maturity of cognitive aspects of preschool children in order to be ready to continue to the next school level.

It will be better if early childhood mathematics learning is connected to the daily activities and experiences of children so that mathematics will be meaningful and of course supported by the help of teachers. Children have a lot of opportunities to learn math through their daily lives (Ashton, 2006). Therefore, mathematical learning for logical mathematical intelligence should be familiarized in conversations and conversations between teachers and children or children and parents. So that there will be a creativity formed as a result of a question that is given to the children. Moreover, children will be creative in answering a question. Sometimes, the children's answers vary greatly depending on what the children think at the moment.

According to Wakefield (2000) mathematics activities for both children and adults are a process of deep thinking that leads to a question, finding solutions, applying math to solve real problems and playing with math. In order to develop logical mathematical intelligence, its teaching should offer a situation which relates closely to the world of play such as provides the challenge and excitement away from the feeling of pressure, especially mingling with their world that is the world of play. Many children feel that math is a scary and boring thing for them or even intimidating (Ginsburg, Greenes, and Balfanz, 2003).

4 CONCLUSION

Logical mathematical learning is a very important thing to teach from an early age because it is related to the power of thought and creativity of children in facing difficulties in everyday life. For example, children learn to solve problems and make decisions appropriately. In this case, logical math needs to be trained and needs to be guided by the teachers and also the parents in the children's daily life. Unfortunately, only some of us know about logical mathematical learning. Thus, the mathematics, according to most people, is just about numbers. There are many speculations that early childhood mathematical is an unimportant thing to be taught.

But actually, it is important to teach mathematics to children since the early age with the way and principles of early childhood learning. Logical math is related to logic. With the hope of logical math can help and hone the children's power of thought and also their creativity as well as the vocabulary of the children. Logical mathematical intelligence includes all the teaching of algebra, classification, comparison, arrangement, patterns, geometry, measurement, data analysis, and probability. Because, if the children have learned mathematical logic, it means they have been thinking logically.

REFERENCES

Ashton, E. (2006). Children's Math Thought. New Brunswick Early Learning and Children Curriculum Care Framework. Retrieved from https://www2.gnb.ca/content/gnb/en/departments/education/elcc/content/curriculum/curriculum_framework.html

Chard, D. J., Baker, S. K., Clarke, B., Jungjohann, K., Davis, K., & Smolkowski, K. (2008). Preventing early mathematics difficulties: the feasibillity of a rigorous kindergarten mathematics curriculum. *Learning Disability Quarterly*, *31*, 11–20.

Ginsburg, H. P., Lee, J. S., Boyd, J. S. (2008). Mathematics education for young children: what it is and how to promote it. *Social Policy Report Giving Child and Youth Development Knowledge Away*,*12*(1), 3–22.

Greenes, C., Ginsburg, H. P., & Balfanz, R. (2004). Big math for little kids. *Early Childhood Research for Quarterly*, *19*(1), 159–166.

Masitoh. (2008). *Strategi pembelajaran TK*. Jakarta: Universitas Terbuka.

Musfiroh. (2010). *Perkembangan Kecerdasan Majemuk*. Jakarta: Universitas Terbuka.

Pound, L. (2008). *Thinking and Learning About Mathematics in the Early years*. London and New York: Routledge.

Said, S., & Budimanjaya, A. (2015). *95 Strategi mengajar Multiple Intelligences*. Bandung: Prenadamedia.

Santrock, J. W. (2004). *Live-Span Development* [Indonesian translation edition]. Jakarta: Erlangga.

Sriningsih, N. (2009). *Pembelajaran Matematika Terpadu Untuk Anak Usia Dini*. Bandung: Pustaka Sebelas.

Wakefield, D. V. (2000). Mathematics as a second language. *The Educational Forum*, *64*(3), 272–279, DOI: 10.1080/00131720008984764.

Yanti, N. K. I.. (2014). Pengaruh Permainan Aktif Kreativitas Pada Penguagsaan Konsep Matematika Awal. *Jurnal Pendidikan Usia Dini*, *8*(1), 1–12.

Zaman, B., & Eliyawati, E. (2010). *Bahan ajar Pendidikan Profesi Guru (PPG) Media Pembelajaran Anak Usia Dini*. Universitas Pendidikan Indonesia.

Early Childhood Education in the 21st Century – Yulindrasari et al. (eds)
© 2020 Taylor & Francis Group, London, ISBN 978-1-138-35203-2

Developing problem-solving skill using project-based learning

O. Setiasih, N.F. Romadona & E. Syaodih
Universitas Pendidikan Indonesia, Jawa Barat, Indonesia

S.M. Westhisi
IKIP Siliwangi, Bandung, Indonesia

ABSTRACT: Problem-solving skill is one of the cognitive aspects which should be enhanced by young children since the obstacles occur in their daily activities. The achievements of this ability depend on teacher's pedagogical skills and teacher's accuracy to apply the teaching methods. The effective approach to develop problem-solving skills to young children is project-based learning. It encourage them to get the opportunity to improve observing, asking, collecting information, analysing information and communicating information. This chapter is a literature review which examined and analyzed the textbooks and journal articles about problem-solving skill and project based learning approach through manual and online seeking. This chapter could be used as a reference to early childhood education teachers in order to develop problem-solving skill for young children.

1 INTRODUCTION

McTighe and Schollenberger (cited in Costa, 1985) believed that three factors which underlie the importance of problem-solving skill for students, namely *"... (1) characteristics of present and future societies, (2) student thinking capabilities, (3) today's teaching methods"*. Firstly, the society in the present and in the future day marked by the rapid and compact information development which affects learning objectives and educational practices directly or indirectly. These require students to learn how to solve the problems which they face. Secondly, in regard to student thinking capabilities, the researches consistently showed that percentage of students who had higher thinking capabilities was low. The result of evaluation related to reading comprehension conducted by *National Assesment of Education Progress* (Costa, 1985) showed that the students used low thinking capabilities to comprehend the text. They satisfied with their superficial interpretation and the response of students was low in regard to problem-solving skill and critical thinking development. Thirdly, teaching methods used by the teachers now tend to not involve the students to be the active learners. The learning process focuses on the teacher, thus the learning activity is likely to make the students sit down nicely, listen to the teacher, write the notes and memorize the learning materials. Most of the teachers give fewer opportunites for students to think, express the ideas and solve the problems.

The phenomenon of problem-solving skills in primary, secondary and higher education is still low due to lack of stimulation in their early years The learning experience in early childhood becomes the foundation to them to go the the higher level of education. National Association for the Educational of Young Children (NAEYC, 2003) stated that the learning experience of the children to solve the problem will facilitate them to enhance their curiousity, patience, accuracy, flexible thinking skill, open-minded, and facilitate them to comprehend the cause and effect of the event. Children will learn how to achieve goals and build confidence through problem solving experiences.

To develop the problem-solving skill properly for young children, it is needed a suitable and relevant learning model by considering their characteristics, children's development, and how they learn well. One of the approaches which is suitable to develop problem-solving skill is "Project Based Learning". It was introduced by John Dewey and Killpatrick (1920) who stated that the children will learn well if they could learn based on their interests. According to Katz and Chard (2000), the implementation of the project approach in learning process will encourage the children's thought through observation and investigation about the selected topics around us.

The outcome of the research conducted by Fry and Addington (1984) showed that the young children who learned by using the project based learning had better problem-solving skill to solve the social problems than the children who used traditional curriculum during two years on their education. Bartelmay et al. (2008) found that Project Based Learning (PBL) model was able to improve grade 4th students' achievement in grade four at Laboratory of Duke University. In addition, Yuen (2009) found that the teachers got benefits from this approach which consist of activity, interest, autonomy, communication skill, and motivation. In line with this, Kimsesiz et al. (2017) believed that project-based learning would improve English vocabulary better than the conventional way. Then, the children who learn by using this approach are able to be an active learner in the classroom.

The results of this research indicates positive impacts of the project based learning model to children achievement in primary school. Those give the sufficient rationale that the project based learning is able to apply in learning process to improve the problem-solving skill.

2 METHOD

The article used a literature study. The literatures were used to give the rich information, among others are: (1) journal articles and (2) reference books. The literatures obtained by seeking manually to some libraries and visited some places where the information could be obtained well. Additionally, seeking information by online with keywords such as problem solving skill, project-based learning, and young children. The analysis used in this article is comparison among theories that become references.

3 PROJECT-BASED LEARNING

Project-based learning or project approach is based on the biggest work of an expert in education and philosophy, John Dewey (1859–1952), who regarded education was reconstruction of experience and he mentioned that *"Knowledge is not absolute, immutable, and eternal, but rather relative to the developmental interaction of man with his world as problems arise to present themselves for solution"* (cited in Clark, 2006). The project was initiated by Dewey in America (1920), and was popularized by his colleague, William Kilpatrick, as a project method in 1922 (cited in Katz & Chard, 1989). Following that, it was implemented in school as a learning model.

There are several reasons why project-based learning was developed. Katz and Chard (1989) believed the following: (1) research related to children's development and learning 20 years ago supported the statement which mentioned that project learning was one of the ways to stimulate and improve intellectual and social development properly for children; (2) there was no evidence which stated that a project approach could pose a risk, particularly in relation to the intellectual and academic development of young children; and (3) project-based learning is a part of curriculum in early childhood education which emphasizes integrated learning and child development

3.1 *Definiton of project-based learning*

A project is an extensive and a meaningful investigation about one particular topic. The students do a project individually, in a small group, or a big group, and the process should be flexible based on time, interest, and ability of child and the school (Katz & Chard (1993:209, 2006:1). The study of the project is an investigation of selected topics which grab children's interest in order to engage them to be active. Henry (1995) believed that six criteria were necessary for a project: (1) a child should choose the topic; (2) a child finds the resource; (3) a child presents a result (a report that is evaluated by the teacher); (4) a child gets a freedom to work; (5) the project should be done over an extended time period; (6) a teacher is a consultant.

According to this statement, project means a learning approach which is child-centered. It can be seen starts from a child picks a topic up to the implementation, whereas a teacher is a facilitator. Project is an approach (Katz & Chard, 1989). It is based on some reasons. Firstly, describing a perspective that project could be included into the early childhood education curriculum in many ways, depend on situation and condition of each institution. Secondly, project is a learning model which refers to the way of learning and teaching, learning materials which are taught and learnt. The implementation should be flexible, in terms of topic, learning activity, time which are adjusted by the condition such as the capability of institution and the environment of the children. The period of time could be extended or shortened, the topic could be expended or narrowed down, the activity could be developed based on children's interests as well.

3.2 *Characteristics of project-based learning*

The project approach has different characteristics from other approaches. Dearden (1983) stated that the main characteristics of project-based learning is a hands-on experience, learning by doing, and spontaneously playing. Moreover, Katz and Chard (1989) believed that the characteristics include: (1) a child's interests and engagement promote effort and motivation, (2) a child is able to choose a collection of activites provided by the teacher; find out a proper level of challenge, (3) a child is an expert and the teacher supports the child's profiences, (4) a child shares responsibility with teacher for learning and achievement. Those become teacher's references to implement project-based learning.

To implement the project-based learning in early childhood education institution, it does not have to replace the curriculum; however, it can be a part of the whole curriculum program of institution. This could be implemented in long, middle or short period.

Selecting a topic is the first step towards implementing the project-based learning. The topic can appear sponteneously from child's interests or it is proposed by the teacher. Afterwards, the teacher cooperates with the students. To achieve meaningful learning, there are some considerations in choosing the topic, for instance compatibility the interest and the need of the child, the environment, and the curriculum. In line with this, Kostelnik et al. (2000) believed that the topic should be adjusted with child's life, it has to give the advantages for a child, it is connected with the program objectives, and it can be done through learning experience. Based on principle child's learning, the implementation of the project in kindergarten should consider the principle of learning through playing since the child's world is playing.

4 RESULTS AND DISCUSSIONS

4.1 *The definition and factors of problem-solving skill to young children*

Problem-solving skill is one of the critical thinking aspects which should be developed by each person. In terms of young children, it is related to how the children develop their thinking or cognitive aspect (Beaty, 1994). Wortham (2006) stated that problem-solving skill for young children is an ability to use the experince in formulating the hypothesis, collecting data, making decision of hypothesis, and making conclusion about information from data obtained by them through scientific process. Furthermore, Brewer and Scully et al. (in Wortham, 2006) mentioned that problem-solving skill for young children consists of skill for observing, classifying and comparing, measuring, communicating, doing experiment, connecting, concluding and using.

A problem-solving skill is a scientific process. It has connections with another scientific processes such as using symbol, observing, describing, predicting, collecting and interpreting data, investigating, classifying, segmenting and blending, and formulating a conclusion since a particular skill of scientific approach has a relation with others. As well as the implementation of learning process for young children. It should be done through integrated approach in any curriculum areas.

According to these statements, the problem-solving skill for young children is a skill used by them to use their experience in developing thinking skill through scientific process. The development of it is different from the development of the problem-solving skill for the higher education level, for instance the primary or secondary student. In line with the children's cognitive development characteristics in which pre-operational level, the development of problem-solving skill should be done through learning experience, started from simple concept up to abstract concept, and emphasize on process rather than outcome.

Wetzel (2008) believed that problem solving is the core of scientific investigation. It emphasizes on scientific process effectively which is done by the children to investigate a particular object or event occured around them. It engages the children to do the action, look for information, compare, differentiate information, categorize and obtain data through various measurement, do a simple experiment, describe the relationship between cause and effect, and share the result of observation to the peers. Moreover, the precious learning experience through playing, doing experiment, finding a new thing, and social interaction will benefit the children's development.

The investigation in this process is a place for children to learn. They need to be pushed to build their own knowledge when the teacher plans the problem-solving activity, for instance providing time, place, tool, and material needed.

There are factors which influence problem-solving skill for young children. Babbington (2006) stated:

> There are a variety of influences and factors that impact upon the development of problem solving skills; the disposition and interests of the child, family and educational environments, the experiences a child may have had as an infant, relationships with caregivers/parents, and whether the child has a disability or special ability in any domain.

In line with this, Britz (1993) mentioned that factors which influence problem-solving skills for young children are disposition of children in family, society type, family structure, residence, experience, social economy status, family stability, health, tension, violence, and inability faced and experienced by the children. Based on that, on planning the learning activity, the teacher should understand and know the development and characteristic of each child, the background and culture of family in order to assist the children to understand surroundings.

4.2 *Teacher's role in developing problem-solving skill for young children*

A teacher plays an essential role in enhancing problem-solving skills for young children. Bloom et al. (cited in Britz, 1993) stated that adult should give example of problem-solving behaviour and the teacher should facilitate it in the classroom. If the teacher reveals a problem in a learning activity, children should discuss solutions, because they must realize that the process of finding out the solution is important. In developing this skill, a teacher has roles such as observing process, giving opportunity and confidence to children, and organize a challenging learning environment for children to solve the problem.

The definition of the teacher's role in this matter is not teacher-centered; this means that the teacher is a facilitator. The learning process should be a child-centered, use countless teaching methods, enhance social experiences, movements, and expressions to facilitate the children in developing problem-solving skill. Moreover, a teacher should identify the children through giving attention, listening to them nicely, asking, sharing knowledge and information, giving an example of problem-solving process as well. If the teacher could design the learning activity properly, means in line with development characteristics and the way the children learn, they will be enthusiastic to follow the learning process by trying to find out a new thing surrounds them. Furthermore, Britz (1993) agreed that by doing problem-solving activity, the children are able to develop emphaty attittude, to share responsibility, and to learn how to take decision as a part of learning to appreciate in democratic society.

The teacher provides time, space, and learning materials which assist the children to do problem-solving process (Britz, 1993). Then, a teacher has a role to assist the children to identify the problems in appropriate by formulating the questions, as follows (Goffin, 1985 in Bullock, 1990): (a) Is the problem meaningful and interesting for children? (b) Can the children solve the problem at various ages? (c) Can a new decision be made? and (d) Can a child's skill be evaluated? In addition, a teacher provides a time in order to make them learn how to make a choice, to take a decision, and to evaluate a mistake in problem-solving process. Providing a space or place is important which assists them easy to move, to communicate, and to work together. Then, providing the learning and tool materials is needed to accommodate this process.

5 CONCLUSIONS

Problem-solving skills should be developed in early childhood through learning models that are suitable with development characteristics and the way children learn. The teacher has an essential role in stimulating and enhancing that skill. Project-based learning could be done in a small group is a potential approach to develop problem-solving skill for young children by doing construction activity, investigation object or event surroundings, and also dramatic play. It gives advantages for children to employ the knowledge and skill of particular topic surroundings.

REFERENCES

Babbington. S. (2003). *Emma's story: A case study of a toddler's problem solving development.* [Online]. Retrieved May 15, 2008 from: http://www.vmcstallite.com/?aid=51733.

Bartelmay. (2008). *Research based project.* Unpublished Master's thesis.

Beaty, J. J. (1994). *Observing development of the young child.* Englewood Cliffs, NJ: Prentice Hall.

Britz, J. (1993). *Problem solvings in early childhoods classroom.* [Online]. Retrieved from: http://www.vmcstallite.com/?aid=51733,30 July 2007. Accessed 15 May 2010.

Bullock, J.R. (1990). Child-initiated activity: its important in early childhood education. *Early Childhood Educational Journal December, 18*(2).

Clark, A. M. (2006). Changing classroom practice to include the project approach. *Early Childhood Research Practice, 8*(2). Retrieved from http://ecrp.uiuc.edu/v8n2/clark.html

Costa, A. L. (1985). *Developing mind: resource book for teaching thinking.* California: ASCD.

Dearden, R. F. (1983). *Theory and practice in education.* London: Routledge & Kegan Paul.

Fry, P. S., & Addington, J. (1984). Comparison of social problem solving of children from open and traditional classrooms: A two year longitudinal study. *Journal of Educational Psychology, 76*(2), 318–329. http://dx.doi.org/10.1037/0022-0663.76.2.318

Henry, J. (1995). *Teaching through project: Open & distance learning series.* London: Kogan Page.

Katz, L. G., & Chard, S. C. (1989). *Engaging children,s mind: the project approach.* Stamford: Ablex Publishing.

Katz, L. G., & Chard, S. C. (2000). *Engaging children's minds: the project approach* (2nd ed.). Norwood, NJ: Ablex.

Kostelnik, M. J., Soderman, A. K., Whiren, A. P., Rupiper, M. L. (1999). *Developmentally appropriate curriculum: Best practices in early childhood education.* Upper Saddle River, NJ: Prentice Hall.

National Association for The Educational of Young Children (NAEYC). (2003). *Helping toddler become problem solvers.* [Online]. Retrieved July 30, 2007 from http://www.naeyc.org/2003.html.

Wetzel, R. D. (2008). *Problem solving and science process skill.* [Online]. Retrieved November 2, 2009 from http;//www.suite101.com/profile.cfm/drwetzel.

Wortham, S. C. (2006). *Early childhood curriculum.* Upper Saddle River, NJ: Prentice Hall.

Yuen, L. H. F. (2009). From foot to shoes: kindergartens', familys, and teacher perception' of the project approach. *Early Childhood Educational Journal, 37*, 23–33. DOI 10:107/s 10643–009322–3

Early Childhood Education in the 21st Century – Yulindrasari et al. (eds)
 ISBN 978-1-138-35203-2

The effectiveness of inquiry-discovery learning and games on mathematical skills

R. Mariyana
Universitas Negeri Jakarta, Jakarta, Indonesia
Universitas Pendidikan Indonesia, Bandung, Indonesia

M.C. Handini & M. Akbar
Universitas Negeri Jakarta, Jakarta, Indonesia

ABSTRACT: Young learners learn to find solutions to problems through self-regulatory activities, such as playing self-regulated game. This can be implemented in mathematical learning activities as it allows the young learners to play while studying as well as learning independently. This study aims to determine the effect of inquiry-discovery learning and games on basic mathematical skills. The study employs quasi-experimental method. The expected outcome of the study includes the identification of games that allow young learners to be active, critical, and creative in learning, determination of game condition, and selection of learning and gaming experiences. Specifically, the study is expected to identify the appropriate learning strategy and game for young learners to learn basic mathematical skills.

1 INTRODUCTION

Children are assets of a nation. They are a nation's long-term investment. A nation can invest in children by making teachers develop active learning activities for young learners using various learning approaches and methods. This will result in the ability of the children to develop their potential and solve their own problems. Furthermore, mathematics has a big influence on people's life as it exists in many aspects of life. That being said, the success of one's mathematical skills is greatly influenced by their mathematical mastery. Therefore, it is important for children to be introduced to mathematics and make it part of their life. It is also important not to force them to learn mathematics but to make it challenging and interesting. In order to achieve this, proper strategies and methods are needed.

Based on research by the Organization for Economic Cooperation and Development (OECD) carried out in the 2012 Program for International Student Assessment (PISA) with children under 15 years old, Indonesian children's mathematical skills were still unsatisfactory. Indonesia was ranked 64 out of 65 countries in mathematics, science, and reading. To address this issue, this study tested the effectiveness of inquiry-discovery learning and games using blocks and flashcards in improving the basic mathematical skills of young learners.

2 LITERATURE REVIEW

Teachers make various efforts to overcome young learners' displeasure and to increase their interest in learning mathematics by including different strategies. A strategy is as "a plan, method, or series of activities designed to achieve a particular educational goal". Soejadi (1999) argues that "A learning strategy is a way to carry out learning activities that aim to change an existing learning condition into an expected learning condition. A variety of learning approaches can be taken to make the changes".

In other words, a learning strategy is a series of procedures for planning activities, which include the use of methods and various resources in a learning activity that is structured to achieve a particular learning goal. This includes specific approaches, models, methods, and learning techniques. Some examples of learning strategies are inquiry learning and discovery learning. With these strategies, young learners learn mathematics by solving problems through block playing, where children are given a stimulus to put triangular, quadrilateral, pentagonal, or circular blocks into their respective places.

Basic mathematical skills are a set of basic skills of young learners in applying mathematical concepts, including the concept of (1) weight, length, size, and height; (2) objects categorization based on shape and color; (3) left-right and top-bottom; and (4) addition and subtraction.

3 METHOD

The purpose of this study is to identify the effect of inquiry-discovery learning and games on basic mathematical skills. The study used quasi-experimental method with 2 x 2 factorial design to test the effect of the learning strategies and games and the interaction effect of the two variables. The data were obtained from young learners through observation using a checklist and analyzed using inferential statistics technique with The hypotheses of the study were tested using a two-way analysis of variance to test the main effect, the interaction effect, and the simple effect of inquiry-discovery learning and games on young learners' basic mathematical skills. Furthermore, the interaction effect of the inquiry-discovery learning and games on the young learners' mathematical skills was calculated using Tukey test.

Table 1. 2x2 factorial design.

A / B		Learning Strategies A	
		Discovery A_1	Inquiry A_2
Games B	Blocks B_1	A_1B_1	A_2B_1
	Flashcards B_2	A_1B_2	A_2B_2
Interaction		A x B	

4 FINDINGS AND DISCUSSION

4.1 *Data normality test*

Table 2. Normality test results with the Kolmogorov-Smirnov Test criteria of H_0 accepted, H_0 rejected if $KS_{count} \leq KS_{table}$ at $\alpha = 0.05$.

Group	Kolmogorov-Smirnov Test					Conclusion*
	Mean	Std. dev.	Var.	KS_{value}	KS_{table}	
A_1B_1	111.625	10.223	3.197	0.085	0.160	Normal
A_1B_2	101.958	10.668	3.266	0.090	0.160	Normal
A_2B_1	109.681	12.278	3.504	0.088	0.160	Normal
A_2B_2	103.903	9.901	3.146	0.077	0.160	Normal
A_1	116.681	10.04	3.168	0.122	0.227	Normal
A_2	106.389	7.404	2.720	0.084	0.227	Normal
B_1	102.5	9.947	3.153	0.096	0.227	Normal
B_2	101.417	11.46	3.385	0.1	0.227	Normal

* Conclusion: Each group of data has a $KS_{count} < KS_{table}$, which is declared normal. Since the overall distribution of samples is normally distributed, H_0 is accepted. Therefore, the normality requirements for testing the hypotheses are fulfilled.

4.2 *Homogeneity test*

Table 3. Summary of homogeneity test results using Bartlett test.

Group	Calculation results		Conclusion*
A1 and A2	$\chi^2_{value} = 0.13$	$\chi^2_{table} = 3.84$	Homogeneous
B1 dan B2	$\chi^2_{value} = 3.26$	$\chi^2_{tabe} = 3.84$	Homogeneous
A1B1 A1B2 A2B1 A2B2	$\chi^2_{value} = 6.94$	$\chi^2_{tabe} = 7.81$	Homogeneous

* Conclusion: Since the results of homogeneity test is $\chi^2_{value} < \chi^2_{table}$ on $\alpha = 0.05$, H_0 is accepted. This means that all groups have the same variance (homogeneous).

4.3 *Hypotheses tests*

(1) There is a difference in the basic mathematical skills between the young learners taught using discovery learning and those taught using inquiry learning.
(2) There is a difference in the basic mathematical skills between the young learners taught using a blocks game and those taught using a flashcards game.
(3) There is an interaction effect of the games and the learning strategies on the young learners' basic mathematical skills.
(4) There is a difference in the basic mathematical skills of the young learners taught using discovery learning with a blocks game and those taught using inquiry learning with a blocks game.
(5) There is a difference in the basic mathematical skills of the young learners taught using discovery learning with a blocks game and those taught using discovery learning with a flashcards game.
(6) There is a difference in the basic mathematical skills of the young learners taught using discovery learning with a flashcards game and those taught using inquiry learning with a flashcards game.
(7) There is a difference in the basic mathematical skills between the young learners taught using inquiry learning with a blocks game and those taught using inquiry learning with a flashcards game.

Statistical factorial was used to test the main effect of hypothesis 1, 2, and 3, while Tukey test was used to test the simple effect of hypothesis 4, 5, 6, and 7.

Table 4. Results of variance analysis using two-way ANOVA.

	JK	df	RJK	F	Sig.
Corrected Model	5359.139[a]	3	1786.380	18.516	.000
Intercept	1642242.250	1	1642242.250	17022.324	.000
strat	3364.000	1	3364.000	34.869	.000
media	1201.778	1	1201.778	12.457	.001
strat * media	793.361	1	793.361	8.223	.005
Error	13506.611	140	96.476		
Total	1661108.000	144			

* Significant level of $\alpha = 0.01$ and $\alpha = 0.05$

After going through the treatment of discovery and inquiry learning and blocks and flashcard games, the data regarding the young learners' mathematical skills were obtained and analyzed.

Table 5. Simple effect.

Games (B)	Learning strategies (A)	
	Discovery (A_1)	Inquiry (A_2)
Blocks (B_1)	A_1B_2 (116,86)	A_1B_2 (102,50)
Flashcards (B_2)	A_1B_2 (106,39)	A_2B_2(101,42)

Below is a simple graph of the interaction effect of the learning strategies (A) [Discovery (A_1) and Inquiry (A_2)] and Games (B) [Blocks (B_1) and Flashcards (B_2)] on the basic mathematical skills (Y) of young learners in Class B

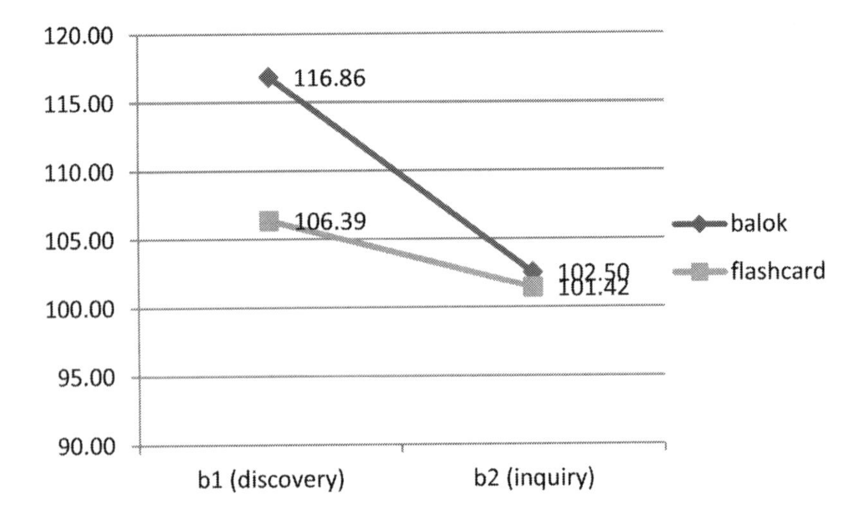

4.4 *Research findings*

Table 6. Summary of research findings.

Number	Hypothesis	Results
1	H_o: $\mu A_1 \leq \mu A_2$ H_1: $\mu A_1 > \mu A_2$ H_o: μDiscovery ≤ μBlocks H_1: μDiscovery-Block > μInquiry-Blocks	F_{value} (34.869) > F_{table} (3.909) on α= 0.05 H_0 is rejected, H_1 is accepted
2	H_o: $\mu B_1 \leq \mu B_2$ H_1: $\mu B_1 > \mu B_2$	F_{value} (12.457) > F_{table} (3,909) on α= 0.05 H_0 is rejected, H_1 is accepted
3	H_o: interaction A x B = 0 H_1: interaction A x B ≠ 0	$F_{value\ interaction}$ (8.223) > F_{table} (3.909) on α= 0.05 H_0 is rejected, H_1 is accepted

(Continued)

Table 6. (*Continued*)

Number	Hypothesis	Results
4	H_o: $\mu A_1 B_1 \leq \mu A_2 B_1$ H_1: $\mu A_1 B_1 > \mu A_2 B_1$ H_o: μ Discovery-Blocks ≤ μ Inquiry-Blocks H_1: μ Discovery-Blocks > μ Inquiry-Blocks	$Q_h = 8.77$ while $Q_t = 3.681$ ($\alpha = 0.05$; $db_1 = 4$; $db_2 = 140$), so $Q_h > Q_t$ H_0 is rejected, H_1 is accepted
5	H_o: $\mu A_1 B_1 \leq \mu A_1 B_2$ H_1: $\mu A_1 B_1 > \mu A_1 B_2$ H_o: μDiscoveryBlocks ≤ μDiscovery Flashcards H_1: μDiscoveryBlocks > μDiscovery Flashcards	$Q_h = 6.397$ while $Q_t = 3.681$ ($\alpha = 0.05$; $db_1 = 4$; $db_2 = 140$), so $Q_h > Q_t$ H_0 is rejected, H_1 is accepted
6	H_o: $\mu A_1 B_2 \leq \mu A_2 B_2$ H_1: $\mu A_1 B_2 \leq \mu A_2 B_2$ H_o: μDiscovery Flashcards ≤ μInquiry Flashcards H_1: μDiscovery Flashcards > μInquiry Flashcards	$Q_h = 3.037$ while $Q_t = 3{,}681$ ($\alpha = 0.05$; $db_1 = 4$; $db_2 = 140$), so $Q_h < Q_t$ H_0 is accepted, H_1 is rejected
7	H_o: $\mu A_2 B_1 \leq \mu A_2 B_2$ H_1: $\mu A_2 B_1 > \mu A_2 B_2$ H_o: μInquiryBlocks ≤ μInquiry Flashcards H_1: μInquiryBlocks > μInquiry Flashcards	$Q_h = 2.375$ while $Q_t = 3.681$ ($\alpha = 0.05$; $db_1 = 4$; $db_2 = 140$), so $Q_h < Q_t$ H_0 is accepted, H_1 is rejected

4.5 *Findings*

1. The basic mathematical skills of the young learners taught using discovery learning (A1) was higher than those taught using inquiry learning (A2)
2. The basic mathematical skills of the young learners taught using a blocks game (B1) was higher than those taught using a flashcards game (B2)
3. There was an interaction effect of the learning strategies (A) and games (B) on the basic mathematical skills (Y)
4. The basic mathematical skills of the young learners taught using discovery learning with a blocks game (A1B1) was higher than those taught using inquiry learning with a blocks game (A2B1)
5. The basic mathematical skills of the young learners taught using discovery learning with a blocks game (A1B1) was higher than those taught using discovery learning with a flashcards game (A1B2)
6. The basic mathematical skills of the young learners taught using discovery learning with a flashcards game (A1B2) is not different from those taught using inquiry learning with group with a flashcards game (A2B2). The difference in the value of Q = 3.037 between the discovery learning with a flashcards game and the inquiry learning with a flashcards game is not statistically significant.
7. The basic mathematical skills of the young learners taught using inquiry learning with a blocks game (A2B1) was not different from those taught using inquiry learning with a flashcards game (A2B2). The difference in the value of Q = 2.375 between the inquiry learning with a blocks game and the inquiry learning with a flashcards game was not statistically significant.

4.6 *Discussion*

Based on the main purpose of discovery learning and the results of this study, it can be said that discovery learning is very effective in improving young learners' basic mathematical skills. This is consistent with Glasson (Santrock, 2009, p. 171) stating, "Students who participate in discovery-based activity learning scored higher on science subject than students in science classes with a traditional direct teaching". Above all, as synthesized from Presseisen

(2001) and Arsyad (2011), which is also supported by Sari and Abdullah (2016), innovative mathematics learning activities, especially in terms of the media, had greater student learning outcomes potential. The advantages of discovery learning make it possible to be used as an alternative in improving young learners' basic mathematical skills in kindergarten because young learners in this stage need guidance and direction in the learning process. In addition to being able to improve their basic mathematical skills, the use of discovery learning can make young learners to be more active in the learning processes, which is appropriate to the development stage of children at an active age, in exploring the environment around them. This is in line with the research results of Yulida et al. (2016, p. 189) on the use of guided discovery learning in improving learning activities and outcomes; Martin, Cirino, Sharp, Carla, Barnes, and Marcia (2014, pp. 12–13) on number skills of kindergarten students, specifically on the ability of kindergarten students to predict mathematical results in first grade; and Istikomah (2014) and Susilowati (2015) on the development of discovery learning devices for science learning in kindergarten. In addition, the research results of Tindangen and Vandalita (2016) on the development of inquiry-based learning tools to improve conceptual understanding, problem solving ability in learning biology describes the main cause of the problem in learning activities that occur in science subject in junior high school and biology subject in high school is that the practicum activities carried out as part of the learning has never been applied as part of the hypothesis verification despite it being one step in the syntax of inquiry learning. Finally, research conducted by Salo (2017) found that discovery learning can improve the activeness of students in learning.

5 CONCLUSION

Based on this discussion, the following conclusions have been drawn: (1) Discovery learning could improve young learners' basic mathematical skill, (2) Blocks games could improve young learners' basic mathematical skill, (3) Discovery learning with blocks games improved young learners' basic mathematical skill, (5) There was an interaction effect of the learning strategies and games on young learners' basic mathematical skill.

REFERENCES

Arsyad, A. (2011). *Media pembelajaran.* Jakarta: Rajagrafindo Persada.

Istikomah. 2014. Pengembangan perangkat pembelajaran metode discovery learning untuk pemahaman sains pada anak TK B. *Nur El-Islam, 1*(52), 81–93.

Martin, R. B., Cirino, P. T., Sharp, C., & Barnes, M. (2014). Number and counting skills in kindergarten as predictors of grade 1 mathematical skills. *Learning and Individual Differences, 34*, 12–23. https://doi.org/10.1016/j.lindif.2014.05.006.

Presseisen, B. Z. (2001). Thinking skills: Meanings and models revisited. In A.L. Costa (Ed.). *Developing minds: a resource book for teaching thinking* (*3rd edition*), (pp. 47–53). Alexandria, VA: ASCD.

Salo, A. Y. (2017). Pengaruh metode discovery learning terhadap keaktifan belajar siswa (studi quasi eksperimen kelas VII SMPNI 6 Banda Aceh). *Jurnal Penelitian Pendidikan, 16*(3), 297–304.

Sari, D. M., & Abdullah, M. H. (2016). Pengaruh permainan balok angka terhadap kemampuan mengenal lambang bilangan pada anak usia dini. *Jurnal PAUD Tambusai, 2*(2), 18–25.

Santrock, J. W. (2009). *Psikologi pendidikan.* Translated by D. Angelica. Jakarta: Salemba Humanika.

Soejadi, 1999. *Kiat Pendidikan Matematika Di Indonesia, Konstanta. Keadaan masa kini menuju harapan masa depan.* Departemen Pendidikan dan Kebudayaan Direktorat Jendral Pendidikan Tinggi.

Susilowati, T. (2015). Kemampuan number sense melalui metode learning by playing. *Jurnal Pendidikan Dasar, 6*(2), 339–350.

Tindangen, M., & Vandalita, V. (2016). Pengembangan perangkat model pembelajaran berbasis inkuiri [Inquiry Based Learning] untuk meningkatkan pemahaman konseptual, kemampuan pemecahan masalah dalam pembelajaran biologi siswa SMP dan SMA di Propinsi Kalimantan Timur. *Proceeding Biology Education Conference* (ISSN: 2528–5742), *13*(1) 2016, 237–248.

Yulida, R., Kausar, K., & Adriani, Y. (2016). Penggunaan model pembelajaran penemuan terbimbing dalam meningkatan aktivitas dan hasil belajar mahasiswa pada matakuliah penyuluhan dan komunikasi pertanian in *MIMBAR PENDIDIKAN.Jurnal Indonesia untuk Kajian Pendidikan, 1*(2), 189–200.

Early Childhood Education in the 21st Century – Yulindrasari et al. (eds)
© 2020 Taylor & Francis Group, London, ISBN 978-1-138-35203-2

The teacher strategy informing discipline on pre-school children in Kendari Kuncup Pertiwi kindergarten

Salwiah & Asmuddin
Halu Oleo University, Sulawesi Tenggara, Indonesia

ABSTRACT: This research describes the teacher strategy informing discipline on pre-school children in Kendari Kuncup Pertiwi kindergarten. The research used qualitative study with ethnographic approach. The informants of the research are teachers and students in Kendari Kuncup Pertiwi kindergarten. The data was collected by using observation, interview, recording, document and photos, then analysed by applying domain, taxonomy and thematic data analysis. The result of the research shows: (1) disciplines applied by the teacher in Kendari Kuncup Pertiwi kindergarten has resulted in children arrive on time at school; wear uniform, pray before and after activities; wash their hands before and after eating; pack up their toys; wait for their turns in taking meals; accept and completed their tasks; and throw rubbish in the proper place, (2) teachers in Kendari Kuncup Pertiwi kindergarten uphold discipline comprehensively throughout the process of learning. The strategy was integrated in every sequence of teaching. The teachers also used teaching media to help children understand the message that they tried to convey.

1 INTRODUCTION

Teaching and learning strategy is the general pattern of the actions of students and teacher in teaching and learning activities. The definition of strategy in this case refers to the abstract characteristics of a series of student teacher actions in an actual teaching and learning event called instructional procedures. Teaching and learning strategies are tools to achieve learning goals. The teacher's strategy in creating a learning and teaching atmosphere that is safe and enjoyable for children by holding good relations between teachers and children so that there is no feeling of pressure on the child or fear of the teacher, then this can cause children to feel at home in kindergarten and want to carry out the tasks given by the teacher.

Kindergarten education's effect on positive behavior can be seen from the results of the main activities, namely: (1) the activity of behavior formation through habits manifested in daily activities in kindergarten, and (2) activities that develop basic abilities. The first activity can be seen in collaboration between friends, neatness, discipline, tolerance, courage, ability to carry out tasks and so on. Whereas in the second activity can be seen in: creativity, language, skills, and physical abilities. Discipline can help a child grow with good self-confidence and control which is demanded by a good awareness of himself and his life and good feelings about him and his sense of responsibility and care for the environment.

The essence of discipline is to teach or someone who wants to follow the teachings of a leader. The close aim of the meaning of discipline is to make children trained and controlled by teaching them appropriate and inappropriate forms of behavior that are unfamiliar to them (Schaefer, 1987, p. 7). The long-term goal of discipline is the development of self-control (self control and self direction), which is where children can direct themselves without influence or outside control. In the case of disciplinary and behavioral formation programs are activities carried out continuously in the daily lives of children in kindergarten, so that it becomes a good habit. Formation of behavior through habituation includes religious faith, emotion, social ability and discipline. While the Basic

Ability Program is a program prepared by the teacher to achieve certain abilities in accordance with the stages of child development, the Basic Ability Program includes creativity, language, thinking, skill and physical. Basically, forming a discipline is a teaching process for us and a learning process for children.

1.1 *The essence of children preschool*

In order to achieve the research objectives, namely to improve the quality of learning in fostering the discipline of preschoolers, especially kindergarteners, it is better to start by looking for the needs of preschoolers (Harianti et al., 1994, p.7). The needs of preschoolers are based on the sequence of stages of their development. In addition, it is also necessary to consider the forms of preschool education in Indonesia with their suitability to develop the ability of preschool children. Every child is unique with individual temperament, learning style, family background, and pattern and timing of growth. At every age, meeting basic health and nutritional needs are essential (http://www.wordbank.org/children/what/stages.html). Thus, children need stimuli that can train their motoric, cognitive, language, and social skills.

1.2 *The essence of strategy*

Purwanto and Suparman stated that teaching strategies can be called a systematic way of communicating the content of the lesson to students to achieve certain learning content (Purwanto & Suparman, 1999, p. 157). Learning contains systematic and directed instructions to help students achieve the learning goals. Gagne and Briggs suggested that learning is an organized plan to develop the optimization of learning outcomes. A good plan must involves experts including teachers, curriculum specialists, learning media developer (Gagne, 1974, p. 19).

Based on the views of Gagne and Briggs (1974), it can be seen that learning must have a specific purpose and that goal is obtained through a process that is planned in advance by experts. According to both of these views, the main focus or purpose of learning lies with students that students can capture the content of learning well (Harefa, 2001). It is intended that in the learning process there is an interaction between the teacher and students.

1.3 *The essence of discipline*

The problem of coaching discipline is a problem in a fairly broad life. In general, discipline is part of inner training and character so that all one's actions are in accordance with the prevailing rules. Therefore, the study of discipline is also a concern of experts. This is in line with what was stated by Wursanto, that discipline is a condition that causes or gives encouragement to children to do and do all activities in accordance with the norms or rules that have been established (Wursanto, 1989, p. 108). Thus the definition above, emphasizes discipline to create a conducive situation, where children can carry out activities that are based on established norms or rules. Discipline is motivated by a sense of confidence in values, aware of his own position, aware of the objectives to be achieved so that he has the ability to live the rules that apply. Such conditions then give rise to rational, conscious, non-emotional and selfless self-respecting behavior. Thus discipline can be said that the ability of a child to live the rules, the rules that apply so that consciously willing to obey certain rules.

Based on this description, it can be concluded that discipline is an inner training and character so that all children's actions are in accordance with the applicable rules so that the child can consciously carry out activities that are based on the norms or rules set. Through teaching, guidance or encouragement conducted by the teacher (instructor) is intended to help children learn to recognize discipline and responsibility oriented to cultural values, religion, or even the value of life that is considered good by the community and is believed to be right for life as social beings that should be upheld in humane, decisive and inner-growth ways: self discipline and self control (self-development: self-discipline and self-control).

2 RESEARCH METHODS

This research is a qualitative research using a naturalistic approach. Especially the ethnographic approach proposed by Spradley. This type of research is a procedure that produces descriptive data in the form of expressions, activities and notes in the process of learning activities involving teachers and students. Data collected as much as possible in accordance with the research focus. The 12 steps put forward by Spradley carried out by researchers include: (1) selecting social situations (selecting social situations) which are the subject of research namely Kendari Kuncup Pertiwi kindergarten. (2) doing participant observation, namely the researcher directly observes the subject of the study. (3) making ethnographic record, namely the researcher carrying out recording in accordance with the focus of observations from below or inductively, (4) making descriptive observation, namely carrying out descriptive observations of what is heard, seen from the actions or words of the subject, (6) making domain analysis, (7) making observation observation, (8) making selective observation, (9) making a theme analysis, (10) taking a cultural inventory, (11) writing the ethnography.

2.1 *Data collection technique*

As a researcher who performs the task of observation in the learning process, he tries to listen carefully to various activities carried out by the subject so that the researcher can capture more information on the focus under study. The objects observed were (1) the steps of teaching activities and teaching methods (2) the concept of discipline, the process of formation of disciplines and obstacles in forming discipline (3) efforts to shape the discipline of preschoolers, which includes the approach to experience used, material selection learning while playing in this case the method of forming the discipline itself. From the results of observations made during the implementation of the teaching with the focus of the rules and strategies applied and the results found in the Kuncup Pertiwi Kindergarten in this case about the formation of discipline in Kindergarten A children, researchers made field notes. Field notes are arranged based on what researchers see, hear, experience and think during the data collection and are reflected.

2.2 *Sampling*

The place and situation of the study in accordance with the priority research focus is the teaching activities of the formation of discipline, although other lessons are also associated with the formation of discipline. In this study, the sample selection was carried out with the aim (purposive sampling) which in its implementation used snowball techniques or the sample chosen was informational in purpose to enrich information. The selected sample limitation in kindergarten TK A1, TK A2, TK A3. and TK A4 in Kendari Kuncup Pertiwi kindergarten, which is the data source grouped on: the learning activities process, which includes activities during the learning process of discipline formation, learning activities in Kendari Kuncup Pertiwi kindergarten students in group A, teachr kuncup pertiwi kindergarten which is carrying out teaching activities in the formation of discipline.

2.3 *Data recording process*

To facilitate data analysis, the writer used the following procedure: data recording is carried out in a field record format which is carried out through stages: preliminary recording, recording during observations/interviews using keywords, expansion which is a form of field notes consisting of notes field consisting of descriptive and reflective notes which are observers' responses and improvements. Make certain instructions (coding) like the following: CL Field Notes, A Paragraph, H Pages, P Observer, W Interviews. Choosing tools that are easy to use in collecting data such as notes, pencils/ballpoints, recording devices and photographing devices (tustels) to be used to take pictures of situations.

3 DISCUSSION

3.1 *Discipline applied in Kendari Kuncup Pertiwi kindergarten*

Analysis of the research findings will be described in this discussion by prioritizing some of the experts' thoughts (tringulation of theory). The changes include the steps that are applied in preparing the teaching of the discipline of preschoolers.

For preschool-aged children, especially the age of the students in kindergarten A, the methods used in schools put pressure on the efforts of teachers to introduce and get used to. In this case the teacher must be able to look for various efforts to implement eight family functions namely religious, socio-cultural, love, protection, reproduction, socialization and education, economy and coaching (Linda & Eyre, 1997, p. xxxii). Linda and Eyre provide the following methods for forming child discipline: scenario play, discussion of concepts, positive praise, rewards, rewards and other forms of distribution, second chance or opportunity to correct mistakes, and channeling positive behaviors (BKKBN, 1996, p. vi).

Schaefer parents can also act to allow children's behavior, ignoring and redirecting. It is normal for children to be noisy, impulsive, demanding and their treatment or actions are often unpredictable. They oppose parents' doubts and order, norms, order, cleanliness and politeness (Schaefer, 1997).

3.2 *Teaching strategy applied by teachers in Kendari Kuncup Pertiwi kindergarten*

Development of teacher teaching strategies can be identified starting from the teaching sequence, teaching methods, teaching media, and teaching time.

3.2.1 *Teaching method*

In the teaching and learning process the main task of the teacher is to create an atmosphere of teaching and learning that can motivate students to always learn well and be enthusiastic. With the challenging teaching and learning atmosphere, competing in a healthy manner is expected to have a positive impact on achieving optimal learning outcomes. To achieve the learning process the teacher should have the ability to choose or determine and use appropriate teaching methods. To teach in kindergarten, teachers must be rich in teaching methods, so that children feel happy and will easily receive lessons. Learning methods are part of an activity strategy. The method chosen by the teacher is the lecture method, question and answer, group work discussion, experimentation, demonstration, and integrated learning unit method. The method is a tool to achieve goals. For example, teachers can introduce alphabets through songs, story telling, pictures, and many more.

3.2.2 *Teaching media*

In the teaching and learning process the teaching media is very important because by using the media the learning process itself will be more effective. In other words, learning media can be used as a tool to succeed in teaching and learning activities in the classroom. The media used in schools include: audio media is the type of media that is heard. This media serves to channel messages from the teacher to students where the way to receive the message students use their sense of hearing. Audio media that is often used in classrooms such as tape recorders, because in addition to this easy-to-use media is also easily controlled by the teacher. Visual media is a type of media whose message can be captured with the sense of sight. Visual media is often used in classrooms, for example: charts, maps, diagrams and charts. This media serves to help students experience the real world that they have not seen directly. Audio-visual media is a type of media whose messages can be received through the senses of sight and hearing and function to overcome the effectiveness of audio or audio-visual media. Multicolored media is an actual objects that teachers can use in the teaching and learning process.

3.2.3 *Teaching time*

In the process of teaching and learning the use of effective time is needed. In order for the subject matter to be accepted by students, the teacher should have a plan to help the child in learning to achieve each goal. When a teacher decides to teach something to his students, he/she should know what he will teach and how he will teach it.

There are three teacher behaviors that have a strong relationship with the effectiveness of teaching, is a learning environment, providing sufficient time in learning activities and the use of appropriate teaching methods. The teaching strategy applied by the teacher in preparing the disciplinary learning of preschoolers is to obtain more satisfying disciplinary learning outcomes in kindergarten A.

Based on this description, it can be said that the strategy is an activity that relates to the order of teaching, methods, media, and time to present teacher teaching can take place properly and systematically in order to improve learning outcomes that is a learning program that has criteria such as attractiveness, usability, and results to be able to support the success of students to achieve the planned goals.

4 CONCLUSION

The teachers taught the children discipline by introducing and enforcing rules. A clear information about rules enable children to understand the rules and follow them without being told by the child to do it on its own as children come to school on time, children dress uniformly, pray before and after activities, wash their hands before and after eating, store and tidy up toys in place, wait their turn when taking food, receiving and completing tasks and disposing of garbage in its place. These stages show that children have understood the meaning of the discipline applied from Kendari Kuncup Pertiwi kindergarten. Likewise, the children simultaneously answer questions from the teacher and questions from friends.

In the teaching and learning process the main task of the teacher is to create an atmosphere of teaching and learning that can motivate students to always learn well and be enthusiastic. With a challenging teaching and learning atmosphere competing in a healthy manner is expected to have a positive impact on achieving optimal learning outcomes. To teach in kindergarten, teachers must be rich in teaching methods, so that children feel happy and will be easy to receive lessons. Methods are part of an activity strategy. The methods to be chosen by the teacher include: lecture method, question and answer, discussion, group work methods, assignments, demonstrations, experiments and learning methods. In teaching and learning process the role of the teaching media is very important because by using the media the learning process itself will be more effective. In other words, learning media can function as a tool to succeed in the stages of teaching and learning in the classroom. In the process of teaching and learning the use of effective time is needed. In order for the subject matter to be accepted by students, the teacher should have a plan to help the child in learning to achieve each goal. This means that the teacher in presenting the subject matter has already had careful planning in accordance with a predetermined time allocation by showing various students' abilities in order to achieve the expected goals. Thus when the teacher decides to teach something to his students, it should be reflected in a process of thinking about what he will teach, what procedures and materials he/she needs to achieve the desired learning outcomes.

REFERENCES

BKKBN. (1993). *Guidelines for implementing child care patterns in 0–5 years of prosperous families*. Bandung: BKKBN.

Gagne, R. M., & Briggs, L. J. (1974). *Principles of Instructional Design*. Exford, UK: Holt, Rinehart & Winston.

Harefa, A. (2001). *Learning the age of all autonomy*. Jakarta: Compass Book Publishers.

Harianti, D., et al. (1994). *Development of children in kindergarten*. Jakarta: Ministry of Education and Culture Center for Development of Education Curriculum and Facilities research report.

Huitt, W. (2001). System model of human behavior overview. *Educational Psychology Interactive*. Retrieved from http://www.edpsycinteractive.org/materials/sysmdlo.html
Linda & Richard, E. (1988). *Principles of instruction design*. Florida: Harcourt Brace Jovanovich Collage Publishers.
Purwanto & Suparman, A. (1999). *Evaluasi Program Diklat*. Jakarta: STIA-LAN.
Schaefer, C. (1997). *How to guide, educate and discipline children effectively*. Jakarta: Restu Agung.
Spradley, J. P. (1990). *Participant observation*. New York: Holt, Rinehart and Winston.
Wursanto, I. G. (1989). *Manajemen kepegawaian 1*. Yogyakarta: Kanisius.

Early Childhood Education in the 21st Century – Yulindrasari et al. (eds)
© 2020 Taylor & Francis Group, London, ISBN 978-1-138-35203-2

How nonverbal communication psychoeducation programs for teachers can improve a preschool's prosocial behavior

A.I.M. Razak, A. Mulyanto & H.S. Muchtar
Early Childhood Department, Universitas Islam Nusantara Bandung, West Java, Indonesia

ABSTRACT: Language is an integral and central part of the importance of communicating and interacting as a social activity and consists of verbal and nonverbal communication (Boyce, 2012). This study aims to train teachers through a non-verbal communication psychoeducation program in the form of (1) facial abilities; (2) gestural abilities; (3) postural abilities; (4) artifactual abilities; and (5) artifactual abilities (Boyce, 2012). A series of psychoeducation procedures are applied to 10 teachers in applying non-verbal communication to children. This study uses a quantitative research design the-pretest-posttest-experimental design (Kumar, 2011), which was tested on 26 children aged 4–6 years involving three forms of prosocial behavior, including (1) sharing, (2) helping, and (3) comforting (Dunfield, 2010). The results showed that in the pre test stage, 11 children were in the low prosocial behavior category, while 15 children were in the moderate prosocial category and no one child was in the high prosocial behavior category. The 19 post-test children were in the moderate category and 5 people were in the high category of prosocial behavior. Hypothesis testing shows that the non-verbal communication psychoeducation programs assist teachers in improving the prosocial abilities of early childhood and proved effective (Asymp.Sig = 0.189 > 0.05).

1 INTRODUCTION

Prosocial behavior is a behavior that benefits others (Froming, Nasby, & McManus, 1998). The strategic and central prosocial behavior is described as the main factor in an individual's success in establishing social relations with others (Koening, 2000). Even babies at the age of 10 months have been able to show prosocial behavior when asking their mother to get them something (McGinley, 2008). Prosocial behavior is one factor of individual success in establishing social interactions, which can be manifested through helpful behavior, sharing, collaborating, providing comfort, being able to play with friends, and providing ideas and ideas in carrying out game activities with friends. Children first learn prosocial behavior from their microsystem environment (McGinley, 2008). Children who live with mothers and fathers, or in extended families, have a different view of prosocial behavior than do children who live in orphanages. At the age of four, on average children expand their microsystem environment through the school environment (Nantel-Vivier, 2010). In preschool children the forms of prosocial behavior are extended again through a friendship environment (Partington, 1980). Prosocial behavior needs to be developed by children so that they can play a role according to the demands of prosocial behavior in each of their environments. There are many programs to improve prosocial behavior in preschool, including those involves constructive play (Partington, 1980), through storytelling activities (McCutchen et al., 2002), through the role of mother and child interaction (Pilgrim, 2006), as well as through peer-playing activities (Bryant & Budd, 1984). The overall effort emphasizes the active interaction between children and their environment through active communication, one of which is the role of language which has a central position in the role of social interaction between humans.

Language is a form of communication between humans to convey messages, both verbal and non-verbal. Verbal communication is defined as communication that arises from a series of letters that form words, then form meaningful and acceptable sentences and respond to the other person (Berk, 2008). Non-verbal communication is interpreted as a form of communication other than generated from words, such as facial expressions, body movements, and distance settings to communicate with the other person (Dodge, Colker, & Heroman, 2009). One form of non-verbal communication that has been applied in learning is gestures or the body language of the teacher when teaching language to students (Roberts & Strayer, 1996). Gesture also has a role in shaping the mindset of students, receiving learning information, and playing an important role in class management (Dodge, Colker, & Heroman, 2009).

Improving the skills of teachers in non-verbal communication is obtained through psychoeducation procedures, namely educational programs delivered by professionals through a psychological approach to improve a skill and change a person's behavior, which can be done on children, adults and parents and carried out in groups or groups. individuals through diverse and explored media (Pilgrim, 2006). In general, the psychoeducation model consists of three types, including: (1) life skills model; (2) developmental model; and (3) helpful model (Berk, 2008). The psychoeducation model applied in this study is a helpful model that is specifically on the ability of the skill training model (Froming, Nasby, & McManus, 1998). The psychoeducation model shows a systematic development of behaviors that are specific to the behavior so that individuals can increase their ability to deal with problems more effectively. In this study, we tried to develop non-verbal communication skills of teachers in order to improve children's prosocial behavior.

2 METHOD

This study aims to examine the research question "how the effectiveness of non-verbal psychoeducation programs on teachers can improve children's prosocial behavior?" The research design used was experimental research with the design of the before-and-after design (Kumar, 2011) without using groups control and experimental group. There were 10 teachers involved in the non-verbal communication psychoeducation program from five schools in Cibeber Village and 26 children were involved. The study employed a non-probability sampling design with purposive sampling technique (Kumar, 2011), and is explained in the following sections.

2.1 *Instruments of prosocial behavior research*

Prosocial behavior tested in this study consisted of three forms of behavior (Dunfield, 2010) including. First, helping, recognizing and responding to another individual's inability to complete a specific goal directed action (Dunfield, 2010), which can be observed through the behavior of asking friends for help, asking the teacher for help, helping friends, and following the teacher's request to help friends and consists of seven item indicators. Second, sharing, recognizing and responding to another individual's lack of a good desired material (Dunfield, 2010), which can be observed through the behavior of giving objects to friends, taking turns, and exchanging objects with friends and consisting of five indicator items. Third, comforting, recognizing and responding to the observation of another individual's negative affective state (Dunfield, 2010), which can be observed through recognizing and responding to feelings of pleasure, sadness, anger, fear, surprise, and disgust and consisting of 11 indicator items.

2.2 *Non-verbal communication psychoeducation training*

We conducted a two-days interactive psychoeducation training consists of five material related to non-verbal communication in children consisting of (1) facial; (2) gestural; (3) postural; (4) proxemic; and (5) artifactual (Boyce, 2012). The psychoeducation training was held for two days divided into three training sessions, as explained in the next paragraph.

First day: Friday, August 31, 2018 at 13.00–15.00. The activity on the first day was given by the teacher about the concept of non-verbal communication in children consisting of facial, gestural, postural, proxemic, and artifactual abilities. The purpose of this training is for teachers to understand forms of non-verbal communication that can be applied when teaching in early childhood. Second day of the first session: Saturday, September 1, 2018 at 08.00–10.00. The activity on the second day of the first session was that the teacher was given an explanation of the practice of non-verbal communication in children which could be applied when the process of teaching and learning activities took place. Second day of the second session: Saturday, September 1, 2018 at 10.00–12.00. The activity on the second day of the second session was to give an explanation of the techniques of measuring children's observation and prosocial behavior interviews consisting of three forms of prosocial behavior (1) sharing; (2) comforting; (3) helping, which is then arranged into 23 indicator items.

2.3 *Data collection method*

Data was collected before (pretest) and after (posttest) a non-verbal communication interventions. The collected data was used to test the effectiveness of non-verbal communication psychoeducation intervention programs on teachers in improving children's prosocial behavior. The study used observational data collection methods with event sampling observation techniques, by counting the behaviors that arose while observations were taking place (Sattler, 2002). Observation data collection methods are applied to prosocial helping behavior (seven indicator items) and sharing (five indicator items), while data collection methods through interview procedures are applied to comforting behavior (11 indicator items). Below are all three forms of prosocial behavior observed,

a. Helping, observed through ± 30 minutes of joint eating activities consisting of 7 indicator items, with scoring techniques as follows:
   Score 1: behavior is expected not to appear
   Score 2: 1–3 expected behavior appears
   Score 3: 4–6 expected behavior appears
   Score 4: all behaviors appear
b. Sharing, observed through ± 30 minutes of joint eating activities consisting of 5 indicator items, with scoring techniques as follows:
   Score 1: behavior is expected not to appear
   Score 2: 1–2 expected behavior appears
   Score 3: 3–4 expected behavior appears
   Schore 4: all behaviors appear
c. Comforing, tested through questions submitted by interviewers during the activity which consisted of 11 indicator items, with scoring techniques as follows:
   Score 1: the questions asked cannot be answered correctly
   Score 2: 1–5 questions can be answered
   Score 3: 6–10 questions can be answered
   Score 4: all questions can be answered

4 IMPLEMENTATION OF INTERVENTION

The implementation of the intervention implementation of non-verbal communication in early childhood was carried out for seven days of teaching and learning activities in schools that were applied to 26 children from five schools, with the intervention program as follows: First, facial: non-verbal communication for forms of prosocial behavior that can be applied by the teacher is always smiling, speaking softly and being friendly. Through this form of communication, children can be stimulated by helping various friends, and the ability to know the emotions of their friends. Second, gestural: gestural non-verbal communication for the form of prosocial behavior can be applied by the teacher is looking down when talking to children, applying eye level when talking to children, and holding the shoulders of the child as a form of care and warmth of an educator. Third, postural: postural non-verbal communication for the form of prosocial behavior can be applied by the teacher is to respond to the child in an enthusiastic way when called upon, when the child asks for help, and when interacting with the child. Fourth, proxemic: proxemic non-verbal communication for children's prosocial behavior can be applied by the teacher when setting close distances to children at 0 cm–15 cm and long distances with children between 15 cm–45 cm. Fifth, artifactual non-verbal communication for children's prosocial behavior can be applied by regulating the appearance of the teacher.

5 RESULT AND DISCUSSION

This study aimed to examine the effectiveness of non-verbal communication psychoeducation programs on teachers in improving children's prosocial behavior. The researcher tested the prosocial behavior of children through two stages. The results showed that at the pre-test of 11 children in the low prosocial category, 15 children were in the moderate prosocial category and the average prosocial behavior was at 56%. Based on the results of the pre-test, it was illustrated that children's prosocial behavior has a diverse score, because prosocial behavior is formed through modeling, learning by doing, and reinforcement (Nantel-Vivier, 2010). Stimulus that is diverse such as prosocial modeling is obtained through parental actions, peer actions, teacher actions and also influences from media such as television, computers, and other information media. Children learn about prosocial behavior through live modeling and the modeling symbol (Berk, 2008).

This can be seen in subjects with names B, Z, G, V, and Y, each of which has a pre-test score of 62, 63, 65, 64, and 68 in the category of moderate prosocial behavior. Based on the results of observations on daily activities carried out by classroom teachers, it shows that research subjects named B, Z, G, V, and Y have previously known and had prosocial behavior. These five subjects have the behavior to be able to play with friends, help friends, and the ability to respond to their friends' emotions and these five subjects are also included in the category of children who are able to control emotions. Secondary data based on interviews with parents indicate that parenting applied to the five subjects has similarities, namely educating children in a gentle way and almost never yell at children, scold children, and also nail children. Regarding the modeling symbol, the five subjects cannot be avoided from technology products. Secondary data from interviews with parents stated that the five subjects in one day no more than 1 hour were given spectacle from both mobile and television products, and all were based offline so that they could be easily controlled. The post-test results showed that the five subjects showed consistency in prosocial behavior and some of them had increased behavior, as can be seen in subject B (62 to 72), Z (63 to 63), G (65 to 65), V (64 to 71), and Y (68 to 70).

Prosocial behavior is also formed through the mechanism of learning by doing, meaning that prosocial behavior can be learned through daily practice learning (Dodge, Colker, & Heroman, 2009). This is shown in subjects who have prosocial behavior in the low category, namely subject X (pre-test = 29), N (pre-test = 34), P (pre-test = 35), and E (pre-test 37). Based on secondary data from observations of the behavior of the four subjects in daily learning shows that the subject was initially reluctant to play with friends, share toys, not yet know the variety of emotional expressions, and even found it difficult to control emotions when a wish cannot be fulfilled. When the teacher applies non-verbal communication to learning activities for 7 days, the behavior of the four subjects begins to show significant changes, the average of the four children has increased scores X (29 to 61), N (34 to 58), P (35 becomes 67), and E (37 to 62), the average of four subjects who have a low category of prosocial behavior can increase significantly by 50% after being given an intervention in the form of teacher non-verbal communication.

Prosocial behavior can then also be formed through a reinforcement mechanism, namely a reward and punishment for a child's behavior (Kusanagi, 2015). The form of reinforcement carried out by the teacher through non-verbal communication is smiling, holding the child's shoulder gently, showing enthusiasm for the child's positive behavior, and showing disagreement to negative behavior with teacher's gestures. This is very effective in reinforcing children's prosocial behavior. In general, the post-test results showed that 19 children were in the moderate prososocial category, five children were in the high prosocial category, and the average prosocial behavior was at 72%. Based on the results of the pre-test and post-test scores, this intervention program can effectively improve prosocial behavior. The basis for decision making is in line with the results of hypothesis testing (Asymp. Sig = 0.189 > 0.05), which indicates that observers state the behavior being tested in accordance with the expected behavioral indicators.

6 CONCLUSION

This study aimed to examine the effectiveness of teacher training through non-verbal communication psychoeducation programs in the form of (1) facials; (2) gestural; (3) postural; (4) proxemic; and (5) artifactual. The intervention has improved three forms of child prosocial behavior, including (1) comforting; (2) helping; and (3) sharing (Dunfield, 2010). The results showed that the teachers were very enthusiastic in participating in the non-verbal communication psychoeducation training, as evidenced by the presence of 100% when training took place. Pre-test results were 11 children in the low prosocial behavior category, while 15 children were in the medium prosocial behavior category, and the prosocial behavior average was 56%. Post-test data showed that there were 19 children in the medium prosocial behavior category, and five children in the high prosocial behavior category; the average prosocial behavior is at 72%, so children's prosocial behavior increases by 16% after intervention through non-verbal communication. The effectiveness test was shown by hypothesis testing, which showed that the non-verbal communication psychoeducation program in the teacher was effective in improving children's prosocial behavior (Asymp. Sig = 0.189 > 0.05).

REFERENCES

Berk, L. E. (2008). *Exploring lifespan development*. Boston: Pearson Education.

Boyce, J. (2012). *Not just talking: Identifying non-verbal communication difficulties: A life-changing approach*. Milton Keynes: Speechmark Publishing.

Bryant, L. E., & Budd, K. S. (1984). Teaching behaviorally handicapped preschool children to share. *Journal of Applied Behavior Analysis*, *17*(1), 45–56.

Dodge, D.T., Colker, L.J., Heroman, C. (2009). *The creative curriculum for preschool*. Washington DC: Teaching Strategies.

Dunfield, K.A. (2010). *Redefining prosocial behavior: the production of helping, sharing, and comforting acts in human infants and toddlers*. Dissertation, Doctor of Psychology. Queen's University, Ontario.

Froming, W. J., Nasby, W., & McManus, J. (1998). Prosocial self-schemas, self-awareness, and children's prosocial behavior. *Journal of Personality and Social Psychology*, *75*(3), 766–777.

Kusanagi, Y. (2015). *The roles and functions of teacher gestue in foreign language teaching*. Nagoya: Temple University Graduate Board.

Koening. A. L. (2000). *Moral development: the effect of childhood maltreatment on prosocial behaviors and transgressions*. New York: University of Rochester.

Kumar, R. (2011). *Research methodology: a step-by-step guide for beginners*. London: SAGE Publications Ltd.

McCutchen, D. Abbott, R. D., Green, L. B., Beretvas, N., Cox, S., Potter, N. S., Quiroga, T., Gray, A. L. (2002). Begining literacy: links among teacher knowledge, teacher practice, and studend learning. *Journal of Learning Disabilities*, *35*(1), 69.

McGinley, M. (2008). *Temperament, parenting, and prosocial behaviors: applying a new interactive theory of prosocial development*. Nebraska: University of Nebraska.

Nantel-Vivier, A. (2010). *Patterns and correlates of prosocial behaviour development*. Montreal: Mcgill University.

Partington, J. W. (1980). *The effects of play materials on the sharing behavior of normal preschoolers*. Tallahassee: Florida State University.

Pilgrim, C. L. (2006). *Afrocentric education and the prosocial behavior of African American children*. New York: Fordham University.

Roberts, W., Strayer, J. (1996). Empathy, emotional expressiveness, and prosocial behavior. *Child Development, 67*, 449–470.

Sattler, J. M. (2002). *Assessment of children*. La Mesa, CA: Jerome M. Sattler Publisher.

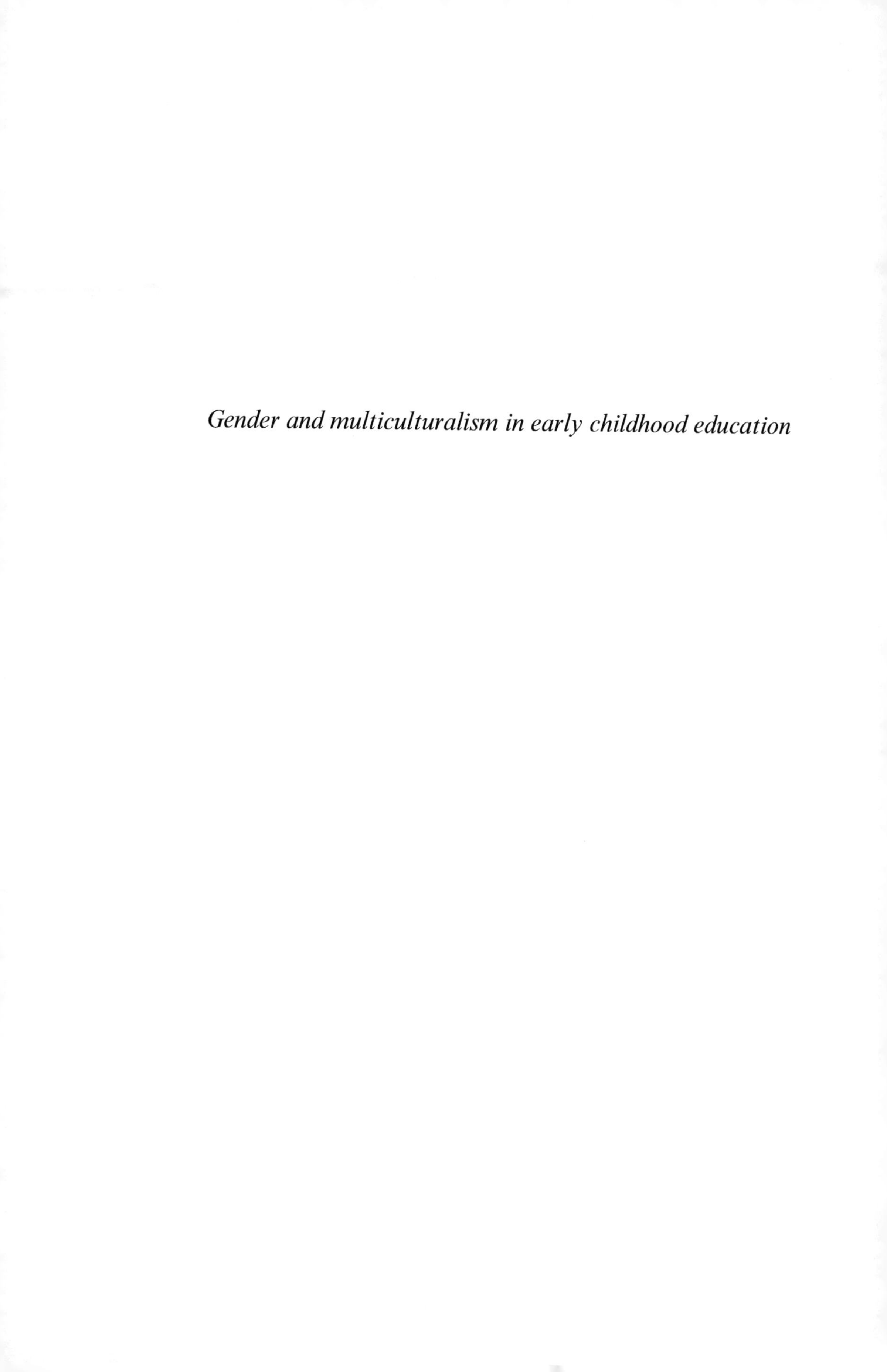

Gender and multiculturalism in early childhood education

Early Childhood Education in the 21st Century – Yulindrasari et al. (eds)
© 2020 Taylor & Francis Group, London, ISBN 978-1-138-35203-2

Leadership in early childhood: Gender and age intersectionality

F. Ulfah, H. Yulindrasari & V. Adriany
Universitas Pendidikan Indonesia, Jawa Barat, Indonesia

ABSTRACT: This research aims to explore the intersectionality between gender and age in leadership at PAUD. The theory of intersectionality helps to understand how children construct leadership in a kindergarten context. We conducted focused ethnography at a kindergarten in Bandung. Based on the findings, children and teachers construct leadership based on gender and age. Gendered perception about leadership results in unequal access to leadership experience for boys and girls. Children's expression of leadership were also hampered by age. Boys and girls cannot express their leadership when there are adults present.

1 INTRODUCTION

Young children are considered too young to understand gender (Tobin, 1997). This argument could be right when referring to Piaget's theory of cognitive development where 4-7 year-old children are in a concrete pre-operational phase (Piaget, 1964). In this phase, learning should be based on hands-on activities or studies that are concrete or real. Hence, it would be considered impossible to learn leadership or other social issues such as gender issues.

Piaget's theory is criticised by post-developmentalism (Adriany, 2018) because it seems to underestimate children's competence and ignore diversity (Donaldson, 1978; Kagitcibasi, 1996). In addition, Piaget's view also ignores children's perspectives and experiences, neglects socio-cultural aspects surrounding them, and universalises children's development (Morrow, 2006).

Children can understand social issues that are relevant to their lives and children are active agents who can build their gender meaning and culture (McNaughton et al., 2007; Dahlberg et al., 1999; Renold, 2005; Corsaro, 2005; Morrow, 2006; Adriany, 2018). Hence, we assume that children can understand leadership and gender issues and they even can construct both of them simultaneously.

The discussion of gender and leadership is not something new (Richardson & Loubier, 2008). There are several studies which focus on the intersection between the leadership style of boys and girls (Mawson, 2009; Lee & Shin, 2005). Mawson (2009) has shown that women's leadership style same as a director who organises and participates in activities. Meanwhile, the term of a dictator who has a role in governing is more appropriately pinned on boys' leadership style. On the other hands, Lee and Shin (2005) reveal that children's leadership characteristics are unique. Both boys and girls are quite powerful in their ways and styles to lead. Moreover, someone considered as a leader is influenced by age differences and class dynamics.

Indeed, the difference in leadership style occurs because of the existing power that is negotiated by girls and boys. Mawson (2011) states that social literacy allows children to negotiate leadership and power in their groups. The children negotiated power in several discourses (Adriany & Warin, 2012). In one discourse, children can be so powerful, but in another discourse, they can be so powerless (Ford, 2006; Adriany, 2018). For example, when a girl acts as a teacher, she can manage her male friends, but when she is playing freely, girls do not have control over their friends, especially her male friends.

Intersectionality approach can be used to identify inequality in the situation mentioned earlier. The term intersectionality is used to unveil the problems related to the multidimensional inequality of individuals or particular groups (Crenshaw, 1991). Also, it not only reveals inequalities but also reveals that oppression has built by the multidimensionality of subjectivity through a hierarchy of powers and privileges (Carastathis, 2014). So it is appropriate to use this approach on social issues research to provide a deeper understanding on racism, sexism, classism, and other social realities which can affect one's life or experience in a specific situation (Jean-Marie, Williams & Sherman, 2009).

Some studies used intersectionality as a frame of analysis, including Hellman et al. (2014) and Taefi (2009). Hellman et al. (2014) conducted their research at two preschools in Sweden. Social norms associate age with ability. When children are not able to behave according to the norm, they are considered not capable or labelled as an immature/childish person. Besides, age and gender interact in the process of gender normalisation which encourages girls and boys to meet gender stereotyping standards. Taefi (2009) concludes that the theory of intersectionality helps to understand the unique position of girls in international law. It helps to unpack why girls are marginalised and neglected. Intersectionality explains that oppression and discrimination are a result of a combination of more than one social conditions, for example, gender and age. Whereas, Christou and Spyrou (2016) in their study concluded that the involvement of children on the "other side" or situations that are different from those of children faced in general provides a great deal of intersectional understanding of themselves and others.

Those studies are mostly carried out in Global North such as Sweden (Helman et al, 2014), New Zealand (Taefi, 2009), and Greece (Christou & Spyrou, 2016), and it is not focused on early childhood leadership. Thus, our study intends to fill in the gaps in the literature on this topic.

In this chapter, we contribute to understanding gender and age leadership and its intersectionality in early childhood education. Postdevelopmentalism theory is used to help in seeing a problem, not from one point of view only. It means that universality does not apply to various problems in the same or different places and it believes that children are an experienced figure (Morrow, 2006).

2 CONCEPTUAL FRAMEWORK

2.1 *Gender and leadership*

Men and women must have equal opportunity and position in society. Nonetheless, the socialisation of gender roles makes men and women different (Kruger, 2008). Therefore, gender is the result of social construction (Poggio, 2006; Adriany & Warin, 2014). For example, leadership is built on masculine terms (Due Billing & Alvesson, 2000).

Leadership in children and adults context is identical to men. In kodrat discourse, women are often in a subordinate position, and it has perpetuated by society through the domestication of women (Warin & Adriany, 2017). Gender stereotypes prescribe women to have feminine qualities such as gentle, caring, graceful and all submissive (Adriany & Warin, 2014; Prentice & Carranza, 2002) while men consider as a strong being with an agency (Prentice & Carranza, 2002).

This perception is one of the factors that hinder women from being able to compete with men in terms of leadership. In general, female leadership is in a feminised sector such as the early childhood education sector, while male leadership is freer to be able to occupy a variety of public domains (Warin, 2013).

2.2 *Gender and age: An intersectionality*

Children can involve in their gender-building process actively (Adriany, 2018; Renold, 2005). In certain situations, being boys and girls are socially negotiated, and they learn about gender and age norms (Hellman et al., 2014).

In this chapter, intersectionality is used to be able to reveal gender and age gaps in leadership since a power dynamic playing on it. When a child becomes a leader, s/he has power over other children, but when dealing with their teacher, the child becomes powerless. Generally, children's leadership over other children depends on the permission of the teacher (Maxcy, 1991).

3 METHODOLOGY

This research is a qualitative focused ethnography. Focused ethnography refers to a study that follows the principles of ethnographic research, but it is conducted in a relatively short time (Knoblauch, 2005). The duration of this study was seven weeks, with 22 meetings and 88 hours.

The research took place at a kindergarten in Bandung, Indonesia. The learning model that is used in kindergarten is a classical approach. This research is part of a Master's thesis that focuses on Early Childhood Leadership and Intersectionality. This research is also part of the Higher Education Excellence Research (PUPT) which is funded by the Ministry of Research, Technology and Higher Education (Kemristekdikti) which focuses on leadership in PAUD, and intersectionality of gender and age. The research participants were 64 children consisting of 31 boys and 33 girls, as well as 6 teachers.

In this study, data was collected through observation using field notes and semi-structured interviews (Higginbottom et al., 2013). We also paid attention to the rights of children by not interfering with their activities (Adriany, 2018). Moreover, the first author made some conversations with children informally and did not force them to respond and protect their rights in terms of maintaining confidentiality. The name of the research participants are pseudonyms (Adriany & Warin, 2014). In this study, the first author was responsible for collecting data in the field. Then the data analysis process is assisted by the second and third authors.

Data analysis uses a grounded theory approach. Charmaz (2006) explains that grounded theory makes it easier for researchers to categorise findings needed by researchers.

4 FINDINGS AND DISCUSSION

The general observation of this research is to describe how the child's position with age and gender in leadership. In this section, we will explain how gender, age and leadership issues are manifested in everyday social interactions in kindergarten.

Findings in this study are based on the results of informal interviews between us as researchers and the children in the kindergarten, interviews with the teachers and field notes.

4.1 *I am a leader: Boys' leadership in kindergarten*

Children can understand about leadership based on their perspectives and experiences. Raihan with his brave acknowledgement claimed himself as a leader. He did it because he could protect her female friends who were disturbed by other boys peer. He also revealed that a leader is someone who protects the surroundings so that it remains safe. It is illustrated in the field notes below.

> When I was sitting, Noah, Raihan, Denis, Fito, and Dinan came over me. Then we talked, starting with Noah's question about what I was writing about. Then, the conversation continues who will become a leader or not. Raihan admitted that he was a leader and Noah's supported him. Raihan explained why he was a leader.
>
> Piet: "Why do you like to lead?"
> Raihan: "I just love to do it".
> Piet: "Why?"
> Raihan: "To make peace, Denis has been fierce with Davina, and I will defend her."
> Piet: "Why do you stand up for her?"
> Raihan: "I punched him back because Davina was punched with Denis".
>
> (field notes on April 18, 2018)

Raihan, as a boy, acknowledges leadership as something that can protect Davina (a girl) by punching Denis (another boy). The dominant characteristic that he shows perpetuates the traditional leadership model, which is masculine leadership (Fletcher, 2004). One characteristic of masculine is to act to be dominant and aggressive (Prenstice et al., 2002). Raihan is competent as a leader among his friends when they are in play activities. It has shown that children's leadership is inseparable from individual experience and other contextual factors (Mawson, 2011).

On the other hand, the next field notes show that the child's power as a leader become powerless. Even though the children dare to lead his friends, the child becomes helpless in the presence of the teacher.

> Noah is in front of the other children who were lining up. The children in the line shouted "Noah, Noah!" Hearing that shout, Noah returned to the line and shouted "Raihan, Raihan!" followed by the others, but Raihan refused to go forward because there was a teacher. Finally, Noah went forward again.
>
> Noah: "Let us all, do gymnastics hahahhaha, follow me.... 1, 2, ..."
> Then Mrs Ratna who led the exercise this morning came and Noah was silent.
> Bradly: That is Noah in the front
> Mrs Ratna: I want to accompany Mrs Ratna, this is a little teacher
>
> Then Mrs Ratna led the gymnastics followed by the children including Noah whose in front with Mrs Ratna, but Noah did not give instructions just following what was instructed by Mrs Ratna
>
> (field notes on April 26, 2018)

In the previous note (April 18, 2018), Raihan seemed so powerful when playing time came. However, based on the note (April 26, 2018), he showed a powerless position because the activities were guided by the teacher who was more powerful than him. It can be seen that power is not centralised but dynamic and can be exchanged between subjects (Foucault, 1971).

The children have shown their courage to become a leader because leadership can emerge in early childhood (Fox et al., 2015). However, their courage has not been facilitated by the teacher and the teacher considers the child to come forward as "the little teacher" and only follows the movements and direction of the teacher. It shows that age, maturity, and competence are positioned children as "other" compared to adults (Shundall, 2012). This othering process turns children into a marginalised figure (Adriany & Warin, 2014).

4.2 *Girls can become a leader in elementary schools*

When girls in kindergarten talk about leadership and becoming a leader, they argued that a leader is a person who leads direct friends to pray and march. Also, based on their understanding, leaders are older than them or people who have a higher level of education, namely the level of Elementary School. The following notes show girls' imagination about leadership.

> Kay and Chia told me that in her class there were no leaders, even though there was a leader, Mrs Eet (headmaster) who likes to lead praying and ceremonies. They told me that a leader is a person who leads sholat (Islamic praying ritual).
>
> Piet: "Oh Mrs Eet is the leader. How about children? is there any of your friends take the lead?"
> Kayla: "No, there is not hahaha".
> Piet: "Ohhh."
> Kayla: "If we are in elementary school, there is a leader who commands "sikaaap!"
> Piet: "Hmmm ... oh, really, Kay?"
> Chia: "My brother is a class leader in an elementary school."
> Kayla: "My brother too."
> Piet: "Ohhh there is a leader in the elementary school".

Chia: "My sister has a friend in her school".
Kayla: "My sister has become a leader once, and her friends had a chance to become a leader as well.
Piet: "What is a leader?"
Kayla: "If anyone wants to pray the leader will say "sikaaap gerak!"
Piet: "Ehmmm, to her/his friends?"
Kayla: "Hmm ... all of her/his friends immediately sat down (she were demonstrating an attitude with folding her hands).

(field notes on April 26, 2018)

The National Association for Gifted Children (NAGC) has defined a leader as a person who can regulate and influence others to achieve the common goals (Fox et al., 2015). Similarly, it was also mentioned in the field notes, when the leader command "*sikap*" then the others follows the leader instruction. It reflects children's leadership. Then, they assume that kindergarten children could not become a leader because of their young age. The finding is in line with what Hellman *et al.* (2014) have stated that age is associated with inability.

While children reveal that in their school there are no children who become a leader, it can be indicated in not only age but also gender which play the role to make them unable to become a leader. One of the reasons is because girls are born into a gendered world where the experience of boys are defined as norms whereas the experiences of girls are devalued and ignored (Taefi, 2009).

5 CONCLUSIONS

Social literacy provides an opportunity for children to interrelate leadership with power. It gives space for children to construct leadership per se. Also, the construction of leadership is related to gender. Boys get more opportunities to be a leader than girls. The more chance that the children get could provide insight into leadership. The boys interpret themselves as a leader who has the power to control the environment.

Apart from gender, there are age factors that influence leadership. Girls think that there are no girls who become leaders unless they are already mature. Whereas, boys cannot express their leadership when there is an adult female figure who becomes a leader. It shows that there is an intersection between gender and age in children's leadership.

ACKNOWLEDGEMENT

We want to thank the Ministry of Research, Technology and Higher Education (Kemristekdikti) for the grants and supporting us to complete the research.

REFERENCES

Adriany, V., & Warin, J. (2012). Gender power relations within the school's space and time: An ethnography study in an Indonesia Kindergarten. In paper presented in Space, Place, and Social Justice in Education Conference, July 13, 2012. Manchester: Manchester Metropolitan University.

Adriany, V., & Warin, J. (2014). Preschool teachers' approaches to care and gender differences within a child-centred pedagogy: findings from an Indonesian kindergarten. *International Journal of Early Years Education*, *22*(3), 315–328.

Adriany, V. (2018). Being a princess: young children's negotiation of femininities in a Kindergarten classroom in Indonesia. *Gender and Education*, 1–18.

Due Billing, Y., & Alvesson, M. (2000). Questioning the notion of feminine leadership: A critical perspective on the gender labelling of leadership. *Gender, Work & Organization*, *7*(3), 144–157.

Carastathis, A. (2014). The concept of intersectionality in feminist theory. *Philosophy Compass*, *9*(5), 304–314.

Charmaz, K. (2006). *Constructing grounded theory: A practical guide through qualitative research*. London: Sage Publications.
Christou, M., & Spyrou, S. (2017). The hyphen in between: Children's intersectional understandings of national identities. *Children's Geographies*, *15*(1), 51–64.
Corsaro, W. A., (2005). The Sociology of Childhood, 2nd ed. London: Pine Forge Press.
Crenshaw, K. (1991). Mapping the Margins: Intersectionality, Identity Politics, and Violence Against Women of Color. Stanford Law Review, 43(6): 1241–1299.
Dahlberg, G., Moss, P., & Pence, A. R. (1999). *Beyond quality in early childhood education and care: Post-modern perspectives*. London: Falmer Press
Donaldson. (1978). *Children's minds* (Vol. 5287). Glasgow: Fontana/Collins.
Foucault, M. (1971). Orders of discourse. Social Science Information, 10(2), 7–30. https://doi.org/10.1177/053901847101000201
Fletcher, J. K. (2004). The paradox of postheroic leadership: An essay on gender, power, and transformational change. *The leadership quarterly*, *15*(5), 647–661.
Ford, J. (2006). Discourses of leadership: Gender, identity and contradiction in a UK public sector organization. *Leadership*, *2*(1), 77–99.
Fox, D. L., Flynn, L., & Austin, P. (2015). Child leadership: Teachers' perceptions and influences. *Childhood Education*, *91*(3), 163–168.
Hellman, A., Heikkilä, M., & Sundhall, J. (2014). 'Don't be such a baby!'Competence and age as intersectional co-markers on children's gender. *International Journal of Early Childhood*, *46*(3), 327–344.
Higginbottom, G., Pillay, J. J., & Boadu, N. Y. (2013). Guidance on performing focused ethnographies with an emphasis on healthcare research. *The Qualitative Report*, *18*(9), 1–6.
Jean-Marie, G., Williams, V. A., & Sherman, S. L. (2009). Black women's leadership experiences: Examining the intersectionality of race and gender. *Advances in Developing Human Resources*, *11*(5), 562–581.
Kagitcibasi, C. (1996). *Family and human development across cultures: A view from the other side*. Mahwah, NJ: Lawrence Erlbaum.
Krüger, M. L. (2008). School leadership, sex and gender: Welcome to difference. *International journal of leadership in education*, *11*(2), 155–168.
Knoblauch, H. (2005). Focused ethnography. In *Forum qualitative sozialforschung/forum: qualitative social research*, *6*(3), Art. 44, http://nbn-resolving.de/urn:nbn:de:0114-fqs0503440.
Lee, S. Y., Recchia, S. L., & Shin, M. S. (2005). "Not the same kind of leaders": Four young children's unique ways of influencing others. *Journal of Research in Childhood Education*, *20*(2), 132–148.
Mawson, B. (2009). Dictators and directors: Leadership roles in children's collaborative play. *New Zealand Research in Early Childhood Education*, *12*, 11–22.
Mawson, B. (2011). Children's leadership strategies in early childhood. *Journal of Research in Childhood Education*, *25*(4), 327–338.
Maxcy, S. J. (1991). *Educational leadership: A critical pragmatic perspective*. New York: Bergin & Garvey.
Morrow, V. (2006). Understanding gender differences in context: implications for young children's everyday lives. *Children & society*, *20*(2), 92–104.
Piaget, J. (1964). Part I: Cognitive development in children: Piaget development and learning. *Journal of research in science teaching*, *2*(3). 176–186.
Poggio, B. (2006) Editorial: outline of a theory of gender practices. *Gender, Work & Organization*, *13*(3), 225–234.
Prentice, D. A., & Carranza, E. (2002). What women and men should be, shouldn't be, are allowed to be, and don't have to be: The contents of prescriptive gender stereotypes. *Psychology of women quarterly*, *26*(4), 269–281.
Renold, E. (2005). *Girls, Boys and Junior Sexualities. Exploring Children's Gender and Sexual Relations in the Primary School*. New York: Routledge.
Richardson, A., & Loubier, C. (2008). Intersectionality and leadership. *International Journal of Leadership Studies*, *3*(2), 142–161.
Sundhall, J. (2012). *Kan barn tala? En genusvetenskaplig underso¨kning av a°lder i familjera¨ttsliga utredningstexter*. Gothenburg, Sweden: Gothenburg university.
Taefi, N. (2009). The synthesis of age and gender: Intersectionality, international human rights law and the marginalisation of the girl-child. *The International Journal of Children's Rights*, *17*(3), 345–376.
Tobin, J. J. (1997). *Making a place for pleasure in early childhood education*. London: Yale University Press.
Warin, J. (2013). The status of care: Linking gender and 'educare'. *Journal of Gender Studies*, *23*(1), 93–106.
Warin, J., & Adriany, V. (2017). Gender flexible pedagogy in early childhood education. *Journal of Gender Studies*, *26*(4), 375–386.

Early Childhood Education in the 21st Century – Yulindrasari et al. (eds)
© 2020 Taylor & Francis Group, London, ISBN 978-1-138-35203-2

Gendering children through song

M.T.R Gunawan, V. Adriany & H. Yulindrasari
Universitas Pendidikan Indonesia, Jawa Barat, Indonesia

ABSTRACT: This chapter explores the process of gender construction through music as a means of learning. By using a post-development perspective, we critically highlight learning tools which are often considered beneficial for early childhood development. In this study, we employed focused ethnography for two months in a kindergarten in Bandung, Indonesia. We show that music as learning tools has constructed gender stereotypes. Consequently, children create and demonstrate their understanding of activities that should be done by boys or girls. It leads to build a gap in stimulating particular skills that both genders should have. According to findings, we recommend that kindergarten teachers need to be aware of promoting learning tools to young children and create the process of learning through tools more gender-equitable.

1 INTRODUCTION

Songs can easily be introduced to young children. Songs are categorised as a verbal symbol and can be used as learning tools so that young children can memorise and absorb the information (Sumiharsono & Hasanah, 2017). In kindergarten, teachers often use it in their learning setting to improve the quality of the education process (Susilana & Riyana, 2008).

As a learning tool, songs can stimulate early childhood development. It helps improve children's music and kinesthetic intelligence through the rhythm of songs (Whidianawati, 2011). Some believe that songs develop children's language skills through its lyrics and it could be able to steer them to do verbal activities (Holden, 2003; Campbell, 2000).

However, in terms of gender in early childhood development, some songs cause some dilemmas. Songs with gender-blind lyrics could situate boys and girl separately. Some children's songs are considered promoting a particular gender ideology. For example, a song about police often denotes *Bapak* in its lyrics. *Bapak*, literally means father or sir, refers to middle age man in Indonesia. It means the songs only associate police profession with male and excludes women in this job. This kind of songs will construct teachers' and young children's perceptions of occupations, toys, and activities regarding their sexes (Adriany & Warin, 2014).

Gender is one aspect of young children development that should introduce in early childhood education. Smith (2017) claims that young children can understand gender. Thus, it is imperative for adults to recognise how to use suitable learning tool that can improve positively for young children's understanding of gender.

Ironically, gender dimensions are often overlooked in early childhood education in Indonesia. Parents and educators are also unaware of children's gender development. The latest curriculum released by the Ministry of Education and Culture, Curriculum 2013 (2014), includes two layers of competencies: core and basic competencies. Gender perspectives are absent in both layers.

Few studies in Indonesia have attempted to links learning tools and gender (Zubaedah, 2010). Some researchers have focused on gender and learning tools such as student's textbooks or images (Emilia, Moecharam & Syifa, 2017; ACDP Indonesia, 2013). To date, specific research related to songs as a learning tool and gender in early childhood education is rare.

We use a post-developmentalism framework in analysing children songs and gender in early childhood education. Within post-developmentalism, children have their agency and power to construct knowledge (Burman, 2008; Adriany, 2016). Children are active subjects who construct and deconstruct meaning for what they are received (MacNaughton, 2000; Gunawan, 2017).

Learning tool, specifically songs, play crucial roles to construct gender in children's mindset. Children also learn gendering things through songs. Based on the narrative, we attempt to ravel how songs as a learning tool can shape gender in early childhood development.

2 METHODOLOGY

Considering our curiosity on specific local culture elements (Denzin & Lincoln, 2009), we used a qualitative approach with a focused ethnographic approach for approximately two months. As the first author, Marina applied a direct observation and recorded in field notes during the fieldwork. Then, Marina shared the notes with co-authors to deeply discuss and analyse it together for better understanding and results.

Since the research was in a pilot kindergarten in Bandung, Indonesia, it is intriguing for us to follow how the kindergarten is inserting the gender values in their teaching activities. We did a preliminary observation for about six months before conducting research. It allows us to understand the routines and condition of the kindergarten that we want to explore. Building trust and familiarity is also vital to allow us to blend in the custom that the kindergarten has constructed.

We collected and analysed it using grounded theory (Charmaz, 1996). We coded the notes and arranged into 204 codes. Then, we focused on coding and reduced some codes into 114 codes. In the next stage, we categorised the codes based on the themes. In this chapter, we focus on only two codes from a total of six codes that we generated in the first author's Master's thesis and our PUPT grant research project funded by the Republic of Indonesia Ministry of Research, Technology, and Higher Education.

3 FINDINGS AND DISCUSSION

3.1 *Gender segregation*

In Indonesia, most of the teachers in kindergarten are female, and usually, children call them *Ibu Guru* or simply *Ibu* (mother). Before the class started, *Ibu* would separate her students into two groups: boys and girls. Then, *Ibu* instructed them to make a small circle and sang an opening song. Thus, this activity segregated girls and boys into two small circles. The separation of boys and girls strengthen a biased understanding of gender, as the following notes taken on March 26, 2018, indicate:

> After snack time, the student enters their class. I followed them into Class B "turtle class". There are ten children with five boys and five girls, and they were in the classroom on this Monday. They began with morning praying and afterwards, Ibu Guru started to sing a "good morning" song. When Ibu Guru sang the song, they automatically gathered based on their sexes separately until it made two a small circle formation.
>
> (6–fieldnotes on March 26, 2018)

These notes showed clear segregation between boys and girls that are directed by a kindergarten teacher. Though the song "good morning", directions are given by the teacher to discipline the children. The children felt that it is necessary to separate themselves and build the formation. Their activities build a routine which reflects on the construction of gender binary in early childhood development (Blaise, 2009; Robinson & Davies, 2010).

Moreover, based on our observation, the gender binary could be found in children's daily activities. They tend to play and do activities exclusively with children of the same sex. It creates a gender stereotype (Juditha, 2015). Gender stereotypes will potentially produce gender inequality, which in turn will affect individual lives. Gender inequality includes marginalisation of women, the subordination of women, labelling of certain groups (groups of boys and girls), violence against women and a double burden on women (Fakih, 1996).

Traditionally, Indonesian gender construction has a clear division of labour between men and women. The discussion of traditional gender will be discussed more deeply in the following section.

3.2 *Traditional Indonesian gender construction*

Gender is a social construction. It is relevant only in a particular geographical area in a specific time. Consequently, the understanding of traditional gender in a region can be different from understanding gender in other regions. Traditional gender contains two parts, traditional masculinity and traditional femininity. Both traditional gender construction will be discussed in the following description.

3.2.1 *Traditional femininity*

According to the observation in the fieldwork, children associated particular activities with their sex. Girls tend to join their girls-peer or choose one activity that is considered feminine by society and in line with the construction of femininity that is developed by the school. The school assumed that girls are inherently different from boys. This also relates to the existence of gender stereotypes that children have learnt. They link those gender stereotypes with activities or existing general objects.

In addition, children tend to associate specific characterisation with femininity discourse. The girls will consider themselves as weak, gentle, sensitive and sympathetic (Hurley, 2005) which is different from boys who are always considered strong, insensitive, and expressive (Warin & Adriany, 2015; Paechter, 2007). Adults also adapted the understanding of traditional femininity as illustrated in the field notes that follow:

> Today Ibu Voni asked her students to practice a dance for performing arts using a contemporary dancing song. They begin the dance practice by following the dancing guide from Ibu Voni. When the session ended, Ibu Voni stopped the practice, and the class are dismissed. Ibu Voni said something while walking out of the class, "The girls should be more sensitive, they know how to dance naturally but when we look at the boys, for example, Kinoa Kino, he is different. He does not have an expression, and his move is too stiff. Ibu Voni continues laughing.

In a dance session, Ibu Voni used a contemporary dancing song that encompasses a value of traditional femininity. The value, then, extends to the daily lives of children. Based on the observation, we found that girls or people around the children both parents and society will use a label of traditional femininity as a measure of a child's ability. As demonstrated earlier, *Ibu* Voni annunciated that the girls are more expressive in dancing than the boys. It indicates that she as a teacher compares the girls has better dancing ability compare to boys who do not express themselves in dancing (Doyle, 1985; MZ, 2013).

Sex does not determine dancing skills, but various intensive dancing exercises do. Associating an activity to sex strengthens gender stereotypes between boys and girls which in turn will limit an individual in developing themselves (Hoorn, 1993).

3.2.2 *Traditional masculinity*

There is not only traditional femininity constructions that we found in the field but also the construction of traditional masculinity embedded in songs. For example, in the national anthem "*Satu Nusa Satu Bangsa*". The anthem demonstrated the construction of traditional masculinity approved by the teacher as we showed in our field note:

> The teacher arranges the students to line up in two lines for each class. The two lines consist of one rows of boys and the other for girls. After the student is in line, the teacher starts playing a national anthem "Satu Nusa Satu Bangsa" and invites the children to

sing. The student sang the song together. When they sang the song, the teacher then asked three boys to come forward to lead other children to sing.

(field notes on March 27, 2018)

These findings show how the teacher manifests traditional masculinity through songs. When the song "Satu Nusa Satu Bangsa" was sung, the teacher asked the boys to come forward and lead other children. Meanwhile, the girls can only follow and sing in the line. It shows that the teacher has traditional masculinity understanding in their perspective. The national anthem is associated with a masculine hero that must be represented by a boy. Being exposed to a masculine male hero, children will stereotype a hero as exclusively male. According to Parsons and Howe (2006), a hero or superhero who is considered strong and invincible is always associated or labelled to a male figure. Even though in Indonesian history, many heroes are women.

4 CONCLUSION

As educators who have learned early childhood education and development, we realised that songs that are designed for early childhood development have neglected gender perspective. We recommend that teachers increase their gender awareness and create learning tools that promote gender equality.

REFERENCES

Adriany, V. (2016). Gender in pre-school and child-centered ideologies: a story of an Indonesian kindergarten. In S. Brownhill, J. Warin & I. Wernersson (Eds). Men, Masculinities, and Teaching in Early Childhood Education (pp.88-100).

Adriany, V., & Warin, J. (2014). Preschool teachers approaches to care and gender differences within a childcentred pedagogy: Findings from an Indonesian kindergarten. *International Journal of Early Years Education*, *22*(3), 315–328, doi: 10.1080/09669760.2014.951601.

Blaise, M. (2009). "What a girl wants, what a girl need": Responding to sex, gender, and sexuality in the early childhood classroom. *Journal of Research in Childhood Education*. 23 (4), 450–460.

Burman, E. (2008). *Deconstructing developmental psychology, second edition*. London: Routledge.

Campbell, D. (2000). *Efek mozart bagi anak-anak, meningkatkan daya pikir, kesehatan, dan kreativitas anak melalui musik (Alex Tri Kantjono Widodo, penerjemah) [Mozart's effects for children, improve children's thinking, health and creativity through music (Alex Tri Kantjono Widodo, translator)]*. Jakarta: PT Gramedia Pustaka Utama.

Denzin, N. K., & Lincoln, Y. S. (2009). *Handbook of qualitative research*. Terj. Dariyatno dkk. Yogyakarta: Pustaka Pelajar.

Doyle, J. (1985). *Sex and gender: the human experience*. New York: McGraw-Hill.

Education Sector Analytical and Capacity Development Partnership (ACDP) Indonesia. (2013). Gender equality in education in Indonesia. *Police Brief*.

Emilia, E., Moecharam, N. Y., & Syifa, I. L. (2017). Gender in EFL classroom: transitivity analysis in english textbook for Indonesian students. *Indonesian Journal of Applied Linguistics*, *7*(1), 206–214.

Fakih, M. (1996). *Analisis gender dan transformasi sosial [Gender analysis and social transformation]*. Yogyakarta: Pustaka Pelajar.

Gunawan, M. T. R. (2017). Learning media and gender in early childhood education. In *The 4th International Conference on Early Childhood Education. Bandung: PAUD SPs UPI.*

Holden, C. (2003). 'Mozart effect' revisted. *Science*. *301*(5635), 914. DOI: 10.1126/science.301.5635.914b.

Hoorn, J.V. (1993). *Play at the center of the curriculum*. New York: Macmillan.

Hurley, D. L. (2005). Children of color and the Disney fairy tale princess. *The Journal of Negro Education*, *74*(3), 221–232.

Juditha, C. (2015). Stereotipe dan prasangka dalam konflik etnis Tionghoa dan Bugis Makassar [Stereotypes and races in ethnic Chinese and Bugis Makassar conflicts]. Jurnal Ilmu Komunikasi, 12(1), 87–104.

MacNaughton, G. (2000). *Rethinking gender in early childhood education*. London: Sage.

MZ, Z.A. (2013). Perspektif gender dalam pembelajaran matematika. *Marwah: Jurnal Perempuan, Agama dan Jender*, *12*(1), 15–31.

Paechter, C. (2007). *Being boys being girls: learning masculinities and feminities*. New York: McGraw-Hill.

Parson, A dan Howe, N. (2006). Superhero toys and boys' physically active and imaginative play. *Journal of Research in Childhood Education*, *20*(4), hal. 287–300.
Robinson, K & Davies, C. (2010). Tomboys and sissy girls: exploring girls' power, agency and female relationships in childhood through the memories of women. *Australasian Journal of Early Childhood*, *35*(1), 24–31.
Sumiharsono, R & Hasanah, H. (2017). *Media pembelajaran [Learning Media]*. Jember: Pustaka Abadi.
Susilana, R., & Riyana, C. (2008). *Media pembelajaran: hakikat, pengembangan, pemanfaatan, dan penilaian [Learning media: nature, development, utilization, and assessment]*. Bandung: CV.Wacana Prima.
Warin, J., & Adriany, V. (2015). Gender flexible pedagogy in early childhood education. *Journal of Gender Studies*. 1–12, doi:10.1080/09589236.2015.1105738.
Whidianawati, N. (2011). Pengaruh pembelajaran gerak dan lagu dalam meningkatkan kecerdasan musikal dan kecerdasan kinestetik anak usia dini. Studi eksperimen kuasi pada anak kelompok bermain Mandiri SKB Sumedang [Effects of motion and song learning in improving early childhood musical intelligence and kinesthetic intelligence: Quasi-experimental studies on playgroup Mandiri SKB Sumedang]. *Doctoral Dissertation*. Bandung: Universitas Pendidikan Indonesia.
Zubaedah, S. (2010). Mengurai problematika gender dan agama [Break down the problems of gender and religion]. *Jurnal Studi Gender dan Anak*, *5*(2), 243–260.

Early Childhood Education in the 21st Century – Yulindrasari et al. (eds)
© 2020 Taylor & Francis Group, London, ISBN 978-1-138-35203-2

Toxic masculinity in sexual violence against children in Indonesia: From a feminist point of view

Rizka Haristi & Vina Adriany
Universitas Pendidikan Indonesia, Jawa Barat, Indonesia

ABSTRACT: The rampant cases of sexual violence against children suggest that the handling efforts taken by the government, community, educational institutions and families have not been fully effective. This ineffectiveness is due to the focus of the carried-out programs which still count heavily on children as target objects, yet the social environment around the children—including family—does not ensure the establishment of an alert and safe environment for them. Therefore, to resolve this issue, the study that may delve the root of the problem is required. One of the views that are capable to provide the different perspective is feminism. By using the study case method through the in-depth interview with three counselors who deal with the cases of sexual violence against children located in Bandung, this study will analyze the issue of sexual violence against children from the point of view of post-structural feminism. The result suggests that there is the contribution of patriarchal culture into the conformation of *toxic masculinity* by which children are being subjected to sexual violence. Thus, by means of this study it is expected that the new alternatives in developing the most effective methods to prevent the sexual violence against children, especially in Indonesia, shall be found.

1 INTRODUCTION

The number of the sexual violence against children cases in Indonesia is still terribly high. According the report of the Indonesia Children Protection Commission (Komisi Perlindungan Anak Indonesia, KPAI), that within the period of five years (2011–2016) the cases of sexual violence against children (rape, molestation, sodomy, etc.) were ranked highest compared to the other cases of violence against children. Hence, the rampant cases of sexual violence that afflict children bring out the certain concern in various parties. On the basis of the very condition, the Government has launched the various prevention programs in hope of reducing the high number of the sexual violence against children cases. Nevertheless, the children-targeted prevention and treatment programs indicate that the approaches taken by the state are still focus on the target as victims (children). For example, with the proliferation of the sexual education aimed at children, in which children are required to be able to take care of themselves. As is known, there are many efforts made by parents and teachers that encouraging children to be able to say “Don’t” to whoever try to approach them. It is related to the assumption that adults, especially men, have the-hard-to-control sexual desire, so that children who are considered targets must be provided by the self-protection method(s) (Paypar, 2002; Lalor, 2004). It is this position that ultimately results in the efforts to control children’s movement and encourage them to be able to fight by reason of avoiding the sexual violence (Irawaty, 2016).

In the contrary to this expectation, the environmental situation surrounding children—and even the people closest to them—especially their own families, cannot be ascertained as an alert and safe environment. A study conducted by Hasanah (2013) point out that 80 percent of the act of sexual violence experienced by children aged 3–6 years old actually occurred in the home environment (Hasanah, 2013). That is, there is an imbalance between the handling carried out with the situation is in effect in the real environment. In this regard, Fairchild

(2016) claims that the countries that hold handling programs that aimed at victims (children) indicate the conformation of social position imbalance. One of them is the form of handling efforts that have been carried out so far put children as a controlled party (*victim blaming*).

Many studies conducted in Indonesia have analyzed various factors of this issue. However, the emphases are generally on the external factors and count heavily on environment. Some of them, namely Seto (2005) and Kurniawan and Hidayati (2017), reveal that the Internet access that freely exposed to pornographic contents is responsible for the occurrence of sexual violence against children. In addition, the other studies suggest that the economic factor takes part as well (Raijaya dan Sudibia, 2017), and Hetinjung (2009) claims that the exposure of private space for children as well as for adults allows adults to commit the abusive acts against children.

Furthermore, the case of sexual violence against children is not the issue that can resolve and corrected solely based on external factors. Feminism offers a different point of view in solving the very issue. Feminism itself emphasizes on the social structure imbalance between men, women and children prevailing in society which may affect the occurrence of sexual violence against children (Seymor, 1998). According to the feminism point of view, the sexual violence against children has its roots in the unequal gender-based power relations in patriarchal system. In other words, feminism may break down the sexual violence against children issue in a wider social context. Accordingly, this study will break down the cases of sexual violence against children that occur using the feminism approach. It is expected that this article may be the new literature that provides the alternative solution to the issue of sexual violence against children in Indonesia.

2 RESEARCH METHODOLOGY

This study uses the qualitative approach with a case study method. Of the three counselors who directly deal with the children victims of sexual violence, their experiences and views were explored using the in-depth interview technique to be further analyzed through the feminism point of view. The data analysis was conducted by the researcher using *Grounded Theory* data analysis (Charmaz, 1996). This study is part of the larger study that is thesis research; hence, out of the three themes that emerge, the one theme would be then discussed, namely the theme concerning patriarchy and *toxic masculinity.*

3 RESULTS AND DISCUSSSION

3.1 *Patriarchy and toxic masculinity*

The result of interviews that have been conducted show that the offenders in the sexual violence against children cases which have been handled by these counselors are fully dominated by men. As explained by the two of them:

> The culprits are of the various kind too. As I told you yesterday, they are raging from age … I handled some boys from age of 12 to … the oldest I've handled is … about 60 year old.
>
> (Respondent I Interview, June 22, 2018)

> They are men on average, including elderly men. They are diverse, different. Some aren't really old, perhaps is still 50 … but in suburb area getting married at young age is common. There are also 17-year-old-perpetrators, there are also those who still in elementary school, junior high school, and even kindergarten.
>
> (Respondent 3 Interview, July 10, 2018)

> Women, huh? It's rare. Never found one. They are all men.
>
> (Respondent 2 Interview, July 10, 2018)

The situation described here proves that men have the very strong position in committing sexual violence against children. The result of interview indicates that men—not only the elderly, but young boys as well—became the offenders. This finding has been in line with the numbers of previous studies which reveal that men as the offenders of sexual violence have a strong dominance over children (Solomon, 1992; Hasssan et al., 2015; Bergstrom et al., 2016; Humaira et al., 2016).

In connection with the earlier finding, the domination of men as offenders indicates the existence of the role of *toxic masculinity* as the result of a firm patriarchal culture. This culture is known as the one that establish the inequalities of the role construction between men, women and children, in which the more mature men possess the highest power compared to others. Fontes and Plummer (2010) and Robertson (2012) promote a claim that one of the factors that became the fundamental of the development of sexual violence against children cases in a certain country is the existence of patriarchal culture. This patriarchal culture put men in a superior position over women and children. This based on the assumption that men have biological traits with greater strength than women and children. Therefore, women and children are positioned as subordinates in society, while men are in the position of superordinate. This situation, which constantly reinforced by various social relations provides a privilege for men. Thus, based on this privilege there is a misappropriation of power and rights in order to exploit others, including sexual harassment against others who considered weaker, namely children. It is also reinforced by the de Young's view that patriarchal culture is able to establish different behaviors on men, and these distinctions affect their sexual patterns toward others.

By relating this view with the study result analysis, it was found that in patriarchal culture men are constructed as tough and strong. It is by this strength men are considered to have the greater libido than women. Therefore, with such a position, an abnormal pattern of the way of thinking (*toxic masculinity*) is formed which consider that sexual activities are a delightful thing for men and require a form of satisfaction. This condition is continually construct through environmental socialization by various social institution and media, that sexual fantasies of men, in which they possess the sexual power over others, are considered as a form of machismo. Consequently, it is normal and considered as appropriate when men seek for sexual gratification objects. Eventually, at the time this situation is accepted and normalized by the environment, as if it were in accordance with nature, it will establishes the misappropriation of the way of thinking that men have an authority in sexual relationship and make the parties who are in a position to represent a weaker parties—that is, women and children—as the object of their power. Dilirio (1989) supports this with his view that in patriarchal culture, a man who became a *hyper-sexual* is appropriately accepted as an expression of masculinity.

The research conducted by Fulu et al. (2013) on adult men in the three areas of Indonesia, that is Jakarta, Purworejo and Jayapura, under a condition in which most of them had ever committed sexual violence against children, it was found that the reason they commit such act was related to their attitude toward gender in equality and masculinity values that really upholds machismo, control and domination over others, especially women and children. The one of these masculinity aspects is the conviction that men are always having rights to receive sexual treatment anytime, anywhere, and from anyone. It is this kind of masculinity concept that allows spreading the culture of sexual abuse.

In addition, the presence of masculinity is also reflected on the forms of sexual abuses received by children. As explained by the counselors:

> There are various forms of sexual harassment that we dealt with, from the genitals touching, the skin doesn't meet the skin, the skin meet panties instead, so the term.
>
> Sodomy, genitals touching, and then ... eeerr ... inserting object or finger into the genitals or into ... eerrr ... rectum. Then, swiping ... if I'm not mistaken, then touching, from head to toe, touching 'that' area, and rubbing as well.
>
> (Respondent 1 Interview, June 22, 2018)

> Some just rubbed it around, no finger inserting. If you just rubbed, well, you just wipe it like that. And for finger inserting, well ... the finger is actually inserted, and the result of

visum et repertum indicates that there is indeed a wound. Then, there is ... eerr, like ... you know that nowadays children of the 3rd grades of elementary school, their breasts have started to grow, huh? Well, they touched, scooped.

(Respondent 2 Interview, June 28, 2018)

The commentary on the forms of sexual violence against children describes that the offenders who are in control, while children are the objects of sexual satisfaction of the perpetrators. The children's bodies are used by the offenders as the object to get the sexual stimulation.

These findings indicate that every form of sexual violence against children is constitute the expression of masculinity of the offenders, who are known are men, in their relation or interaction with the-considered-weaker party, that is children, both girls and boys. As stated by Kageha and Mayer (2013), that, for men, the domination ability in sexual activities is one of the important points of reference of their masculinity. When adult women cannot fulfill the offenders' sexual wants, the possible alternative is children—regardless of gender—as the subordinates. It is reinforced by the viewpoints of Deering and Mellor (2011) that children have a great chance of being the victims of sexual violence, for, basically, in patriarchal culture children are positioned as the weak, vulnerable group, either physically, psychologically, socially, economically or sexually, so that they are in the vulnerable situation to being the object of sexual violence.

While on the other side, children are considered to be more easily to being controlled and deceived, hence the older offender can easily vent their lust. It is in line with the view of McElvaney et al. (2012) that the availability of children and the assumption that children are easily deceived allow the defender to commit sexual violence toward them.

4 CONCLUSIONS

Based on the findings of this study, it can be concluded that the occurrences of sexual violence against children are strongly affected by the presence of patriarchal culture which normalized the masculinity hegemony. Compliance to hegemonic masculinity increases the risk of toxic masculinity. Consequently, either men or adults, regardless gender, may commit fraud to exploit the subordinates, that in this case are children, in order to carry out the acts of sexual violence as the expression of masculinity. By means of this research, it can be obtained the input in designing the gender-friendly prevention and children sexual educational programs. It is expected that children should be introduced to the concept of masculinity appropriately as early as possible.

REFERENCES

Azzopardi, C., Alaggia, R., & Fallon, B. (2018). From Freud to Feminism: Gendered Constructions of Blame Across Theories of Child Sexual Abuse. *Journal of Child Sexual Abuse*, *27*(3), 254–275.

Bergström, H., Eidevald, C., & Westberg-Broström, A. (2016). Child sexual abuse at preschools–a research review of a complex issue for preschool professionals. *Early Child Development and Care*, *186*(9), 1520–1528.

Charmaz, K. (1996). The search for meanings-grounded theory. *Rethinking Methods in Psychology*, 27–49.

Deering, R. & Mellor, D. (2011). An exploratory qualitative study of the self-reported impact of female-perpetrated childhood sexual abuse. *Journal of Child Sexual Abuse*, *20*(1), 58–76.

deYoung, M. (1987). Toward a theory of child sexual abuse. *Journal of Sex Education and Therapy*, *13* (2), 17–21.

Dilorio, J. A. (1986, 1989). The social construction of masculine sexuality in a youth group. K. Richardson & V. Taylor (eds). *Feminist frontiers II: Rethinking sex, gender, and society*. New York: Random House.

Fairchild, K. (2016). *But look at what she was wearing!: Victim blaming and street harassment*. New York: Routledge.

Fulu, E., Jewkes, R., Roselli, T., & Garcia-Moreno, C. (2013). Prevalence of and factors associated with male perpetration of intimate partner violence: findings from the UN Multi-country Cross-sectional Study on Men and Violence in Asia and the Pacific. *The Lancet Global Health, 1*(4), e187–e207.

Fontes, L. A. & Plummer, C. (2010). Cultural issues in disclosures of child sexual abuse. *Journal of Child Sexual Abuse, 19*(5), 491–518.

Gusti, A. K. M. R. I. & Ketut, S. I. Faktor-faktor sosial ekonomi penyebab terjadinya kasus pelecehan seksual pada anak di kota Denpasar. *Piramida, 13*(1), 9–17.

Hassan, M. A., Gary, F., Killion, C., Lewin, L., & Totten, V. (2015). Patterns of sexual abuse among children: Victims' and perpetrators' characteristics. *Journal of Aggression, Maltreatment & Trauma, 24* (4), 400–418.

Hasanah, H. (2013). Kekerasan terhadap perempuan dan anak dalam rumah tangga perspektif pemberitaan media. *Sawwa, 9*(1), 159–178. http://journal.walisongo.ac.id/index.php/sawwa/article/view/671/609.

Hertinjung, W. S. (2009). *The dynamic of causes of child sexual abuse based on availability of personal space and privacy*. Surakarta: Fakultas Psikologi Universitas Muhammadiyah Surakarta.

Humaira, B. D. (2016). Kekerasan seksual pada anak: telaah relasi pelaku korban dan kerentanan pada anak. *Jurnal Psikoislamika, 12*(2), 5–10.

Kageha, I. E. & Moyer, E. (2013). Putting sex on the table: Sex, sexuality and masculinity among HIV-positive men in Nairobi, Kenya. *Culture, health & sexuality, 15*(sup4), S567–S580.

KPAI. (2017). Data Kasus Perlindungan Anak Berdasarkan Lokasi Pengaduan dan Pemantauan media se-Indonesia tahun 2011–2016. Retrieved from http://bankdata.kpai.go.id/tabulasi-data/data-kasus-se-indonesia/data-kasus-perlindungan-anak-berdasarkan-lokasi-pengaduan-dan-pemantauan-media-se-indonesia-tahun-2011–2016.

Kurniawan, D., & Hidayati, F. (2017). Penyalahgunaan seksual dengan korban anak-anak (Studi Kualitatif Fenomenologi Terhadap Pelaku Penyalahgunaan Seksual dengan Korban Anak-Anak). *Empati, 6*(1), 120–127.

Laror, K. 2004. Child sexual abuse in Sub-Saharan Africa: A literature review. *Child abuse and neglect, 28*, 439–460.

McElvaney, R., Greene, S., & Hogan, D. (2012). Containing the secret of child sexual abuse. *Journal of interpersonal violence, 27*(6), 1155–1175.

Price-Robertson, R. (2012). Child sexual abuse, masculinity and fatherhood. *Journal of Family Studies, 18*(2–3), 130–142.

Seymour, A. (1998). Aetiology of the sexual abuse of children: An extended feminist perspective. *Women's Studies International Forum, 21*(4), 415–427.

Solomon, J. C. (1992). Child sexual abuse by family members: A radical feminist perspective. *Sex Roles, 27*(9–10), 473–485.

Early Childhood Education in the 21st Century – Yulindrasari et al. (eds)
© 2020 Taylor & Francis Group, London, ISBN 978-1-138-35203-2

Space and gender: Playing in early childhood education

S. Ramdaeni, V. Adriany & H. Yulindrasari
Universitas Pendidikan Indonesia, Bandung, Indonesia

ABSTRACT: The chapter aims to explore gender relations at the play space in kindergarten. By using feminist poststructuralism, it helps to reveal that the playing space promotes gender neutrality, but it strengthens the dualism idea of femininity and masculinity instead. This study is based on two months of focused ethnographic research in one of the kindergartens in Bandung, Indonesia. This research shows that the play space is often gendered. Children's and teachers' lack of gender awareness potentially excludes young children to access certain spaces just because of their sex.

1 INTRODUCTION

The purpose of this chapter is to exhibit gender relations in early childhood play space. Gender construction plays a significant role in who dominate the space and how. Consequently, girls and boys have different opportunity to access certain space. Eventually, this segregation will affect the development of their skills. It will also confine children's understanding of gender relations (Chase, 2009). MacNaughton (2000) shows that girls and boys control the play space differently. Children also maintain their domination in play spaces and tend to exclude other girls and or boys in playing activities (MacNaughton, 1997, 2000; Azzarito, Melinda & Louis, 2006; Martin, 2011).

Construction of masculinity broadens opportunity for boys to expand their space in the play space to stimulate their physical strength and ability (Chase, 2009; Maccoby, 1988; Paechter, 2003). Azzarito et al. (2006) also argue that girls are often subordinated because of the dominance of boys in physical activities. So it is not surprising that boys have more developed physical skills than girls (Lappalainen, 2004). However, Connell's (1996) study shows that boys experience difficulties in reading and language because they are more encouraged to do physical activities such as sports. It makes boys spend less time on soft motoric activities, such as language and art while girls have less time in developing their hard motoric play (Early et al., 2010).

Meanwhile, the construction of femininity provides more opportunities for girls to develop their language skills. Bhana (2003) states that girls have superiority in language development and speech compared to boys. This is indicated by the opportunity of girls who are often encouraged to do house play and parenting play (Paechter, 2003), and to play dolls (Maccoby, 1988). It has given a chance to stigmatise dolls which become women's play space as feminine space (Adriany, 2018). Consequently, as Adriany and Warin (2014) has shown, boys are not allowed to play princesses, play dolls, or even play Barbie though this stigmatisation. Therefore, boys often avoid dolls (Messner, 2000). Boys with non-conforming gender expression, such as boys who approach feminine space, will be prone to bullying and pressure from their peer (Paechter, 2006).

Lack of teacher's awareness of gender issues blindsight the gendered play space. Teachers generally claim that the play space is gender neutral (Arlemalm-Hagsér, 2010). In this context, the teacher thought that gender segregation towards different play spaces by children is considered natural (Chase, 2009). For example, the teacher's belief that boys have a stronger physique than girls (Chapman, 2015). In addition, the concept of developmentalism strengthens the teachers' assumption in providing opportunities for children to play without any intervention even though the playing activities often affirm traditional gender (Adriany, 2013; Warin & Adriany, 2015).

Although several studies have examined the importance of gender issues in children's play spaces (MacNaughton, 2000; Martin, 2011), research in Indonesia is still rare to examine the gendered playing space in early childhood education. Applying a feminist poststructuralism perspective helps us to see the power and gender relations in the children's play space. This chapter focuses on the play space in the playground and seeks to reveal that the playground is gendered. Therefore, we expect that the results will help teachers to open up their understanding of gender issues and its urgency.

2 APPLYING POSTSTRUCTURALISM: GENDER AND POWER RELATION IN EARLY CHILDHOOD STUDIES

Feminist poststructuralism is used in this study to discover to what extent the concepts of masculinity and femininity can be potentially noxious for young children. Some studies explicate that the ideas of gender binary are promoted in kindergarten (Adriany, 2018; Warin & Adriany, 2015; MacNaughton, 200). The teachers mostly assume that the gender binary is something natural (Maccoby, 1988).

Feminist poststructuralism is an approach which examines how knowledge and power are interrelated. Paechter (2006) explained that knowledge and power on the construction of gender control what it means to be "a real man" or "a real woman" in society. Such categorisation is often considered natural as one of the examples of how knowledge and power work to maintain the differences between men and women through social influences (Paechter, 2007). Gavey (1989) also has the same argument that knowledge and power are constructed socially so that feminist poststructuralism is one approach that challenges the idea of dualistic gender.

As in the earlier explanation, feminists poststructuralism discusses the relation of power and gender. Poststructuralists understand power as control and domination which becomes a disciplinary system that limits a person to behave and act (Connell, 1996). Gender is constructed through the disciplinary power that controls what it means to be masculine and feminine in society. The disciplinary power inevitably significantly influence on how young children understand and construct gender (Paechter, 2007).

This study adopts the definition of gender as a social construction of masculinity and femininity (Adriany, 2018; MacNaghton, 2000; Paechter, 2007). Thus, gender ideals are not fixed; it changes across context, time and places. For example, gender construction in Indonesia during the New Order (1967–1998) situated women as housewives and caretaker for their children, whereas men considered as breadwinners (Yulindrasari & McGregor, 2011).

3 METHODOLOGY

We conducted focused ethnography for two months. This research uses focused ethnographic that is carried out for two months and spent about 108 hours in kindergarten. Knoblauch (2005) explains that ethnographic focus applies short-term research but still uses ethnographic principles.

The research was conducted in one of the kindergartens in Bandung, Indonesia. We observed 86 children consist of 40 boys and 46 girls and eight teachers. We also used observation and informal interviews. We did the observations from the beginning of the class started until the class finished, but we focused on playing activity that children did in kindergarten. We conducted informal interviews with all of the teachers and children. The reason for choosing informal conversations is to allow us to establish a connection and to interact with young children so that they can easily understand the language that we used. As with research carried out by Adriany (2018), simple conversations allow children to be able to answers the questions.

During the fieldwork, we tried to establish a relationship with children by playing together and blend in every conversation by greeting and smiling. Building closeness with children is so important to build trust and freedom of children towards us as researchers (adults). In this chapter, we used pseudonyms as an effort to protect informants and protect their confidentiality. In conducting this research, the first author has a significant role in researching the fieldwork, while the second and third authors help the analysis process of the research.

We used a grounded theory that produced 164 codes that had been reduced from 189 codes. Then, we divided the codes into several themes. This chapter discusses one of the themes that is "space and gender in the playground." Three themes emerge: namely, the construction of masculinity of a "fortress" in the playground, construction of femininity of a "house" in the playground, and traditional gender resistance in the playground. This research is part of the master's thesis and is also part of PUPT's research.

4 RESULTS AND DISCUSSION

4.1 *Construction of masculinity through a "fortress" in the playground*

Boys construct their masculinity in the playground. It is indicated by the dominance of boys in specific game tools. They seem often defend some spaces:

> In the playground, there were three girls and three boys fight to play circled monkey bars, and one boy shouted out "this our fortress". Then, three boys threw girls' bags out.
>
> (field notes on April 17, 2018)

Based on these field notes, it is explained that boys maintain their masculinity through the domination in monkey bars games. Circled monkey bars is a "fortress", where they can show their masculine power. Hence, the boys oppose the girls by throwing out their bags. It has assured that makes girls are excluded from boys' space.

On the other hands, girls made some efforts to resist masculine power through the fortress. It further strengthens gender segregation to maintain dominance in the playground. Consequently, the playing room is contested by both boys and girls as has shown in notes given here.

> In teeter-totter game, a group of boys and girls were fighting against one and another by showing who can lift the weights so that the group is the winner. Initially, a group of girls could maintain the balance, but all the boys who played the ball were arrived to join the other boys and help them in the game. After all, the boys won the game. A Group of boys quipped, "wuuuuu . . . !",
>
> "Cemen (means a coward)!" group of girls
>
> Vino shouted "boys come here to help . . . help" then the boys gathered
>
> The group of boys said to the girls"cemen!" by showing thumbs down to girls
>
> The group of girls shouted out, "aaaakkkk" while closing their ears because of the whoop of boys,
>
> Group of boys called out the girls "cemenn cemenn!"
>
> Then, the girls one by one went to leave the teeter-tooter.
>
> "Cemen . . . cemen . . . cemen . . . cemen . . . cemen . . . cemen!" yelled the boy in unison, one girl then cried.
>
> At first, the teacher only saw what the children were playing, until one girl started to cry to calmed down one of the girls. One girl named Mahira still tried to push down one side of the teeter-tooter but the boys continued to maintain their weight, and they shouted out to her "wuuuuuuuuu". Mahira shouted out them back and left the game until the game left no girls.
>
> (field notes on Thursday, April 3, 2018)

At the playground, there is a struggle to claim power between boys and girls. It happens in the teeter-totter game where all of in children in the kindergarten want to sustain their dominance. As has illustrated above, the boys win because they got help from other boys who are in the pavilion and the opponents were not balanced and lost by the boys. Furthermore, it reinforces the masculine and feminine binaries in kindergarten, so children tend to strengthen even beyond gender constructions that exclude girls.

Also, boys exclude girls using verbal aggression by saying "*cemen*" (coward), showing thumbs down, shouting and whooping. The girls choose to leave the game because of that. A girl who kept trying to maintain the game cried because of continuous verbal attacks from the boys.

This discussion reveals that the play space in the kindergarten is a boy's space. The boys are considered naturally in dominating playground equipment that is believed that improve their physical skills. Some studies have displayed that boys often get benefits from the assumption that boys are physically strong, so they have more access to games that increase their physical strength (Early et al., 2010; Chase, 2009; Azzarito, Melinda & Louis, 2006; Maccoby, 1988). Therefore, ignorance of boys' power and domination in the play space is normalised because of the gender label developed in an early childhood education setting.

4.2 *Construction of femininity through a "houses" in the playgrounds*

Girls also maintain their power through dominance in the playground in some spaces through particular games. A box-shaped monkey bar becomes a space of femininity for the girls as indicated in this excerpt from the field notes:

> Then came the two boys from the A2 group namely Vino and Azzam. Vino then went to a box monkey bar they called "Our Home". When he entered, Mahira immediately shouted out "aaaa . . . not here". The two boys looked surprised, then left the monkey bars.
>
> (field notes on May 30, 2018)

There are three monkey bars in the kindergarten. One is in a circled-shape, the other is combined with a slide, and the last is boxed-shaped with semi-circular-shaped on its top which is often illustrated as a "house" for girls. Based on our observation above, it shows that girls use "home" to maintain their power in the monkey bars. Indeed, "house" is a very feminine space, especially in Indonesia. Yulindrasari and McGregor (2011) argued that gender knowledge is built through the role of being a mother who has a large space in the domestic area. The construction of gender becomes ideal for the girls who play in "house" in the playground. It makes that the girls maintain their space by using the designation of the house for their feminine power.

4.3 *Traditional gender resistance in the playground*

In the playground, the children also reproduce traditional gender construction. At the same time, however, they can reconstruct the traditional gender idea. Both cases are shown in the following conversation:

> In the small-sized monkey bars, two girls did not climb it. They just sat and made houses from barbie pictures. Then I came, the two daughters invited me to play, but I served as a house guard.
>
> I: "Why it should be guarded?"
> Rosyida: "We worried that there are any boys who come in."
> I: "What if there are boys enter this house?"
> Rosyida: "Can't fight."
> I: "Who can't fight?"
> Rosyida: "Me"
> Me: "What if there is a boy come in, Rosyida?"

Rosyida: "If a boy truly comes, I have a friend of mine whose name is Vanya, she can hit a boy with small energy, and also I can hit them with higher energy than Vanya to help him, that is how it is."

Rosyida: "Men should be strong if women ask for help from men. They should act nicely to women and love us too."

(field notes on April 16, 2018)

The conversation describes how the girls feel unable to fight back the boys, so they do not want the boys to enter the play space they call "home". However, on the other hand, Rosyida stated that she was strong, even though the boys were stronger than her. Rosyida tried to reconstruct traditional femininity by assuming that girls can also be strong as the boys although the masculinity stigma is still embedded in her thought. In addition, her further explanation confirms that boys should be good and caring for girls. This statement allows boys and girls to play together equally.

5 CONCLUSION

This study shows that the playground is not gender-neutral. Our findings are consistent with MacNaughton (2000) that free play is influenced by the discourse of masculine and feminine where boys and girls control space differently. It allows boys and girls to be excluded from certain spaces.

The role of teachers is important in tackling gender inequalities that occur in kindergarten. Teachers' involvement will assure that children's play activities are a necessity to enable blocking masculine and feminine binaries that could exclude girls or boys in certain spaces. Martin (2003) revealed that the absence of teacher in supervising those activities makes children free to develop their own rules such as boys who often play physically and aggressively when the teachers left them. It will increase the risk of developing behaviour problems. So, teachers require gender-sensitive qualities that make them aware that children's activities related to the spaces that are often gendered.

Developing the teacher's knowledge about gender equality requires support from governmental institutions such as the education agency and universities. Gender mainstreaming should also touch early childhood curricula (MacNaughton, 1997). Also, teachers training curriculum must include gender awareness (Adriany, 2018; Warin & Adriay, 2015). We recommend that teachers and adults in ECE settings should equip themselves with knowledge about gender equality band equity to provide a non-bias learning environment for young children.

REFERENCES

Adriany, V. (2013). *Gendered power relations within child-centred discourse: an ethnographic study in a kindergarten in bandung, Indonesia*. PhD dissertation, Lancaster University, UK.

Adriany, V. (2018). Being a princess: young children's negotiation of femininities in a Kindergarten classroom in Indonesia. *Gender and Education*, 1–18.

Adriany, V., & Warin, J. (2014). Preschool teachers' approaches to care and gender differences within a child-centred pedagogy: findings from an Indonesian kindergarten. *International Journal of Early Years Education*, *22*(3), 315–328.

Arlemalm-Hagsér, E. (2010). Gender choreography and micro-structures–early childhood professionals' understanding of gender roles and gender patterns in outdoor play and learning. *European Early Childhood Education Research Journal*, *18*(4), 515–525.

Azzarito, L., Solmon, M. A., & Harrison Jr, L. (2006). "... If i had a choice, i would ..." a feminist poststructuralist perspective on girls in physical education. *Research Quarterly For Exercise and Sport*, *77* (2), 222–239.

Bhana, D. (2003). Children are children: gender doesn't matter. *Agenda*, *17*(56), 37–45.

Chapman, R. (2016). A case study of gendered play in preschools: how early childhood educators' perceptions of gender influence children's play. *Early Child Development and Care*, *186*(8), 1271–1284.

Chase, C. M. (2009). Children's play: The construction of gender roles. *Young Children*, 1(21), 83–86.

Connell, R. W. (1996). Teaching the boys: New research on masculinity, and gender strategies for schools. *Teachers College Record*, 98(2), 206–235.
Early, D. M., Iruka, I. U., Ritchie, S., Barbarin, O. A., Winn, D. M. C., Crawford, G. M., ... & Bryant, D. M. (2010). How do pre-kindergarteners spend their time? Gender, ethnicity, and income as predictors of experiences in pre-kindergarten classrooms. *Early Childhood Research Quarterly*, *25*(2), 177–193.
Gavey, N. (1989). Feminist poststructuralism and discourse analysis: Contributions to feminist psychology. *Psychology of Women Quarterly*, *13*(4), 459–475.
Knoblauch, H. (2005). Focused Ethnography [30 paragraphs]. *Forum Qualitative Sozialforschung/Forum: Qualitative Social Research*, 6(3), Art. 44, http://nbn-resolving.de/urn:nbn:de:0114-fqs0503440.
Lappalainen, P. A. (2004). They Say it's a Cultural Matter: gender and ethnicity at preschool. *European Educational Research Journal*, 3(3), 642–656.
Maccoby, E. E. (1988). Gender as a social category. *Developmental Psychology*, *24*(6), 755.
MacNaughton, G. (1997). Feminist praxis and the gaze in the early childhood curriculum. *Gender and Education*, *9*(3), 317–326.
MacNaughton, G. (2000). *Rethinking gender in early childhood education.* London: Paul Chapman Publishing
Messner, M. A. (2000). Barbie girls versus sea monsters: Children constructing gender. *Gender & Society*, 14(6), 765–784.
Martin, B. (2011). *Children at Play: Learning Gender in the Early Years.* Trentham Books Ltd. Westview House 734 London Road, Oakhill, Stoke-on-Trent, Staffordshire, ST4 5NP, UK.
Paechter, C. (2003). Masculinities and femininities as communities of practice. *Women's Studies International Forum*, *26*(1), 69–77. https://doi.org/10.1016/S0277-5395(02)00356-4
Paechter, C. (2006). Power, knowledge and embodiment in communities of sex/gender practice. *In Women's Studies International Forum*, *29*(1), 13–26. https://doi.org/10.1016/j.wsif.2005.10.003
Paechter, C. (2007). *Being boys; being girls: Learning Masculinities and Femininities: Learning masculinities and femininities.* New York: McGraw-Hill Education.
Warin, J., & Adriany, V. (2015). Gender flexible pedagogy in early childhood education. *Journal of Gender Studies*, *26*(4), 375–386.

Early Childhood Education in the 21st Century – Yulindrasari et al. (eds)
 ISBN 978-1-138-35203-2

Multiculturalism knowledge construction of kindergarten teachers in Indonesia

G.A. Putri & E. Kurniati
Universitas Pendidikan Indonesia, Bandung, Indonesia

ABSTRACT: As one of the dimensions of multicultural teaching competency, knowledge is important to be achieved by kindergarten teachers. Moreover, the construction of multicultural knowledge will give more understanding in which aspect they concern. This study was conducted to explore the construction of multicultural knowledge and used qualitative method with case study design and interview technique. Three teachers from different religious backgrounds were selected. The results of this study show that teachers constructed their multicultural knowledge from the concept of diversity, religious aspects, and experiences with children. Therefore, teachers and educational institutions can put more emphasis on these aspects.

1 INTRODUCTION

As one of the biggest multicultural countries, Indonesia still has problems related to multicultural issues. In addition, multiculturalism is a national identity of Indonesia as a country with more than 300 ethnics, around 370 languages, 125 faith (there are 125 faith as local religion based on the traditional animism and dynamism value), and six religions acknowledged by the government (Kementerian Sekretariat Negara Republik Indonesia, 2013). In Indonesia, the term of multicultural, emphasized on national symbol called, *'Bhinneka Tunggal Ika'* or Unity in Diversity.

The multicultural problems that occur in Indonesia not only affect adults but also children. This result might lead to discrimination problems at schools. Because of those problems, children need to develop their understanding of multiculturalism. Therefore, they will learn things like tolerance and respect for each other. The early understanding of multiculturalism is important because young children already notice racial differences by the age of 3 (Arau, 2011). In addition, they learn to treat others differently through the process of interaction with other people (Abdullah, 2009).

Moreover, children are easily influenced both at home, neighborhood, and school, because adults around them such as parents and teachers may have various values about diversity (Carlson et al. 2008; Yusof et al. 2014). The implementation of multicultural education is needed to develop young children's understanding of the diversity in cultures, ethnics, religions, and customs around themselves. Multicultural education provides an equal education for all children with diverse races, ethnics, languages, and cultural groups (Banks 2010), and it would help to construct unbiased knowledge and understanding when they face diversity (Vold, 1992).

As a multicultural country, Indonesia already promotes multicultural education that is included in Indonesian Legislation on National Education System Moreover, through pre-service training, kindergarten teachers receive competency of multicultural education so that they understand how to implement the multicultural education in their classrooms.

However, the study conducted by Formen (2011) says that only 50% of Islamic kindergarten teachers in Indonesia were confident to teach multicultural education. Those previous studies lead to an assumption that there is a different perspective from different religions in multicultural education. Moreover, this study was conducted in order to explore kindergarten teachers' multicultural knowledge construction, so that the aspect behind teachers' understanding about multiculturalism will be better explained.

2 LITERATURE REVIEW

Humans are born with imperfect minds. Children at an early age are great learners, and this is a crucial time for their emotional, moral, and intellectual development (Hutcheon, 1999). By the age of 3, children are aware of the diversity around them and use physical markers to identify the differences because of their understanding had been developed (Arau, 2011). Moreover, children learn to treat others differently from race, gender, age, ability, religion, and cultural heritage (Abdullah, 2009). Because of that, if differences which are recognized by children are not directed to positive things, then those differences will be embedded as negative perspective by the time that they become adults.

Because of the influence from their environment, children learn to treat others differently from the very young age, so that multicultural education is essential to be taught in early childhood education. Multicultural education also become a challenge for teachers in introducing multiple perspectives, cultures, beliefs, and it encourage them to develop their awareness (Potter, 2008).

The importance of multicultural education in the early years of a human being cannot be separated from the methods or techniques that are implemented by teachers during learning, playing, or doing other activities in school. Furthermore, teachers as educators also need to recognize this importance by having a strong foundation in early childhood education quality (Abdullah, 2009).

3 KNOWLEDGE CONSTRUCTION

In order to understand teachers' mindset about multicultural education, the understanding how they construct their own knowledge about multicultural is needed. Banks (2010) said that the multicultural knowledge construction is one of the ways how someone describe or explain the reality.

The multicultural knowledge construction in this study become basic knowledge of perception and culturally responsive attitudes of the teachers in order to implementing multicultural education, especially in early childhood classroom. The knowledge as explained is one of teacher multicultural teaching competency which are describing the culturally responsive teaching and strategy of the teachers. Furthermore, the knowledge construction also become a theoretical concept which are related to influence perception, especially in social work studies (Boggs, 2004).

In this study, the teachers as participants came from different backgrounds. They had different subjective values such as religion values, culture, ethnicity, and so forth. Furthermore, because the multicultural knowledge not only constructed by the text resources but also their own experience with the children (Belenky, 1986), it was important to also focused on teachers' reflexivity.

This study used Belenky's (1986) theory of knowledge construction as a framework to explain knowledge construction. There are five steps: (1) Silence, (2) Received Knowing, (3) Subject Knowing, (4) Procedural Knowing, and (5) Constructed Knowing. Those steps explain that the construction need the teachers and students reflection. Moreover, teachers need to introduce different perspectives, culture, and beliefs to children (Potter 2008, p. 84). Because of that, it can be said that young children will be influenced by their teacher's knowledge about multiculturalism (Carlson et al. 2008; Yusof et al. 2014)

Moreover, the first multicultural teaching competency is knowledge described by Sue and Sue (2012) as understanding the worldview of culturally diverse students. To have a grip on this competency, teachers need to see and accept the other perspectives of worldviews and not judge others' cultural backgrounds, daily life experiences, hopes, fears, and aspirations.

4 METHOD

This study used qualitative design with interview technique in order to seek more detail information about kindergarten teachers' multicultural knowledge construction. The participants in this study were three kindergarten teachers from different types of school. Moreover, the teachers in this study also come from different religion-based schools.

The first participant is a teacher from Islamic-based school, the second participant is a teacher from Catholic-based school, and the third participant is a teacher from a general school without a religion base. All of the teachers are female and have experience as teacher in their school for more than five years.

Data collection was done through interview with each teacher at different times, and all the data was recorded before it was written down. The interviews used a semi-structure method and most of the questions focused on the teachers' perceptions and aspirations about multicultural education in their own school.

For the data analysis, this study used grounded theory (Charmaz, 2006) in order to seek the relevant code and categorized the results into themes. Three steps of coding were conducted from initial coding line by line, focused coding, and axial coding. The results found three themes about multicultural knowledge construction of kindergarten teachers, as explained in the results and discussion section.

5 RESULT AND DISCUSSION

The analysis of this study revealed three themes of the teachers' multicultural knowledge constructions. The first theme was related to the diversity concept, the second theme was related to religious values, and the third theme was related to experiences with children.

5.1 *Diversity concept*

When talking about multiculturalism, all of the teachers described it with different things such as culture, ethnicity, languages, and so forth that were brought by the children. These kinds of things included as diversity concept which is seen by the teachers in their classroom. One of the teachers, Mrs. Yanti from the Islamic-based school, felt that multicultural education is something new, but when she looked up to the words, she found that multi means variety. Then she described it as variety of culture, backgrounds of the children because in Indonesia, they have many diversities.

The same concept was also described by Mrs. Dita from the Catholic-based school. However, in addition, Mrs. Dita add some word, there was inclusive education that facilitated the children with special needs. In Indonesia curriculum, the inclusive education is education for all children, including children with special needs and they intend to become the part of classroom activities (Herawati, 2010). In this case, all of the teachers also brought up the children with special needs as a part of diversity that they saw in the classroom.

Moreover, when looked up the description of multicultural education by Banks (2010), it said that multicultural education is an education for all children from different backgrounds such as gender, culture, ethnicity, languages, social-economy status, and children with special needs. However, from the result of the interview, most of the teachers did not touch the gender aspect as a part of diversity that they saw in the classroom.

It can be said that teachers did not include all the problems related to gender diversity into multicultural issues. This result also made assumption that there are might be unheard voices of the children related to their gender because teachers not seem this issue as something that they need to understand.

5.2 *Religious values*

The second construction came from religious values of the teachers. In this case, the interview questions focused on the meaning of multiculturalism from their religion perspective. As results, the teachers wrapped up their answer into two religion perspectives, which are Islam and Catholic. However, these two religions speak the same values, such as humanity, tolerance, and merciful.

Mrs. Yanti from the Islamic-based School said about the value of *silaturahmi* (means humanity) that leading her way to teach in the classroom when there are different backgrounds of children. This value also one of determined in her teaching style. Moreover, she added the important of merciful, so that children will love each other. The other teachers also said the same things, even Mrs. Dita from the Catholic-based school. She also mentioned how the merciful aspect was determined in her classroom. In addition, Mrs. Dita also mentioned about the universal norm in Catholic values like humanity.

The religious values that were mentioned by the teachers then became the foundation of their multicultural knowledge. Moreover, these values also integrated into their daily life activities, not separated in one day activity only. This is because to develop the paradigm of multiculturalism in religious aspect, the universal values in religion need to develop first (Muliadi, 2012). These universal values then became basic knowledge when teachers faced children from various backgrounds.

The similar religious values that mentioned by the teachers in this study also showed that in their knowledge construction, they have similar perspective about multiculturalism. Formen's (2011) study showed how kindergarten teachers in an Islamic-based school did not have much competence on multicultural teaching. However, the result of this study show that the problems might not in their perspective of multiculturalism but in the implementation of those values.

5.3 *Experiences with children*

The last construction made from teachers' experience with children. The results of this theme came up from the reflection that teachers showed when answering the questions. This is because the multicultural knowledge construction not only gained by the text resource, but also from the experience with people from different backgrounds (Belenky et al. 1986). The classroom experience with the children showed up some basic construction, especially in children way of thinking about diversity.

They were agreed about the importance of multicultural education for early children, because it will lead the children to have more experience with the diversity around them. They also had similar perspective about the children way of thinking, which are children understand about diversity but not seeing it as a problem. *"Children loved to play with their classmate from different background, including their friend with special needs,"* said Mrs. Yanti from the Islamic-based school, when describing how children face multiculturalism in their classroom. Moreover, Mrs. Lanlan from the general school said that in her classroom, the conflicts between children from different religions or cultures was rare.

Those explanations showed that teachers also construct their multicultural knowledge from children's perspective. The children characteristic who are not seeing the diversity as problem also kept by the teachers as their basic knowledge. This is important because, as Banks (2010) said, teachers need to distribute their own multicultural knowledge to the children. If teachers did not pay much attention into children's characteristic and their perspective about multiculturalism, there might be an inequality and lead others serious problems into children's mind about diversity.

6 CONCLUSION AND RECOMENDATION

The results of this study showed three aspects that construct kindergarten teachers' multicultural knowledge: diversity concept, religious values, and experience with children. Those three aspects were showed up from teachers' reflection of their daily activities in the classroom.

From the first aspect, most of teachers said about the different background of religion, culture, languages, including the special needs of children. They also mentioned about the inclusive school that leading them to provide facilities for all the children. However, regarding the definition of multiculturalism by Banks (2010), the teachers did not mention about the diversity of gender. This might because teachers were not given much attention on gender problems.

The second aspect is religious values, that all the teachers, whether they have different religion belief, they still keep similar values when facing the multiculturalism. The main values that teachers mentioned were humanity and mercifulness. This construction aspect also explained the previous study that the problems of multicultural teaching competency might occurred because of their knowledge, but the problems in the implementation.

The third aspect showed that teachers experience with the children also constructed their multicultural knowledge. In this case, teachers were aware that teaching multiculturalism is important for the children. Moreover, they also kept thinking about children's mind when facing the diversity. *Children are different with adults*, said teachers. This means that do children not see multiculturalism as a problem.

Moreover, because this study only seeing the construction of knowledge from teachers' perspective, it became the limitation to explain more about the reality of their implementation, Therefore, as recommendation, teachers and education institution can put more emphasis on these aspects.

REFERENCES

Abdullah, A. C. (2009). Multicultural education in early childhood: issues and challenges. *Journal of International Cooperation in Education, 12*(1), 159–175.

Arau, S. B. (2011). Early education for diversity: starting from birth. *European Early Childhood Education Research Journal, 19*(2), 223–235. https://doi.org/10.1080/1350293X.2011.574410.

Banks, J. A. (2010). *Multicultural education issues and perspective 7th edition*. John Wiley & Sons Inc: New Jersey

Belenky, M. F., Clinchy, B. M., Goldberger, N. R., & Tarule, J. M. (1986). *Women's ways of knowing: the development of self, voice, and mind*. Basic Books: New York

Boggs, J. P. (2004). The culture concept as theory, in context. *Current Anthropology, 45*, 187–209.

Carlson, S. A., Fulton, J. E., Lee, S. M., Maynard, L. M., Brown, D. R., Kohl, H. W., & Dietz, W. H. (2008). Physical education and academic achievement in elementary school: data from the early childhood longitudinal study. *American Journal of Public Health, 98*(4), 721–727.

Charmaz, K. (2006). *Constructing grounded theory*. London: Sage Publications Ltd.

Formen, A. (2011). In between Islam and nationalism what may Indonesian early childhood education learn?. *Contemporary Issues in Early Childhood*, 1–9.

Herawati, N. I. (2010). Pendidikan inklusif [Inclusive education]. *EduHumaniora, 2*(1).

Hutcheon, P.D. (1999). *Building character and culture*. London: Praeger Westport.

Kementerian Sekretariat Negara Republik Indonesia. (2013). *Geografi Indonesia [Indonesian geography]*. Retrieved from https://indonesia.go.id/kementerian-lembaga/kementerian-sekretariat-negara-republik-indonesia.

Muliadi, E. (2012). Urgensi pembelajaran pendidikan agama Islam berbasis multikultural di sekolah [The urgency of multicultural-based Islamic education learning in schools]. *Jurnal Pendidikan Islam* (Islamic Education Journal), *1*(1), 55–68.

Potter, G. (2008). Sociocultural diversity and literacy teaching in complex times. *The Early Childhood Education, 2*, 84.

Sue, D.W., Sue, D. (2012). A tripartite framework for understanding the multiple dimensions of identity. *Counseling the Culturally Different: Theory Practice*, 36–48.

Vold, E. B. (1992). *Multicultural education in early childhood classroom*. (in E. B. Vold, Ed.). West Haven, CT: National Education Association of the United States

Yusof, N. M., Abdullah, A.C., & Ahmad, N. (2014). Multicultural education practices in Malaysian preschools with multiethnic or monoethnic environment. *International Journal of Multicultural and Multireligious Understanding, 1*, 12–23.

Early Childhood Education in the 21st Century – Yulindrasari et al. (eds)
© 2020 Taylor & Francis Group, London, ISBN 978-1-138-35203-2

Re-conceptualization of character values in multicultural education

N.A. Rosfalia & V. Adriany
Univesitas Pendidikan Indonesia, Bandung, Indonesia

ABSTRACT: The purpose of this chapter is to find out the extent to which the implementation of character values in multicultural education can be a solution to eliminate the practice of intolerance in early childhood education. This chapter will try to re-conceptualize multicultural education, which has neglected the value of tolerance in early childhood education. This chapter uses the literature review research method in analyzing previous studies. The results of previous research indicate that the character values of multicultural education still ignore tolerance values such as addressing disrespectful actions such as insulting, denouncing and other negative behaviors. Therefore, this chapter is expected to be a reference for early childhood educators as a solution to the elimination of intolerance practices in early childhood education.

1 INTRODUCTION

Parekh (1997) argues that multiculturalism includes three things: (1) cultural diversity; (2) the diversity that exists; and (3) regarding specific actions in response to diversity. Multicultural as an understanding, appreciation, and assessment of one's culture, as well as a respect and curiosity about other people's ethnic cultures. Multiculturalism includes an assessment of other people's cultures, not in the sense of agreeing to all aspects of the culture but trying to see how a genuine culture can express values for its own members (Blum, 1991).

The attitude of mutual acceptance, respect for values, culture, different beliefs will not automatically develop on their own, because in a person there is a tendency to expect others to be themselves (Ibrahim, 2008). The most important thing that must be done by an educator in multicultural education is able to instill the values of multicultural education, such as democracy, humanism, gender justice, ability to disagree, and cultural pluralism (Hanum, 2009).

Character cultivation in children of this age will process children's experiences and have implications for children's behavior. Therefore, it is appropriate to instill character as early as possible, especially the character of tolerance, because gradually if the child has grown up or stepped into the level of life, the sense of responsibility and respect for others will be an important point (Wibowo, 2012). The low cultivation of tolerance character values will cause various impacts on subsequent lives. If the planting of character values is not resolved well from an early age, children tend to behave negatively. For example, if the child is low in character of tolerance, the child will grow up to be a generation that cannot respect others (Mulia & Aini, 2013).

This chapter tries to fill the gaps in the literature and previous research on character and multicultural values in Early Childhood Education in Indonesia. The results of this study are expected to help teachers understand character values from a multicultural perspective to avoid the negative side of multiculturalism. Thus, it is also expected that teachers and adults can become facilitators to make children better understand and apply character values of tolerance in daily activities regardless of the differences that exist.

2 PROBLEMS WITH EXISTING DEFINITIONS

Multicultural education refers to learning the right knowledge, attitudes and skills related to respect and appreciation from various cultures, and other differences that include race, ethnicity, religion and others. Gollnick and Chinn (1990) explain the issues of multicultural education including the value of cultural diversity, emphasis on human rights and respect for those who are different, acceptance of alternative life choices for society, social justice and equality for all people, and an emphasis on equal distribution of power and income between groups. DES (1985) also explains the issue of multicultural education that emphasizes respecting ethnic, cultural and other people's differences.

Lynch and Hanson (1998) explain the application of multicultural education can be done from an early age. Young children can learn new cultural patterns and differences in certain ways. Some researchers (Aydin, 2013a; Banks, 2009) in the field of multicultural education research show that multicultural education is a type of education that helps all students to develop knowledge, abilities and behaviors needed to participate effectively in a democratic society. In addition to these benefits, multicultural education is a field of study that aims to help students to respect each other from various races, ethnicities, social classes, and cultural groups (Aydin & Ozfidan, 2014; Nieto, 2000).

The facts of previous studies (Glover, 1996; Palmer, 1990; Ramsey & Myers, 1990) have shown that children see differences such as skin color, eyes and hair. As part of their socialization, children develop their identity by comparing themselves with others. Young children observe how other people around them react and respond to these differences, they see what is valued and what is not. They begin to develop positive or negative feelings about observed differences. For example, children show preference for the same race when choosing puppets, refusing to hold the hands of children of different races (Glover, 1991). Other research explains that there is still the practice of intolerance in Indonesia, namely concerning religious freedom, ethnocentrism, and other form of intolerance toward differences (Menchik, 2014).

3 RECONCEPTUALIZATION OF CHARACTER VALUES IN MULTICULTURAL EDUCATION

The reconceptualization of character values in multicultural education explains that character values are not only individual but contribute to others, namely the character of tolerance that must be instilled in every young child. The character of tolerance is expected to be a solution to the practice of intolerance that still occurs in early childhood education.

Koesoema (2007) says that character is a human anthropological structure. Character consists of values about human relation to God, to other human, and to non-human environment. Character education assists children to instill the social and religious values needed to be respectful individuals.

According to Noah (2010), character formation needs start at an early age. If the character has been formed at an early age, it will not be easy to change one's character. An understanding of the character of tolerance is very important for every child to understand because Indonesia has a diversity of races, cultures, ethnicities and others that children should understand and appreciate. The development of character values is generally associated with morality and very few develop it by binding to the characteristics of a multicultural Indonesian nation.

4 CONCLUSION

The differences that occur in multicultural education are very important education. Differences in race, culture, religion, language, and others are still a problem, where children make multicultural differences in a problem. Teachers and adults often explain the theory of multiculturalism but have not fully applied the meaning of multicultural education. As a result, there is often the practice of intolerance in multicultural education that occurs in early childhood education.

Parents, schools, communities, and the government should work together in character education. The many problems of intolerance associated with character should be taken seriously. Intolerant behaviors, including insulting others, disrespecting others and additional negative behaviors, are a problem that arises because of the lack of tolerance. The importance of character education is a homework that must be completed by the government considering that character is the foundation for children to develop themselves in life in the future. Early age is seen as the right time to begin to instill a positive character. Character values can be inserted in the learning process in school and become a habit in the child's life. The multicultural program is one program that aims to instill positive character in early childhood.

Thus, the results of this study indicate that teachers and adults must understand the character of children through using a multicultural perspective. This is important to make children more flexible by providing opportunities and space for all of their friends without being limited by all the differences that exist. The results of this study also implicitly show that children are often excluded from certain types of games because of multicultural differences.

REFERENCES

Aydin, H. (2013). *Dünyada ve Türkiye'de cokkültürlü eğitim tartışmaları ve uygulamaları [Discussions and practices of multicultural education in Turkey and the world]*. Ankara, Turkey: Nobel Academic Press.

Aydin, H. & Ozfidan, B. (2014). Perceptions on Mother Tongue (Kurdish) based multicultural and bilingual education in Turkey. *Multicultural Education Review, 6*(1), 21–48.

Banks, J.A. (Ed.). (2009). *The Routledge international companion to multicultural education.* New York: Routledge Press.

DES. (1985). *Education for All: report of the Committee of Inquiry into the Education of Children from Ethnic Minority Group (The Swann Report)*, Department of Education and Science, London, HMSO.

Glover, A. (1996). Children and bias. In B. Creaser & E. Dau (Eds.). *The Anti-Bias Approach in Early Childhood.* Sydney: Harper Educational.

Gollnick, D. & Chinn, P. (1990). *Multicultural education in a pluralistic society* (3rd edn.). New York: Macmillan.

Hanum, F. & Si, M. (2009, December). Pendidikan Multikultural Sebagai Saran membentuk Karakter Bangsa (Dalam Perspektif sosiologi Pendidikan). Paper presented at Regional Seminar DIY-Jateng Sociology Education Students Association, Universitas Negeri Yogyakarta.

Koesoema, D. (2007). *Pendidikan Karakter, Strategi Mendidik Anak di Zaman Global* (rev. edn.). Jakarta: PT Gramedia Widiasarana Indonesia.

Lynch, E. & Hanson, M. (Eds.) (1998). *Developing cross-cultural competence: A guide for working with children and their families* (2nd ed.). Baltimore: Paul H. Brookes.

Menchik, J. (2014). Productive intolerance: godly nationalism in Indonesia. *Comparative Studies in Society and History, 56*(3), 591–621.

Mulia, & Aini. (2013). *Karakter manusia Indonesia.* Bandung: Nuansa cendekia.

Nieto, S. (2000). *Affirming diversity: The sociopolitical context of multicultural education* (3rd ed.). New York: Longman Press.

Palmer, G. (1990). Preschool children and race: an Australian study. *Australian Journal of Early Childhood,* 15(2), 3–8.

Ramsey, P. & Myers, L. (1990). Salience of race in young children's cognitive, affective and behavioral responses to social environments, *Journal of Applied Behavioral Psychology*, 11, 49–67.

Wibowo, A. (2012). *Pendidikan karakter usia dini.* Yogyakarta: Pustaka belajar.

Early Childhood Education in the 21st Century– Yulindrasari et al. (eds)
© 2020 Taylor & Francis Group, London, ISBN 978-1-138-35203-2

Multiculturalism in early childhood education: Literature review

R.O. Marisa & V. Adriany
Universitas Pendidikan Indonesia, Indonesia

ABSTRACT: The purpose of this article is to explain why multiculturalism must be learned in early childhood education. The existence of ethnic groups, races, languages, gender, religion and others in Indonesia and the differences there of absolutely deserves to be studied in early childhood educational settings. The aim is to help children develop identities, increase their knowledge and understanding of their unique culture. Multiculturalism as a thought and movement consists of elevating values in the form of respect for differences in terms of ethnicity, race, language, gender, religion, and more.

1 INTRODUCTION

Multiculturalism describes the nature of human societies that vary culturally, linguistically, religiously and socio-economically, racially, and so on. To strengthen democracy, the education system needs to consider the multicultural character of society, and aim to actively contribute to peaceful coexistence and positive interactions between different cultural groups (UNESCO, n.d.; Woodhead 2006). Traditionally in the education system the previously mentioned can be achieved through a multicultural education approach and intercultural education. Multicultural education uses learning about other cultures to produce acceptance, or at the very least tolerance of these cultures whereas intercultural education is aimed to go beyond passive coexistence, to achieve a shared and developing sustainable way of life in a multicultural society through the creation of understanding, respect and dialogue between different cultural groups (UNESCO, n.d.). Multiculturalism education includes ideas or concepts, educational reform movements, and processes which recognize the political, social and economic realization experienced by each individual in complex and culturally diverse human meetings (Banks, 2013; Mahfud, 2011). This is intended to foster social equality and justice regardless of race, ethnicity, gender, and social class.

In fact, countries around the world are said to be pluralistic even though the practice of multiculturalism does not always run smoothly. As for example in various countries there are social, religious, political and other differences reflecting that diversity or diversity in many areas is still a trigger for certain class conflicts and discrimination (Miller, 2017; Case & Ngo, 2017; Anastasiou, Kauffman, & Michail, 2016; Lubis, 2015; Parsons, 2005). On the other hand there is also a compulsion to accept the beliefs of the dominant group and the concept of universality, and shows the practices of monoculturalism (Nagayama & Barongan, 2004). Another example that shows problems of the practice of multiculturalism is superiority or the dominance of one particular group over others resulting in other groups becoming minorities (Solehuddin & Adriany, 2017).

The discussion on multiculturalism also addresses racial based discrimination experienced by community groups as there is an assumption that color teachers are considered less competent in teaching, even though they have proven otherwise it is still not recognized (Burciaga & Kohli, 2018). Color teachers also experience structural suppression of work, which is systematically searching for ways for them to be excluded from the profession (Kohli, 2016). In Zimbabwe, primary school teachers are considered to be less understanding of the concept of multicultural education due to the majority of teachers not being able to provide a comprehensive definition of multicultural education (Mapuranga & Bukaliya, 2014). The same occurrence happened to black youth who were marginalized in political decisions and management (Parsons, 2005).

From these phenomena, multiculturalism emerged as a form of criticism both in political philosophy and education which tended to consider the lack of the ability of other cultural differences or parts of cultural diversity (Koppelman & Goodhart, 2011). Multiculturalism is also used as a critique of racist practices in the United States for non-native women where these Asian women find it difficult to position themselves and interact with others (He, 2018). Other criticisms were also made such as criticism of white teachers who were considered too racist to their students who were color (Au, 2017). Multiculturalism also appeared to criticize the practice of monotheism where in certain regions the community must embrace only one dominant religion and keep the community away from other ideologies that can disrupt the harmony of the dominant religion (Abdou & Chan, 2017).

Previous research also shows how various countries in the world have tried to accommodate multicultural based curricula in their education. For example in the United States in International Schools multiculturalism is implemented in the curriculum in order to facilitate the diverse needs of students (Hayden & Thompson, 2008). In Turkey the education system is transforming the curriculum into a multicultural structure with the aim of preventing students who have different cultural features from being exposed to marginalization in the educational environment and through influencing their learning positively by changing their cultural differences into advantages (Demir & Yurdakul, 2015). In Indonesia multiculturalism is implemented in education which aims to build knowledge and peace amongst high school students (Lie, 2000). However, unfortunately this implementation feels difficult because multicultural issues must be addressed in professional development so that teachers can learn how to recognize and accept differences while providing a common set of norms and values to bind students who have various backgrounds (Lie, 2000).

In the context of the ECE itself, discussions on multicultural curricula are still rare. So far, research on the implementation of multicultural curriculums has been carried out in Finnish settings where schools provide strategies and practices for inclusive education for education and social life because Finland has an increasing number of children and parents with multicultural backgrounds. Due to this, the task of developing this field is becoming increasingly important (Ojala, 2010). This practice focuses on preschoolers and their transitioning to elementary school. The development activities in multicultural early childhood education in Finland are focused on four key elements namely value, learning environment, support for children's growth and development, and cooperation and support for parents (Kuusisto, 2010). When working and developing early childhood education, it must always be understood that preschool education plays an important role in preventing social exclusion in education and encourages the success of long-term school students (Ojala, 2010).

Basically the implementation of a multicultural curriculum in the context of its own education aims to establish awareness of the other tasks of teachers who play an important role in schools that act as anti-racists (Coe, 2017), forming the character of students from diverse cultural, religious, racial, social groups (Chiu et al., 2017), and also allows children to be netted with racial communities from different groups (Flynn, 2017). This is intended to create a safe and supportive learning space to become a school culture that is socially just and respects differences.

As stated above, talking about multiculturalism in the end intends to foster social justice. However, in reality studies that try to link multiculturalism with social justice are still very limited. As can be seen in previous studies, the discussion on multiculturalism has focused on the cultural context, even though talking about multiculturalism also speaks of social justice. So the purpose of this article is to attempt to show a clearer relationship between the two, and how multiculturalism can be implemented more broadly from a social justice perspective.

2 THE CONCEPT OF SOCIAL JUSTICE

Social justice is the basic principle for peaceful and prosperous coexistence in community life. The United Nations (UN) upholds the principles of social justice by promoting gender equality or the rights of indigenous and migrant people, to promote development and human dignity (United Nations, 2006). In Indonesia itself the concept of social justice is found in Pancasila as the philosophy of life of the Indonesian nation in the fifth principle which reads "social justice for all Indonesian people", in which one of them is practicing a fair attitude towards others, respecting the rights of others, doing activities in order to realize fair and socially just progress (Majelis Permusyawaratan Rakyat Repubik Indonesia, 2003). Fair in question is not generalization but removes obstacles faced by a person due to gender, age, race, ethnicity, religion, culture or disability (United Nations, 2006). Disguising it would mean benefiting the dominant group, as in the research of Solehuddin dan Adriany (2017) almost all the teachers in his study defined social justice as equal treatment given to all children in school. As a consequence of this orientation, much attention has been given to recognizing diversity among students in relation to their social, cultural and economic background by generalizing rather than refracting differences.

3 MULTICULTURALISM TEACHES ABOUT SOCIAL JUSTICE

Talking about justice in the frame of multiculturalism in early childhood education is often the material to be discussed. This is due to the extent of the concept, commitment and practice of justice being accommodated by the various parties involved in early childhood education. Current studies discuss how justice is interpreted as 'equivalent' which is interpreted as access to high-quality early education that promotes the same results across economic groups to increase the level of educational games for children (NCTE, 2016; Solehuddin & Adriany, 2017). Social justice has been diverted to justify policies and standardization (Luke, Woods, & Weir, 2013) in a way that does not always position the needs and rights of our most vulnerable children as a priority (Woods, Mackenzie, & Wong, 2013).

For many ECE teachers and researchers, commitment to social justice is at the core of their work (Woods et al., 2013), but often in practice teachers still cannot interpret justice as something that does not mean 'equal/same'. As in the research of Solehuddin and Adriany (2017) where auditors desired to teach similarity by using the majority norms that are privileged, so ethnic minorities are forced to be the same as the ethnic majority. These studies illustrated the misconception of the concept of justice, instead of generalizing it as fair but what happens is the existence of other parties who are harmed. In another study, as reported by NCTE (2016) in early childhood education, one example of unfair practice is the disproportionate number of colored boys in the United States referred for disciplinary action. In this case, teachers and administrators were involved in underlying racial bias towards children of color which can create aggressive attitudes that can lead to discriminatory behavior.

The practice of this injustice also discusses one's leadership within the scope of auditing. Like the study of Nicholson et al. (2018) where teachers and administrative staff in the early childhood education sector are compared to other fields (eg. public school administrators and business executives), through marginalized discourses and deficits that disguise their funds, knowledge and strengths. This was also emphasized in the study (Whitebook, Phillips, & Howes, 2014) that there had been inequities of poor working conditions and a lack of compensation related to work in the early childhood education environment.

Socially just education is about creating an educational context consisting of educators who refuse to accept unfair institutional, systemic and social relations (Woods et al., 2013). Fair early childhood education is achieved when children's strength-based views are basic; when local knowledge and family are respected; when children are assessed in an authentic and reasonable amount; and when differences between race, ethnicity, language, gender, religion, class, sexual orientation, family structure, physical/mental abilities are recognized, understood and utilized; and respect for children's voices in conversations surrounding justice, hope and practice of reconciliation (NCTE, 2016). The field of early childhood education requires more leadership theory that illustrates the purpose of working for change including reducing oppression and bringing greater justice to children, families, communities and laborers in the field of early childhood education (Nicholson et al., 2018). Rethinking the concept of equality referring to the basis of justice, because equality can be described as the abolition of privileges, oppression, differences, and losses which historically exclude those belonging to certain groups (NCTE, 2016). Education that is socially fair is all about creating a challenging educational context and the refusal to accept institutional relations, both systemic and socially unjust, that continues adversely benefit some groups and not others (Kathy & Bettez, 2011; Woods et al., 2013).

4 CONCLUSION

Multiculturalism as thoughts and movements elevates values in the form of respect for differences. Social, religious, political and other differences reflect that diversity in many places is still a trigger for conflict and discrimination. Through multiculturalism, differences are not perceived as obstacles or sources of problems but rather as wealth that beautifies life, is not removed but maintained, not curbed but given space. Therefore, education for early childhood must be able to explain the concept of social justice through multiculturalism education. In the case of early childhood education, it needs to be well understood that early childhood education plays an important role in the future life of children at least children must possess basic life skills and can work with peers and adults, be autonomous, creative, solve problems, respect differences and also have just social attitudes.

REFERENCES

Abdou, E. D., & Chan, W. Y. A. (2017). Analyzing constructions of polytheistic and monotheistic religious traditions: a critical multicultural approach to textbooks in Quebec. *Multicultural Perspectives, 19*(1), 16–25. https://doi.org/10.1080/15210960.2016.1263961.

Anastasiou, D., Kauffman, J. M., & Michail, D. (2016). disability in multicultural theory: conceptual and social justice issues. *Journal of Disability Policy Studies, 26*(1), 3–12. https://doi.org/10.1177/1044207314558595.

Au, W. (2017). When multicultural education is not enough. *Multicultural Perspectives, 19*(3), 147–150. https://doi.org/10.1080/15210960.2017.1331741.

Banks, J. A. (2013). Multicultural education: characteristics and goals.pdf (pp. 3–21). Retrieved from http://www2.humboldt.edu/education/images/uploads/documents/3._Multicultural_Education_Characteristics_and_Goals.pdf.

Burciaga, R., & Kohli, R. (2018). Disrupting whitestream measures of quality teaching: the community cultural wealth of Teachers of Color. *Multicultural Perspectives, 20*(1), 5–12. https://doi.org/10.1080/15210960.2017.1400915.

Case, A., & Ngo, B. (2017). "Do we have to call it that?" the response of neoliberal multiculturalism to college antiracism efforts. *Multicultural Perspectives, 19*(4), 215–222. https://doi.org/10.1080/15210960.2017.1366861.

Chiu, C. L., Sayman, D., Carrero, K. M., Gibbon, T., Zolkoski, S. M., & Lusk, M. E. (2017). Developing culturally competent Preservice Teachers. *Multicultural Perspectives, 19*(1), 47–52. https://doi.org/10.1080/15210960.2017.1267515.

Coe, C. A. (2017). Hearing the difference: mockingbird novel inspires student-led anti-racism video project hearing the difference: Mockingbird Novel Inspires Student-Led. *Multicultural Perspectives, 19*(4), 239–243. https://doi.org/10.1080/15210960.2017.1373572.

Demir, N., & Yurdakul, B. (2015). The examination of the required multicultural education characteristics in curriculum design. *Procedia - Social and Behavioral Sciences, 174*, 3651–3655. https://doi.org/10.1016/j.sbspro.2015.01.1085.

Flynn, J. E. (2017). Speaking up and speaking out? long-term impact of critical multicultural pedagogy. *Multicultural Perspectives, 19*(4), 207–214. https://doi.org/10.1080/15210960.2017.1365611.

Hayden, M., & Thompson, J. (2008). *International schools: growth and influence. UNESCO: International Institute for Educational Planning*. Paris: UNESCO.

He, M. F. (2018). Multiracial/mixed narrative of lives in-between contested race, gender, class, power, and place. *Multicultural Perspectives, 20*(1), 53–55. https://doi.org/10.1080/15210960.2018.1408349.

Kathy, H., & Bettez, S. C. (2011). Understanding education for social justice. *Educational Foundations, 25*, 7–24.

Kohli, R. (2016). Behind school doors : the impact of hostile racial climates on urban teachers of color. *Urban Education*, 1–27. https://doi.org/10.1177/0042085916636653.
Koppelman, K. L., & Goodhart, R. L. (2011). *Understanding Human Differences: Multicultural Education for a diverse America, 3rd edition* (Myeducatio). Boston: MA Pearson.
Kuusisto, A. (2010). *Cultural, Linguistic and Ideological Diversity in Daycare Center: Challenges and Possibilities.* Helsingin Kaupungin Sosiaalivirasto: Tutkimuksia.
Lie, A. (2000). The multicultur education for peace and development. In *the 35th Southeast Asian Ministers of Education Organization Regional Language Centre (SEAMEO RELC) Conference* (pp. 81–102). Singapore: SEAMEO Jasper Monograph Series.
Lubis, A. Y. (2015). *Pemikiran kritis kontemporer: dari teori kritis, culture rtudies, feminisme, postkolonial hingga multikulturalisme [Contemporary critical thinking: from critical theory, culture studies, feminism, postcolonial to multiculturalism]*. Jakarta: Rajawali Pers.
Luke, A., Woods, A., & Weir, K. (2013). Curriculum design, equity and the technical form of the curriculum. in *curriculum, syllabus design and equity : a primer and model* (pp. 6–39). New York: Routledge, Taylor & Francis Group. Retrieved from http://purl.org/au-research/grants/arc/LP0989526.
Mahfud, C. (2011). *Pendidikan multikultural [Multicultural education]*. Yogyakarta: Pustaka Pelajar.
Majelis Permusyawaratan Rakyat Repubik Indonesia. (2003). *Peninjauan terhadap materi dan status hukum ketetapan majelis permusyawaratan rakyat sementara dan ketetapan majelis permusyawaratan rakyat Republik Indonesia tahun 1960 sampai dengan tahun 2002 [Review of the material and legal status of the provisional decree of the provisional people's consultative assembly and the decree of the people's consultative assembly of the Republic of Indonesia year 1960 up to 2002*]. Indoncsia. Retrieved from http://www16.plala.or.jp/bouekitousi/tap_mpr_no_1_2003pdf.pdf.
Mapuranga, B., & Bukaliya, R. (2014). Multiculturalism in school: an appreciation from the teachers perspective of multicultural education in the Zimbabwean school system. *International Journal of Humanities Social Sciences and Education (IJHSSE)*, *1*(2), 30–40.Retrieved from www.arcjournals.org.
Miller, A. (2017). Dismantling cultural beliefs, policies, and practices that criminalize and dehumanize Black Girls in U.S School. *Multicultural Perspectives*, *19*(4), 247–249. https://doi.org/10.1080/15210960.2017.1373574.
Nagayama, G. C., & Barongan, C. (2004). *Multicultural psychology*. United States: Prentice Hall PTR.
NCTE. (2016). *Equity and arly childhood education: reclaiming the child.* West Kenyon Road, Urbana. Retrieved from http://www.ncte.org.
Nicholson, J., Kuhl, K., Maniates, H., Lin, B., & Bonetti, S. (2018). A review of the literature on leadership in early childhood: examining epistemological foundations and considerations of social justice. *Early Child Development and Care*, *0*(0), 1–32. https://doi.org/10.1080/03004430.2018.1455036.
Ojala, M. (2010). Developing multicultural early childhood education: education in a finnish context. *International Journal of Child Care and Education Policy*, *4*(1), 13–22.
Parsons, E. C. (2005). From caring as a relation to culturally relevant caring: a White Teacher's bridge to Black Students. *Equity & Excellence in Education*, *38*(1), 25–34. https://doi.org/10.1080/10665680390907884.
Solehuddin, M., & Adriany, V. (2017). Kindergarten teacher's understanding on social justice: stories from Indonesia. *SAGE Open*, *7*(4), 1–8. https://doi.org/10.1177/2158244017739340
UNESCO. (n.d.). *UNESCO guidelines on intercultural education*. France.
United Nations. (2006). *The International Forum for Social Development - Social Justice in An Open World: The Role of the United Nations*. New York.
Whitebook, M., Phillips, D., & Howes, C. (2014). *Worthy work, still unlivable wages: the early childhood workforce 25 years after the national child care staffing study*. Berkeley. Retrieved from www.irle.berkeley.edu/cscce.
Woodhead, M. (2006). *Changing perspectives on early childhood: theory, research and policy*. United Kingdom.
Woods, A., Mackenzie, N. M., & Wong, S. (2013). Social justice in early years education: practices and understandings. *Contemporary Issues in Early Childhood*, *14*(4), 285–289.Retrieved from www.wwwords.co.uk/CIEC.

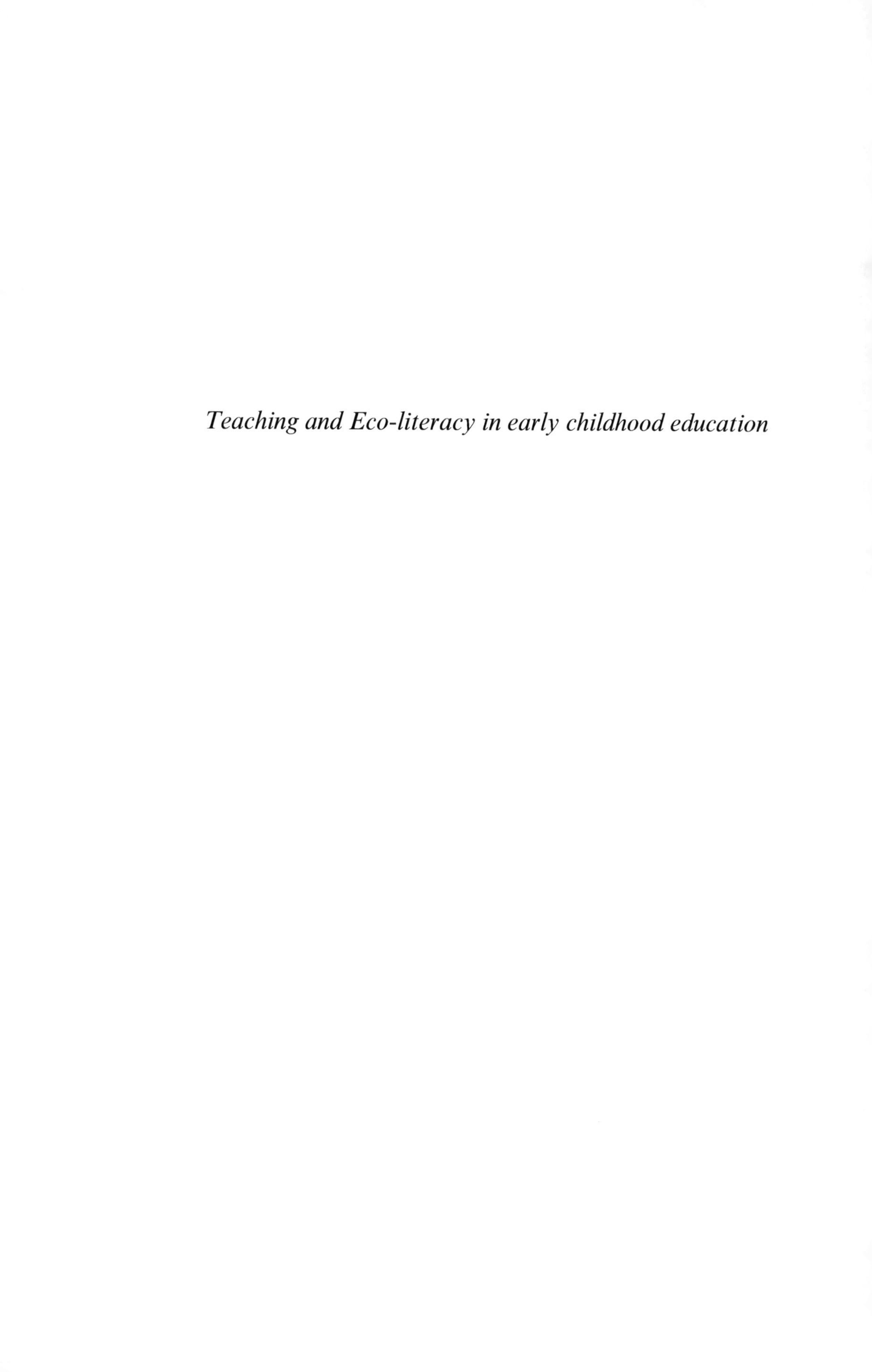

Teaching and Eco-literacy in early childhood education

Early Childhood Education in the 21st Century – Yulindrasari et al. (eds)
© 2020 Taylor & Francis Group, London, ISBN 978-1-138-35203-2

The implementation of scientific approach in early childhood curriculum

S. Wahyuni & H. Djoehaeni
Universitas Pendidikan Indonesia, Bandung, Indonesia

ABSTRACT: The curriculum serves as a reference for teachers to facilitate children in carrying out learning activities. An appropriate learning approach will help achieve the expected educational objectives. The scientific approach is ideally implemented in various early childhood learning experiences. In early childhood the introduction of the scientific process is done by involving the child directly in the activities of doing and experiencing the search for information by asking questions. In order to implement a scientific approach optimally, a thorough understanding of teachers is required so that every stage of scientific approach such as observing, questioning, gathering information, reasoning and communicating can be done optimally and in accordance with the needs and characteristics of the child. This article aims to find out more about the implementation of the 2013 curriculum-based approach in early childhood learning using various literature that relevant to the study. The results of this study are expected to provide an overview of the implementation of a scientific approach in early childhood learning so that it can become a reference for teacher in facilitating learning that suits the needs and characteristics of early childhood.

1 PRELIMINARY

Early childhood education programs are the foundation for developing individual characters in the future. Many experts state that early education is a very basic stage for further development and education. The Victoria Environmental Education Council states that learning experiences at an early age are the foundation for subsequent learning experiences (Djoehaeni, 2017). The curriculum is an educational program that serves as a general guideline in implementation of the education system. The curriculum contains an outline of the program of activities that must be carried out in each education implementation, including the purpose of education as a target that must be sought to be achieved or realized, the main points of the material, the form of activities, and evaluation activities (Nugraha, 2013).

The implementation of the 2013 curriculum is a curriculum actualization in learning and the formation of competencies and character of students. This requires teacher activeness in creating and growing various activities according to the programmed plan. The teacher must master the principles of learning, the selection and use of instructional media, the selection and use of learning methods, the skills of assessing the learning outcomes of students, and choosing and using strategies or approaches (Mulyasa, 2015). One of the characteristics of the 2013 Curriculum for Early Childhood Education is the use of thematic learning with a scientific approach in providing educational stimuli (Permendikbud, 146: 2014).

Learning is a scientific process. Therefore, the 2013 curriculum mandates the essence of the scientific approach to learning and the scientific approach is believed to be the golden mark of the development and development of students' attitudes, skills, and knowledge (Suryana, 2017). This will have an impact on children's thinking and insight when they continue their education to a higher level. The scientific process that can be done by use of scientific approach.. Using a scientific approach, children can integrate children's language, literacy, mathematics, and science development (Hope, Rachel and Barbara, 2013). Through a scientific approach, experience gives information and builds on one another to enhance learning in all aspects of a child's development.

This chapter aims to find out more about the implementation scientific approach in early childhood learning. The results of this literature review will provide an overview of the implementation of the scientific approach in early childhood learning so that it can become a teacher's reference in facilitating learning that is appropriate to the needs and characteristics of early childhood.

2 METHOD

The method used is Literature Review. The authors review and analyze the various sources related to the theme of writing. Topics related to the implementation of the review of scientific approaches in early childhood curriculum in 2013.

3 DISCUSSION

What is a scientific approach?

The scientific approach is a learning process that is designed in such a way that students actively build attitudinal competencies, knowledge and skills through the stages of observing, asking questions, gathering information, reasoning and communicating (Kemendikbud, 2014). A scientific approach is not interpreted as learning science but instead uses scientific processes in learning activities. Bredecamp and Cople (in Masitoh, 2012) suggested that "the environment must enable children to carry out the learning process. The environment is not only the physical environment, but also the psychological environment. In order for children to learn optimally, an environment that can stimulate children is needed to carry out various activities so that children can develop their new understanding through observing or participating with teachers and other children (Masitoh, 2012).

The learning process uses an approach scientifically directed towards achieving spiritual attitude competencies, social attitudes, knowledge and skills and involves six aspects of development in an integrated manner (Kemendikbud, 2014). Competence in PAUD is defined as the ability expected to be achieved through learning that is adjusted to the level of child development. In activities AU: please clarify: I am unsure of your meaning here, it always involves various aspects of development as well (Ministry of Education and Culture, 2015). Mursid (2015) revealed the benefits of learning science in aspects of child development, as described here.

3.1 *Fine and rough motor development aspects*

Various studies indicate that play allows children to move freely so they are able to develop their motor skills. The child learns to climb, step, jump, and so on.

3.2 *Aspects of cognitive development*

Horn's research (in Mursid, 2015) shows that play has a very important role in developing the ability to think logically, creatively and imaginatively. When learning science, children have the opportunity to know the characteristics of object by observing, touching, kissing, and listening. From the sensing the child obtains new facts, concepts, and information that will be compiled into a knowledge structure and used as a basis for thinking.

3.3 *Social development aspects*

When children are playing science, children can learn to socialize in group so that it opens opportunities to interact with other children or people. This will gradually reduce a child's sense of selfishness and help to develop social skills.

3.4 *Language development aspects*

When children are playing and learning science, they are trained to provide answers, which means that children practice using language to communicate and express their ideas or thoughts. Thus, playing will train children's language development.

3.5 *Aspect of moral development*

By playing, children train to be aware of the rules and the importance of obeying regulations. This is the initial stage of moral development.

The learning process in PAUD in the 2013 curriculum emphasizes the scientific approach or scientific approach. The scientific approach in learning as referred in Permendikbud (2013) includes observing, asking questions, gathering information/trying, associating, and communicating. The scientific approach is the process of asking and answering questions with a specific set of procedures. Gerde (2013) states that this process can be used as a guide for making comprehensive and meaningful experiences for children. in line with the 2013 Curriculum PAUD, Gelman and Brenneman (in Garde, 2013) also state that scientific methods include the process of observing, asking questions, producing hypotheses and predictions, experiments, summarizing to draw conclusions, communicate, and identify new questions. Therefore, the end result is an increase and a balance between the ability to be a good human being (soft skills) and a human who has the skills and knowledge to live properly (hard skills) from children which includes aspects of competency attitudes, skills, and knowledge (Daryanto, 2013).

3.6 *What is the process of implementing the scientific approach?*

The Scientific Approach Process is an activity carried out at the stage of the core activity. The series of stages of this scientific approach are:

Observe. Viewing means any activity using all the senses (sight, hear, smell and touch, and taste buds) to recognize an object observed. With this observing activity children can define objects and describe what they are observing, so that children can develop new concepts and vocabulary (Copple and Bredekamp in Garde, 2013). Marchman and Fernald (in Graaf, 2018) said that children with larger vocabulary are better to interpret feedback to improve their next experiment. The teacher helps children define and describe what is being observed (Gerde, 2013).

Ask. Asking is a process of thinking that is driven by the child's curiosity about an object or event. The questioning activity in learning activities as stated by Permendikbud, is to ask questions to get information that is not understood from what is observed to obtain information (Daryanto, 2014). Gerde (2013) said that teachers function as children's language scaffolds and help them develop their ideas and help them make questions.

Collecting Information (Collecting). Gathering information is done through various ways, for example by trying, discussing, reading books, asking questions, and concluding results from various sources. Collecting data is a process that children are very interested in. In this process the child conducts activities that make it possible to experiment and explore answers to questions so that the teacher has the task of organizing the experiment and exploration so that the child is involved in the activity for research questions (Gerde, 2013).

Associating. The association process is a further process in which the child begins to connect the knowledge he already has with the new knowledge he has acquired or that is around him. Here the teacher has associated or connected new knowledge with the surrounding environment, so that the child can build a new understanding of the world around him.

Communicating. Communicating is an activity to convey things that have been learned (Garde 2013). Children share their findings with others so that teachers can provide various methods or media for children to tell about what they are learning.

This scientific approach is in line with Bloom's Taxonomy model, but Bloom's Taxonomy consists of six levels, namely knowledge, understanding, application, analysis, synthetics, and evaluation. Bloom's Taxonomy model has been revised but still has six levels but different places and uses verbs, starting from remembering, understanding, applying, analyzing, evaluating, and creating (Suyadi, 2015). The scientific approach is very relevant to three learning theories namely Burner learning theory, Piaget theory, and Vygotsky theory (Daryanto, 2014). The theory adopted in this article is Vygotsky's theory because in his theory, Vygotsky states children development can be enhanced through scaffolding provided by adults.

Children have cognitive needs that motivate them to recognize themselves and the world around them actively (Garson, in Trnova, 2015). Abstract scientific concepts cannot be fully shaped in childhood due to the gradual development of the brain (Trnova, 2015). The basic method of forming these concepts is investigation through all their senses. Consciously or unconsciously children often make When forming scientific concepts, observation and experimentation play an important role (Bilek in Trnova, 2015). When observing and experimenting, children immediately recognize the characteristics of objects, such as shape and size. At the same time, they began to create the first idea (a simple concept) of space, thanks to the position of each other. This is needed to examine how children conceptualize the shape and size of objects.

Scientific approaches can be implemented in play and games such as Trnova's 2015 research on the formation of science concepts in pre-school, they found a set of four types of simple experiments with toys that are implemented in play and games. The combination of simple experiments in play and games can support the formation of formal and informal pre-school children's science concepts. They are able to develop simple experiments implemented in play and games for the formation of concepts related to the characteristics of substances and natural law. Therefore, it is necessary to apply established methods and tools to pre-school education. To assess children's abilities, early childhood teachers need to practice how to write behavioral indicators from the inquiry process skills.

It is important for teachers to be able to help children understand modes of inquiry and to encourage their inquiry-based skills. To meet current expectations in science education, teachers must provide and integrate ways to improve inquiry skills, apply inquiry-based education, and assess children's abilities (Lee, 2008). The most important thing in implementing this scientific approach is to foster children's understanding. Children have a natural tendency to learn, experiment, and explore so that they develop and expand children's interests (Browman, Donovan & Bums in Kongpa, 2014) as a way to develop future scientific thinking habits (Kongpa, 2014).

The aim of science is to understand the world of nature through a process known as scientific inquiry. scientific knowledge helps us explain the world around, for example, the trees around them. For example, Kongpa's research (2014) examined the scientific concepts of kindergarten and skills in the Tree Unit.. Children can learn scientific concepts from school and everyday life. Some of them can build meaning from concepts about the type and classification of trees. They can also hold back their theory to explain the nature of trees and leaves. This shows that children can be enhanced by scientific concepts through interactions between concepts, scientific reasoning, the nature of science, and the science of social activities (Kongpa, 2014).

4 CONCLUSION

The scientific approach as a learning approach in 2013 Curriculum for Early Childhood can be done by observing, asking questions, gathering information, associating, and communicating with the guidance of the teacher. In the scientific learning model, students are directed to find out their own facts, build concepts, and new values needed for their lives. The focus of the learning process is directed at developing to develop children's attitudes, knowledge and skills. Science learning that needs to be developed includes the need for careful planning, choosing the right material, selecting the right media or tools according to the material being taught, making the learning process always fun, and need an evaluation to find out how much the child is capable of, and to correct the shortcomings of money there and then find a better solution.

REFERENCES

Daryanto, (2014). *Approach to the 2013 scientific learning curriculum*. Yogyakarta: Gava Media.

Gerde, H. K., Schachter, R. E., & Wasik, B. A. (2013). Using the scientific method to guide learning: An integrated approach to early childhood curriculum. *Early Childhood Education Journal, 41*(5), 315–323.

Graaf, J. V. D., Eli, A. S., & Verhoeven, L., (2018). The kind of thinking in kindergarten. *Learning and Instruction. 56*, 1–9.

Kongpa, M., Jantaburom, P., Byne, D., Obmasuy, N., & Yuenyong, C. (2014). Kindergarten's Scientific Concepts and Skills in the Tree Unit. *Procedia-Social and Behavioral Sciences, 116*, 2120–2124.

Lee, J., & Yoon, J. Y. (2008). Teaching Early Childhood Teacher Candidates How to Assess Children's Inquiry Skills in Science Learning. *Contemporary Issues in Early Childhood, 9*(3), 265–269.

Masitoh, et al. (2012). *TK Learning Strategy*. Open University: Jakarta.

Ministry of Education And Culture. (2015). *2013 PAUD Curriculum Educator Guidebook For Children Aged 4–5 Years*. Indonesia: Ministry of Education and Culture.

Mulyasa, (2015). *Development and implementation of 2013 curriculum*. Bandung: PT Remaja Rosdakarya.

Mursid, (2015). *Development of ECD learning*. Bandung: PT Remaja Rosdakarya.

Nugraha, A. et al. (2013). *Kindergarten curriculum and learning materials*. Open University: Jakarta.

Minister of National Education Regulation Number 146 of 2014 concerning 2013 Curriculum for Early Childhood Education. Jakarta: Ministry of Education and Culture.

Suryana, D. (2017). Integrated Thematic Learning Based on Scientific Approach in Kindergarten. *Early Childhood Education Journal*, 11. doi: https://doi.org/10.21009/JPUD.111.

Suyadi. (2015). *2013 PAUD Curriculum educator guidebook for children aged 4–5 years*. Bandung: PT Remaja Rosdakarya.

Trnova, E., Trna, J. (2015). Formation of science concepts in pre-school science education. *Procedia - Social and Behavioral Sciences, 197*, 2339–2346.

Early Childhood Education in the 21st Century – Yulindrasari et al. (eds)
 ISBN 978-1-138-35203-2

Effective communication in child-friendly schools

H. Djoehaeni, M. Agustin, A.D. Gustiana & N. Kamarubiani
Universitas Pendidikan Indonesia, Jawa Barat, Indonesia

ABSTRACT: Teachers have a significant role in creating child-friendly schools by providing a pleasant environment free from pressure or acts of violence. A safe and comfortable environment can be created if the teacher performs his role well. One of the abilities teachers should possess is the skill to communicate. This ability becomes very important because it can help children express their ideas and feelings, stimulate children to respect their friends and bridge various conflicts that arise in various activities. The ability of teachers to communicate with students is expected to help create a child-friendly school. This study intends to reveal more about the ability of teachers to communicate to create child-friendly schools through appropriate literature searches. This study provides a reference for teachers and early childhood education (ECE) institutions to create friendly schools through effective communication.

1 INTRODUCTION

Hate-related incidents massively occur in almost all parts of the world, including in Indonesia. Hatred manifests in various forms, and lately, a lot has happened, especially hatred circulated in social media. In 2016 the Ministry of Communication and Information received 1,769 negative content reports on Twitter, Facebook, and YouTube from January 2016 to early December 2016. Furthermore, the increasing number of sites and social media accounts that spread hatred and fake news is quite worrying.

In terms of culture, hatred originates from a strong stereotypes and prejudice. The forms of stereotypes and prejudices are generally based on differences in ethnicity, gender, religion, politics, aggression, and sex (Sarwono, 2006; Gordijn, Koomen, & Stapel, 2014). Stereotypes and prejudices are both interconnected and mutually reinforcing. Stereotypes and prejudices fuel social conflict and harm, either individually or inter groups, and in some cases perpetuate prolonged wars that cost many lives (Falanga, De Caroli, & Sagone, 2014; Amodio, 2008).

In the context of learning in schools including early childhood education, stereotypes and prejudice can be based on ethnicity, religion, gender and socio-economic class. Stereotypes and prejudice can manifest into bullying, discrimination, marginalisation, mockery and hate speech (Cottrell & Neuberg, 2005; Sarwono, 2006; Olson, 2005; Amodio, 2012).

A solution or an appropriate alternative is needed to handle the negative impact of stereotypes and prejudices in children's learning activities at school. One alternative that can be applied is to implement a child-friendly school-based learning model. This learning model considers the rights of children as well as principles including fulfilling children's educational needs in order to develop their potential, and children's right to grow and develop normally without intimidation, discrimination and violence (UNICEF, 2009; Ahmad, 2016).

Teachers have a significant role in creating a conducive learning or school environment. Related to their role as a facilitator, teachers have a role in making it easy for the children to interact with their learning environment. A conducive learning environment is an environment that provides a safe and comfortable feeling for children to move and learn. There is no discrimination, intimidation and various treatments that discredit children. This paper will examine teachers' communication skills needed to create a child-friendly school.

2 METHOD

The method used in this chapter is the library research, a series of studies relating to the method of data collection library, or research object of research explored through a variety of literature (books, encyclopaedias, journals, newspapers, magazines, and other documents). The research literature or review of the literature is research that examines or critically review the knowledge, ideas, or findings contained in the body of academic-oriented literature. The data used in this research is mailnly based from literatures review.

3 DISCUSSION

3.1 *Child-friendly schools*

Child-friendly schools are schools that consciously strive to guarantee and fulfil children's rights in every aspect of life in a planned and responsible manner. The central principle in creating Child-friendly schools is non-discrimination of interests, rights to life and respect for children. Child-friendly schools are schools that openly involve children to participate in all activities, social life, and encourage child growth and welfare (Solihin, 2015). Furthermore, Solihin stressed that Child-friendly schools are safe, clean, healthy, green, inclusive and comfortable schools/madrasas for physical, cognitive and psycho-social development of girls and boys including children who need special education and/or service education special.

Child-friendly schools are schools that consciously strive to guarantee and fulfil children's rights in every aspect of life in a planned and responsible manner. The central principle is non-discrimination of interests, rights to life and respect for children. As stated in Article 4 of Law No. 23 of 2002 concerning child protection, states that children have the right to be able to live, to grow, develop, and participate fairly according to human dignity and dignity, and protected from violence and discrimination. The right of participation involves the right to have an opinion and to participate in the decision-making process concerning the child's life. Child-friendly schools are schools that involve children to participate in all activities, social life, and encourage child growth and welfare.

The study by Kusdaryani, Purnamasari, and Damayani (2016) emphasises the need for friendly schools through the creation of a culture of friendliness and respect. Furthermore, learning in child-friendly schools has the following indicators that are inclusive, proactive, healthy, safe. Protective climate involves active community participation, child-centred learning, and gender equality. The application of learning also implements activities that invite children to be more active, innovative, creative and fun that are carried out jointly by all learning citizens.

In order to create a conducive atmosphere in learning at child-friendly schools, several aspects need to be considered, notably: (1) appropriate school programs; (2) a supportive school environment; and (3) aspects of adequate infrastructure. (Solihin, 2015). The teacher has a significant role in creating a supportive school environment. One aspect that plays a vital role in creating a supportive school environment is effective communication between teachers and students in the learning process.

3.2 *Effective communication in child-friendly schools*

Educational communication and, more specifically, instructional communication is one aspect of the educational function, whereas interpersonal communication is a form of communication that proceeds from the existence of ideas or ideas of someone's information to others. Effendy (1986) says that interpersonal communication is the process of delivering a message by a communicator to the communicant to change his attitude, views and behaviour. The definition of interpersonal communication based on components is revealed by De Vito (1997: 231) which is the delivery of messages by one person and the reception of messages by other people or a small group of people, with various impacts and with the opportunity to make immediate feedback.

The communication process runs effectively when the communicant well understands the message conveyed by the communicator (Tubbs & Moss, 1996). Furthermore, Tubbs and Moss (1996) suggest that there is five measure of effective communication: understanding, pleasure, influence on attitudes, better relationships and actions.

Effective communication requires a communication competency that is an ability to communicate effectively. This ability includes knowledge about the role of the environment (context) in influencing the content and the form of communication messages, knowledge of procedures for non-verbal behaviour such as touch propriety, loud voice, and physical closeness (De Vito, 1997). Rakhmat (1998) argues that interpersonal communication is declared effective if communication meetings are fun for the communicant. The process of social communication is closely related to the interest that arises among the communicators.

Regarding interpersonal attractions, Barlund (in Rakhmat, 1998) argues that knowing the lines of attraction and avoidance in the social system means being able to predict where the message will appear, to whom the message will flow and moreover how to order will be accepted. In simple language, it means that by knowing who is interested in who or who avoids whom, we can predict the flow of interpersonal communication that will occur. The more interested we are in someone, the higher our tendency to communicate with that person.

De Vito (1997) argues that a pragmatic approach to interpersonal effectiveness sometimes called a competency model focuses on specific behaviours that must be used by communicators to get the desired results. This model offers five qualities of effectiveness namely: confidence (confidence), Unity (immediacy), management interaction (interaction management) power of disclosure (expressiveness) and orientation to other parties (other orientation). Furthermore, De Vito (1997) revealed that, effective communicators have social self-confidence with behavioural characteristics that do not show anxiety, feel comfortable with other people and communication situations in general, are relaxed, not rigid, flexible in sound and gestures, not fixated on a certain tone of voice, controlled, not nervous or awkward. Unity (immediacy) refers to the creation of a sense of togetherness and unity between the speaker and listener. De Vito further revealed that in non-verbal unity communicated by maintaining eye contact, a physical closeness that describes psychological closeness, as well as direct and open body figure (De Vito, 1997).

McCroskey et al. (1995) revealed that immediacy is a behaviour that describes closeness and nonverbal interaction between communicators. These nonverbal behaviours include eye contact, gesture, movement, vocal variations, smile, body position, and touch. The results of his research show that vocal variation, relaxed body position, eye contact and smile are non-verbal aspects of immediacy that get high scores in students' assessment of their teacher. Within effective interaction management, no one feels neglected or important. Each party contributes to overall communication (De Vito, 1997). Furthermore, De Vito (1997) states that the power of disclosure or power of expression (expressiveness) refers to the skill of communicating sincerity in interpersonal interactions. The power of expression is equal to openness with an emphasis on involvement. Such as the expression of responsibility for thoughts and feelings, talking and listening. The power of expression is accounts through the use of variations in speed, volume tones, and sound rhythms to signal involvement and attention. Also, gestures are crucial in showing involvement. Regarding orientation to other parties, De Vito (1997) states that this orientation refers to the ability to adjust to the other person during interpersonal encounters. This orientation includes communicating attention and interest in what the other person is saying.

The teacher is a significant figure in learning. The 2003 Indonesian law of teachers and lecturers states that the teacher should have a set of competencies including pedagogic, professional, social and personal competence. These four competencies are an essential part of strengthening the role of teachers as facilitators in the learning activities. The ability of teachers to communicate became one of the factors supporting the creation of a conducive learning atmosphere, safe, fun and far away from the pressure. Early childhood is the individual that has unique characteristics. Communicating with early childhood learners requires many skills, including greeting, smile, touch and affection.

During the preschool years, children acquire the language and communication skills necessary to express their needs, thoughts, and feelings in social interactions, and they learn to respond appropriately to others. Through effective communication, they also learn to be socially competent individuals, building respectful, positive interactions and relationships with others. According to Erbay, Omeroglu and Cagdas (2012) one of the essential factors for preschool education is teachers with effective communication skills. Effective communication between educators and children influences children's adjustment, development, and their future relationship with teachers.

4 CONCLUSION

Child-friendly schools are learning environments that provide a sense of comfort and security free from discrimination. Child-friendly schools also fulfill the students' needs according to their circumstances. Including a conducive environment so that students especially early childhood will be stimulated all aspects of their development perfectly.

The role of the teacher is critical in creating a child-friendly school. The teacher's ability to manage conducive learning is inseparable from his communication skills. Teachers must be able to give equal attention to all students, foster mutual respect for others, foster motivation and build empathy.

Some of the skills of teachers in communicating with early childhood to create child-friendly schools include placing children as parties who must be respected, respected and treated equally by maintaining eye contact and maintaining an equal body position. Body language in communication must reflect closeness and warmth. Speak in a simple language that is easy to understand and be a good listener.

REFERENCES

Ahmad, D. (2016). *Konsep Sekolah Ramah ANAK Sebagai Penangkal Kekerasan di Dunia Pendidikan [The Concept of Child Friendly Schools as an Antidote to Violence in Education]*. Yogyakarta: Indonesia Education Studies and Research Forum. Retrieved from: http://iesryogyakarta.blogspot.co.id /2016/01/konsep-sekolah-ramah-anak-sra-sebagai.html.

Amodio, D. M. (2008). The neuroscience of prejudice and stereotyping. *Journal of Personality and Social Psychology*, *91*(4), 652–666.

Amodio, D. M. (2012). The social neuroscience of intergroup relations. *Journal of Personality and Social Psychology*, *88*(5):770–789.

Cottrell, C. A. & Neuberg, S. L. (2005). Different emotional reactions to different groups: a sociofunctional threat-based approach to "prejudice". *Journal of American Psychologist*, *61*(8), 741–756.

DeVito, J. A. 1997. *Komunikasi Antar Manusia: Kuliah Dasar [Inter-Human Communication: Basic Lecture]*. Jakarta: Professional Book.

Effendy, O. U. 1997. *Ilmu Komunikasi Teori dan Praktek [Theory and Practice Communication Science]*. Bandung: CV Remaja Karya

Erbay, Omeroglu, & Cagdas (2012). Development and validity-reliability study of a Teacher-Child communication scale. *Educational Sciences: Theory and Practice*, *12*(4), 3165–3172.

Falanga, R., De Caroli, M. E., & Sagone, E. (2014). The relationship between stereotypes and prejudice toward the Africans in Italian University students. *Journal of Personality and Social Psychology*, *56*(1), 5–18.

Gordijn, E. H., Koomen, W., & Stapel, D. A. (2014). Level of prejudice in relation to knowledge of cultural stereotypes. *Procedia-Social and Behavioral* Sciences, *159*. Retrieved from: http://internetsehat.id/ 2017/01/kemkominfo-kembali-blokir-11-situs/.

Kusdaryani, W., Purnamasari, I., & Damayani, A. T. (2016). Penguatan Kultur Sekolah Untuk Mewujudkan Pendidikan Ramah Anak. *Cakrawala Pendidikan Jurnal Ilmiah Pendidikan. Retrieved from* http://journal.uny.ac.id/index.php/cp/article/view/8383.

McCroskey, J. C., Richmond, V. P., Sallinen, A., Fayer, J. M., Barraclough, R. A. (1995). A cross-cultural and multi-behavioral analysis of the relationship between nonverbal immediacy and teacher evaluation. *Communication Education, 44.*

Olson, M. A., & Fazio, R. H. (2006). Reducing automatically activated racial prejudice through implicit evaluative conditioning. *Journal of American Psychologist, 12*(6), 1273–1280.

Rakhmat, J. (1998). *Psikologi Komunikasi.* Bandung: PT. Remaja Rosdakarya.

Sarwono, W. S. (2006). *Psikologi Prasangka Orang Indonesia Kumpulan Studi Empirik Prasangka Dalam Berbagai Aspek Kehidupan Orang Indonesia.* Jakarta: PT Raja Grafindo Persada.

Solihin, A. (2015). *[Introducing and Developing Child Friendly Schools].* Retrieved from https://visiuniversal.blogspot.co.id/2015/09/mengenal-dan-mengembangkan-sekolah.html.

UNICEF. (2009). *Model Gerakan Sekolah Aman, Sehat, Hijau, Inklusif, Ramah Anak dan Menyenangkan [The School Movement Model is Safe, Healthy, Green, Inclusive, Child Friendly and Fun].* Indonesia: Yesforsaferschool.

Early Childhood Education in the 21st Century – Yulindrasari et al. (eds)
© 2020 Taylor & Francis Group, London, ISBN 978-1-138-35203-2

Enhancing children's ecological literacy skill through traditional games: An overview from Indonesian kindergartens

E. Silawati, D.N. Muliasari & W. Ananthia
Universitas Pendidikan Indonesia, Cibiru Campus, Indonesia

ABSTRACT: This chapter is an exploratory study that describes some Indonesian traditional playing as learning strategy in kindergarten setting to develop ecological literacy skill of children. Ecological literacy skill is crucial ecause it is believed to be one of the solutions to overcome global warming. The ability to understand the basic principles of ecology and to live accordingly should be developed from early years as children have the potential skill of curiosity and affinity for nature. Play-based learning is assumed to be a suitable learning strategy as children love playing. The learning is implemented in a form of games. Furthermore, Indonesia is a rich country with its natural resources and cultural values in which hundreds of tribes are live and has many different kinds of traditional games that suitable for children. Some of the games have natural content values that promote ecological literacy skills to children. Therefore, this study explores its implementation through qualitative approach to analyze its potential as learning strategy in Indonesian kindergarten setting. The study involved two kindergartens. The analysis identified that some Indonesian traditional playing namely *icikibung, babalonan, leuluetakan, huhuian* and *orayorayan* have potential as learning strategies in developing some aspects of ecological literacy skills in children. The five traditional games were then implemented in the kindergartens as learning strategies. The result shows that the five traditional games that have been implemented in the kindergartens are effective in developing ecological literacy of children in terms of understanding key concepts and ecological connectivity, appreciate the links between human action and the environment, and introducing the children about how to care and share natural resources, such as water, air, land and energy. However, the element of thinking scientifically about ecological issues has not fully explored yet by the using of traditional games as the learning strategy. Therefore, further research is needed to investigate this matter.

1 INTRODUCTION

One of the most urgent issues today is saving our environment from destruction. Global warming and pollution causes ecological destruction (Goleman, Benett & Barlow, 2012). Indonesia is tropical country and it faces environmental problems including deforestation, air pollution, water pollution, soil pollution, and other problems that will affect the environment globally. Therefore, we should take some actions to overcome this problem before environmental systems spin out of control and crack.

There are a lot of strategies that can be implemented in conserving the environment. One of the strategies is the implementation of environmental education to develop the ecological literacy skills of students. Moreover, UNESCO (1976) has stated concept of environmental education which include ecological literacy as one its skill elements Ecological literacy is the ability to understand the basic principles of ecology and to live accordingly (Capra, in Stone & Barlow 2005). These skills should be promoted in a child's early years, because children have potential of affinity for nature, known as "biophilia" (Callenbach, 2005). When people have skills in ecological literacy, natural sustainability will hopefully increase. Therefore, in the coming decades, the survival of humanity will depend on human's ecological literacy (Capra, 2002).

Consciousness of the importance of nurturing values and attitudes in young children which relate to sustainability has occurred as awareness and concern grows in response to climatic and ecological changes. In this case, some countries have conducted environmental education programs, for example; China's Green Schools Project, Sweden's Green School Award Programs, Australian Sustainable Schools Programs and New Zealand's Enviro-Schools Programs (Maxwell & Mawson, 2015).

In Indonesia, the program of Go Green School starts growing but but mostly, it is - applied in the level of primary school to high school. Consequently, the development of educational environment learning strategy for children of early years in Indonesian context which has their own characteristic in learning strategies is also needed to be developed.

Children love playing and the basic principle of early childhood education (ECE) learning strategy is playing (Dockett & Fleer, 2000). Playing has become an important topic in ECE because playing is the important phase of optimizing development of cognition, affection, emotional personality and sensory of perception (Dockett & Fleer, 2000). There are efforts to explore objects, people, events, space and time when playing to develop children's imagination and creativity (Tedjasaputra, 2002). Playing is a mean of transferring values as well as sustainability from adult to children (Alif, 2013). Moreover, the environmental and cultural sensitivity skills of children can be enhanced through playing (Alif & Retno, 2009; Milfont & Duckit, 2009; Nitecli & Chung, 2016), specifically traditional playing. Some findings have proposed that traditional playing, which is part of local culture, gives positive influence to developmental aspects of the children (Silawati & Ardiyanto, 2014). Through playing, children will gain some concepts of knowledge automatically. Traditional playing shows some cognitive, motoric, social and emotional practices (Muliasari, 2017) subconsciously. Thus, traditional playing has the potential to be developed as a learning strategy and it has local culture learning elemants that are able to develop eco-literacy in children.

In this study, the play-based learning stategies are implemented in the form of games, as these have sets of rules. Henceforth, the five traditional playing are referred to as games. The research explored five Sundanese traditional games: *icikibung, babalonan, leuluetakan, huhuian* and *orayorayan*. Sundanese is one of the tribes in Indonesia. The five Sundanese traditional playing games were implemented as a learning strategy for children in kindergarten to develop their ecological literacy.

Hence, the study is important to be investigated as the need to preserve the environment is urgent and it has more impacts on children.

2 METHOD

Qualitative approach is employed in this research. It described in detail the implementation of the five traditional games in the learning process (Creswell, 2009). The study involved two kindergartens in Bandung, Indonesia, with two teachers and 30 students. Both kindergartens were chosen as they have a local culture program on their curriculum and they were willing to participate in the research. The data is gathered through observation and document analysis. The document analysis was conducted through focus group discussion. The purpose was to identify the traditional games that carried the ecological issues and was suitable for children. The focus group discussion involved two experts on Indonesian traditional playing: Mr. Zaini Alif and Mr. Solihin Ichas Hamid. As a result, *icikibung, babalonan, leuluetakan, huhuian* and *oray-orayan* were selected for this study.

Observation was undertaken to obtain data for the following purposes: (1) to understand key concepts and ecological connectivity, (2) to think scientifically about ecological issues, (3) to appreciate the links between human action and the environment, and (4) to introduce the children to the concepts of sharing and caring about natural resources like water, air, land and energy (Reynolds & Lowman, 2015; Dasgupta, 2011). In addition, a thematic approach was used to analyze the data, identifying patterns and grouping them into particular themes (Braun & Clarke, 2006). The themes were selected based on the four elements of ecological literacy skill.

3 RESULTS AND DISSCUSION

3.1 *Developing children's ecological literacy through Indonesian traditional games: An analysis*

Indonesia has hundreds of traditional games with each version based on different regions. H. Overback identifies 690 Indonesian traditional playing (Suyanto, 2005). It was recognised that most traditional games characterized as girls' games. The first stage of this project is analyzing Indonesian traditional games that suitable for developing ecological literacy of the children. The data is gathered through literature study and focus group discussion. Focus group discussions involved two experts of Indonesian traditional playing and the analysis of traditional playings based on ecological literacy elements proposed by Reynolds and Lowman (2015) and Dasgupta (2011).

The findings show that there were 24 games that have -potential to be used to develop ecological literacy skill of children. Most of the games have one or two elements of ecological literacy. All of the elements of ecological literacy are potentially covered by the implementation of Indonesian traditional games. However, none of the games were used by the teachers to develop the children's scientifical thinking of ecology. In fact, some of the games could be implemented in enhancing the children's awareness about the ecological issues. For example, in the game *Ngala Papatong* (catching dragonflies), children could learn that dragonflies live in fresh and non-polluted air. When they could not find a dragonfly, children learned that the air is not as clean as it should be. Consequently, this will raise their awareness of the issues of air polution. The next step, children could be asked for their participation in maintain clean water and fresh air

Based on time consideration and games that facilitated the five elements of ecological literacy skill, the researcher specified five gamesto be implemented on learning activities, The local values carried in the playings were related to the children's daily life and would definitely made the learning meaningful. Below is the description of each game.

3.1.1 *Icikibung*

Icikibung is a game that usually played by children who live near a river. *Icikibung* means a sound produced from when children's palms and elbow hit water. The talented children, who has musical sense could make musical harmony from the sound of water. Through this game, children are introduced to water and its characteristics (Alif & Sachari, 2016). This interaction is believed can build children's environmental sensitivity to water conservation.

In a kindergarten setting, children did not only play *Icikibung* at the river but also played it in some big buckets. In the first activity, children playing water in the big buckets. Some children played freely and some of them played based on their teacher's instruction to make musical harmony. This activity used clean water. Next, children were taken to the river nearby. Unfortunately, the water of the river was polluted so the children could not play *icikibung*. They asked what had happened with the water: why was the colour dirtywhy was the smell bad. The teacher explained that the water had been polluted by textile factories which were located around the river. From this activity, children could differentiate between clean water and polluted water. Hopefully this will lead the children to develop environmental sensitivity and learn to conserve clean water.

3.1.2 *Babalonan*

Babalonan is a game that uses traditional fabric (sarong) as the media and usually this game is played on a field. There are two playing strategies. First, the sarong is thrown into the air so that it forms a balloon. Second, one side of sarong is tied to a player's waist and another side is held up by their hands, and then the players run so that the sarong forms a balloon.

In a kindergarten setting, this game was used as strategy to introduce air and its characteristics. This activity also promoted the functions and advantages of air. Children conducted two strategies of playing at the yard near the school. They could feel air pressure when they ran with the sarong at their back. The teachers explained the characteristics, functions and advantages of air after the activity.

3.1.3 *Leuluetakan*

Leuleutakan is a game that facilitates interaction between children and soil. This game is usually played in an unplanted rice field. There are no rules in this game; children just play freely at the rice field, which full of mud. This game has two goals such as for having fun an for competition (Alif & Retno, 2009).

In a kindergarten setting, children did not go to the rice field to play the game, but teachers prepared some mud and clay in the classroom. Some activities arranged by the teachers, such as playing freely with the mud or making some forms from the clay. through this game, children have experience to explore the mud and soil. In the last part of the activity, teacher explained that soil is part of the ecosystem that should be protected. It is the place for growing the food that we consumed. If the soil is polluted so the vegetables or fruit will be also polluted. From this exploration, children will have effort to protect the environment.

3.1.4 *Huhuian*

Hui means sweet potatoes. In this game, some children make a line while sitting and -hugging each other to pretended as they were sweet potatoes that was grew by a farmer. A child is chosen to become *ucing* (main player). His/her main aim is to separate each child from the group one by one. Usually the game is played at field. While playing, the children usually sing a Sundanese song:

> Kelenang-keleneng samping koneng
> Keledat-keledut samping butut
> *(pointed child asks question):* *saha eta?* (who is that?)
> Answer: *Perdadi (*landlord)
> Question: *arek naon*(what do you want?)
> *Answer:* *menta hui* (I want sweet potatoes)
> Question: *kekakarek macul*(the ground just start preparing)
> (because it still hacking the ground using hoe the pointed child is return again while singng the song)
> *Kelenang-kelening gogorosak* (because the ground just cultivating, so *ucing* starts go around and asking the same question)
> Question: *saha eta?*

The question and answers is started from cultivating process of sweet potatoes, then growing the leaves, growing the seed until the sweet potatoes are ready to be harvested Then one child is pulled away from the line. If the main player succeeds in taking a child, then other child will become the main player.

From the lyrics of the song, children can learn the process of planting sweet potatoes, starting from preparing the ground through the harvest. This game promotes the system of life cycles, one of the important factors of sustainability. The children also learn that there is a process of grow food. They will not perceive anymore that sweet potatoes come from market or a refrigerator. Furthermore, teachers gave reinforcement by explaining that to get the good quality of sweet potatoes, we need to keep the soil fertile and keep it away from the pollution.

3.1.5 *Oray-orayan*

The principle of this game is answering questions between players. This game is played by 7 to 20 players at field. The rule is like London Bridge gameIn which there are children (usually eight or more), two of whom join hands high to fom an arch (the gridge). The other players march under the bridge, each holding onto the waist of the player in front. This game does not need any material or tools.

This game was started by making a line of the players. Children put their hand on other shoulders. The front child pretend as the head of the snake,and the backmost child pretend - as the tail of the snake. After line is ready, all the children walk sway just like the real snakes and sing an answer-question song. The player who pretend as head of the snake attempts to eat the player that pretend as tail of the snake. The children in the middle of line try to adjust the movement. When the line breaks, children try to have it back. When the head succeeds in bitting the tail, the child at the back, as the tail of the snake, has to stop playing. But he still can participate by singing the song. When the line gets shorter, the game is over. There are many version of songs for this games, such as Indonesian or javanesse version but the most famous song of this game is from Sundanese, arranged by an expert in Sundanese culture, Mang Koko Koswara:

Y: oray-orayan luar leor kasawah,
X: entong ka sawah parena keur sedeng beukah,
Y: oray-orayan luar leor kakebon,
X: entong kakebon, di kebon loba keur ngangon
Y: mending ge teu leuwi,
X: di leuwi loba nu mandi,
Y: saha nu mandi,
X: Nu mandina pandeuri...
Kok...kok...kok....

Through these lyrics, children learn about the ecosystem of snakes and the consequences if humans disrupt it. This game promotes understanding key concepts and ecological connectivity as well as appreciating the links between human action and the environment.

In a kindergarten setting, children play this game at the yard next to the school. After playing three rounds, teachers explained that snakes is part of our ecosystem that must be maintained

3.2 *Indonesian traditional games as learning strategy in kindergarten: The implementation*

3.2.1 *Planning the learning*

In implementing the playing, teachers used some stages in learning planning, namely (1) determining developmental stages in particular ages, (2) determining developmental indicator that will be developed, (3) analysing learning materials that will be introduced to the children, (4) determining learning themes, (5) arranging learning activities, and (6) developing learning media, as prescribed by Dirjen PAUD (2012). Those processes were conducted by the teachers before the learning activity. The teachers arranged traditional games as the main activity. The playing is adjusted with the kindergarten's learning themes.

Meanwhile, a lesson plan in the context of ECE is very important in order to ensure that all developmental aspects is developed in every daily activities through students' active involvement (Curtis & O'Hagan, 2003). Teachers are expected to ensure that the planned activities are going to go smoothly through well-organised preparation. The ability to design activities is one of the indicators this.

3.2.2 *Learning activity*

There are three stages of learning activity: opening activity, main activity and closing activity. In the opening activity, the teachers prepared an ice-breaking game related to the learning theme and then explained the game that was to be played. The main activity was started by teachers' explanation of the rules of the game to be played. Then children played the game, guided by the teacher. In the end activity, children shared their experiences of the game and teachers gave some reinforcement related to environmental topics.

3.2.3 *Learning evaluation*

The evaluation process used authentic assessment approach, because it is a true test that requires students to demonstrate their deep understanding, higher-order thinking, and complex problem solving through the performance of exemplary tasks (Koh, 2016). The data for evaluation process was gathered from children's responses, performances and products. The output of learning evaluation was also discussed in the focus group discussion. The teachers confirmed that the five Indonesian traditional games that were implemented as the learning strategies are able to develop children's ecological literacy in terms of the elements proposed by Reynolds and Lowman (2015) and Dasgupta (2011), namely understanding key concepts and ecological connectivity, appreciating the links between human action and the environment, and introducing the children about how to care and share natural resources, such as water, air, land, and energy. However, the teachers found difficulties in exploring the second element: thinking scientifically about ecological issues. They think that the second element is the biggest challenge to be communicated to and with the children.

4 CONCLUSION AND RECOMMENDATION

Traditional games have the potential to be used as learning strategies in promoting ecological literacy skills to children. Based on data analysis, it is found that there are 24 traditional games that could be effective in developing children's ecological literacy skills. Five, *Leuleutakan, Icikibung, babalonan, Huhuian*, and *Oray-orayan*, were modified and implemented. This study reveals that games can be effective in facilitating children' ecological literacy skills. However, teacher have difficulties in developing the sub-skills, namely thinking scientifically about ecological issues. In fact, all of these five Indonesian traditional games have the potential to develop all of the elements of ecological literacy skills. Therefore, it is recommended that teachers are given training to sharpen their sensitivity to the current issues related to environment and nature.

REFFERENCES

Alif, Z. (2013). *Pendidikan karakter dalam mainan dan permainan tradisional jawa barat*. Bandung: Pemprov Jabar.

Alif, Z. & Retno. (2009). *Permainan rakyat jawa barat dalam dimensi budaya*. Bandung: Pemerintah Provinsi Jawa Barat.

Braun, V. & Clarke, V. (2006). Using thematic analysis in psychology. *Qualitative Research in Psychology, 3*(2), 77–101.

Capra, F. (2002). *The Hidden Connection, Integrating the Biological, Cognitive and Social Dimension of Life into a Science of Sustainability*. New York: Doubleday.

Callenbach, E. (2005). *Values Ecological literacy, educating our children for a sustainable world*. San Francisco: Sierra Club Books.

Creswell, J. W. (2009). *Qualitative, quantitative, and mixed method approaches* (3rd edn.). Thousand Oaks, CA: Sage Publications.

Curtis, A., & O'Hagan, M. (2003). *And education in early childhood, a students' guide to theory and practice*. London: Routledge Falmer.

Dasgupta, S. (2011). *Environmental education-a book of activities*. New Delhi: Center for Science and Environment.

Dockett, S., and Fleer, M. (2000). *Play and pedagogy in early childhood: Bendinng the rules*. Sydney: Harcourt.

Goleman, D., Bannett, L. & Barlow, Z. (2012). *Eco literate, how educators are cultivating emotional, social and ecological intelligence*. San Fransisco: Jossey-Bass.

Koh, K. H. (2016). Authentic assessment. *Oxford Research Encyclopedia of Education*.[online]. Available at: http://education.oxfordre.com/view/10.1093/acrefore/9780190264093.001.0001/acrefore-9780190264093-e-22. [Accessed 12 Februari 2018]

Maxwell, A., & Mawson, B. (2015). A kindergarten's journey into sustainability through the enviroschools in the early years programme. *New Zealand Journal of Teachers' Work, 12*(1), 14–29.

Milfont, T. L. & Duckitt, J. (2010). The environmental attitudes inventory: A valid and reliable measure to assess the structure of environmental attitudes. *Journal of Environmental Psychology, 30*(1), 80–94.

Muliasari, D. N. (2013). Promoting critical thinking through children's experiential learning. *Advances of Social Science, Education, and Humanities (ASSEHR), (58), 3rd International Conference of Early Childhood Education.*

Nitecli, E., & Chung, M-H. (2006). Play as place: a safe space for young children to learn about the world. *International Journal of Early Childhood Environmental Education, 4*(1), 25–26.

Reynolds, J. A., & Lowman, M. D. (2015). Promoting ecoliteracy through research service-learning and citizen science. *Ecoliteracy*. The Ecological Society of America.
Silawati, E., & Ardiyanto. (2014). *Sundanesse Traditional Playing as Learning Strategies in developing children's language skills*. A paper presented at the 6th UPI-UPSI International Conference on Teacher Education. A Proceeding.
Suyanto, S. (2005). *Dasar-dasar pendidikan anak usia dini*. Yogyakarta: Hikayat Publishing.
Stone, M. K., & Barlow, Z. (2005). *Ecological literacy, educating our children for a sustainable world*. San Francisco: Sierra Club.
Tedjasaputra, M. S. (2001). *Bermain, mainan dan permainan*. Jakarta: Gramedia.
United Nations Educational, Scientific, and Cultural Organization and United Nations Environment Programme (UNESCO-UNEP). (1976). The Belgrade Charter. Connect: UNESCO-UNEP Environmental Education Newsletter 1, 1–2.

Early Childhood Education in the 21st Century – Yulindrasari et al. (eds)
© 2020 Taylor & Francis Group, London, ISBN 978-1-138-35203-2

The identification of eco-literacy practices in early childhood education

I. Anggraeni & B. Zaman
Universitas Pendidikan Indonesia, Bandung, Indonesia

ABSTRACT: The purpose of this paper is to introduce eco-literacy education practices in early childhood education. This is based on being able to increase the chances of children to understand the crisis that exists in the future, by being positive about ecological issues, and helping them to make the transition to a sustainable way of life. In this study, a literature study of various sources was conducted related to eco-literacy learning in children of early age. The results of the literature study show that learning in early childhood can be used as a first step to introduce eco-literacy by using approaches that are appropriate to the stages of children's cognitive development, namely through habituation to daily activities, and through real applications in the surrounding environment. The results of this paper can be used as study material and an information source on how to practice eco-literacy education

1 INTRODUCTION

Environment is where all living things coexist. If one of these components is damaged, the environment of living things in it will be affected (Irwan, 2010). For example, forests and mountains that experience deforestation, logging and destruction, also damage the system and place of storage (and filtering and retaining) of water, resulting in environmental damage such as landslides and soil erosion.

The natural damage phenomenon that occurs is not new. This situation is getting worse with the ignorance of our nation towards its environment (Supardi, 1994). The attitude of ignorance is generally caused by nature/individual talents themselves and the insistence of needs beyond their abilities, both in the economic field and in the socio-cultural field. Humans tend to exploit natural resources haphazardly without regard to the effects they cause (Rusmana & Akbar, 2017). Progress in development that is not accompanied by awareness of the environment is a clear example of human attitudes that exploit natural resources in a haphazard manner. Because of that, humans have a role in maintaining the environment so that the survival of the next generation remains safe (Juhriati, 2016).

Jucker (2012), Gaziulusoy & Carol (2012) refer to the above phenomenon as an ecological crisis. Where this ecological crisis shows unsustainable community behavior. The emergence of a sustainable concept is a form of understanding that brings people as agents of change (Duhn, 2012). This sustainable concept is a key concept in the ecological movement, where awareness arises to safeguard the surrounding environment and an understanding of the insights of ecological principles called eco-literacy. Eco-literacy needs to be developed/grown so that people understand how the ecosystem on earth supports human life in meeting their daily needs, so that people's lives can be sustainable. Unfortunately the eco-literacy attitude is increasingly eroded. This can be seen from a series of global problems that endanger the biosphere and human life (Capra, 2002).

Education has a major role in helping people to make the transition to a sustainable way of life (Elliott & Julie, 2009). One way to overcome ecological challenges is through education (Hattingh, 2001; Sterling, 2008; Manteaw, 2012; Swilling & Annecke, 2006).Environmental Education in Indonesia has been sought by various parties since the early 1970s (Landriany,

2014). So far, the implementation of Environmental Education has been carried out by each of the environmental education actors separately.

The concept of continuing education (sustainable) or initial planting program (love the environment), should be promoted at all levels of education. However, it is generally applied during years of elementary to high school education (Juhriati, 2016). As some of the results of eco-literacy research by Prasasti (2017), Widiani (2017), and Mamu (2017) in Indonesia, revealed that activities such as farming, utilization of plastic waste, and eating healthy activities can increase the eco-literacy of elementary students, various aspects of knowledge, awareness and skills through project learning.

While in Indonesia alone, research on eco-literacy in the ECE sector is still relatively small. In line with Elliott and Julie (2009) in his research revealed that the early childhood education sector experienced delays in being involved in continuing education (planting environmental programs). There is little research in addressing how the concepts of sustainability in Early Childhood Education (ECE) may be understood differently in different social and cultural contexts (Kim, 2016). In the social and cultural view children are seen as less competent in solving problems and making decisions in their own environment. Even though children have strong curiosity about everything around them and have an attitude of adventure and strong interest in observing the environment.

Pangeti (2013) in her research revealed that learning in early childhood can be used as a first step to introduce the environment by respecting and loving the environment or eco-literacy, and increasing their chances of understanding the future crisis by being positive about ecological issues. Planting environmental love from the beginning is needed because children are in the golden age, where the development of intelligence at this time has increased to 50% (Jalal, 2002). This period is a sensitive period for laying the initial foundation in developing physical, cognitive, language, art, social emotional abilities, self-discipline, religious values, self-determination and independence (Sujiono, 2004). In the context of child development, early childhood is not only in a period of rapid physical growth, but also a time of extraordinary mental and cognitive development. Cognitive development has a very positive effect on the future. In addition, children can understand the issues that are relevant to their lives and children become active agents who can build their own meaning and identity. One of the issues is the issue of ecology, thus childhood is an important period to develop basic habits, norms, dispositions, values, and lifestyle (Davis, 2008). Based on the explanation above, the author intends to know deeply about how eco-literacy education practices in ECE.

2 METHOD

The approach used in this study is a qualitative review approach that is literature review, to analyze trend issues or research topics that have been predetermined and analyzed according to relevant scientific developments (Kuang & Maya, 2015). The steps that can be carried out are as follows: 1) used textbooks at this stage an assessment of issues related to eco-literacy is carried out. 2) related research is an effort to find research results related to the issues to be studied by reading the latest research journals and relevant both national and international peer-reviewed journals that are accessed manually or by online aceses. The journal came from two studies conducted in Indonesia (Rusmana & Akbar, 2017; Setyowati, 2013) and five studies abroad (Carr & Luken 2014; Johnson, 2014; Pangeti, 2013; Reynolds, & Lowman, 2013). 3) state-of-the-art research, namely analyzing theories obtained from book sources, the results of research both from journal sources and proceedings to be synthesized (Cresswell, 2014).

3 RESULTS AND DISCUSSION

The results of the literature review show two main themes namely eco-literacy learning through habituation to daily activities, and eco-literacy learning through real applications in the surrounding environment.

3.1 *Eco-literacy learning through habituation to daily activities*

The development of education regarding ecological literacy begins with the cultivation of environmental awareness in children. Growing ecological literacy in children can be realized through environmental care in children. Eco-literacy in children needs to be grown from an early age. This can be done in everyday life. The attitude of eco-literacy begins with the emergence of environmental awareness in children, after which behavior arises that is responsible for maintaining the environment around children (Johnson, 2014). The following are indicators of environmental care in early childhood, are 1) always maintaining the preservation of the surrounding environment; 2) not taking, cutting or uprooting plants along the road; 3) not scribbling, writing inscriptions on trees, rocks, roads or walls; 4) always throw garbage in its place; 6) carry out cleaning activities; 6) hoard used goods; 7) clean up the garbage that clogs the waterways (Nenggala, 2007). In addition, Hyun (2000) explained that the easiest example we can do is to maintain a clean environment, not littering, not damage trees carelessly, inviting children to start loving plants with gardening. Give them the responsibility to look after the plants they plant themselves. All of that is an effort to plant the love of the child's environment.

Instilling the love of the environment for children from the beginning, we will indirectly also contribute to saving the environment on earth so that it does not become more damaged and polluted. So that one day later it will be a healthy generation and a comfortable environment (Johnson 2014). Children who have an ecolithic attitude expected to have awareness, sensitivity (awareness) that the environment needs to be maintained, managed and utilized not only for now but for future generations who have the right to enjoy it too. Therefore it must be accustomed as early as possible the things that make our children to be literate to be ecological. For example, recycling waste around us, such as used cartons, mineral water products, instant noodles, milk cans, plastic bottles, plastic packaging into reusable items, planting ornamental plants, planting live pharmacies in pots, For example, recycling trash around us, such as used cartons of mineral water products, instances of noodles, milk cans, plastic bottles, plastic packaging into items that can be reused, planting ornamental plants, planting live dispensaries in pots or planting vegetables in the yard (Setyowati, 2013). Based on the description above, the eco-literacy attitude in children further needs to be applied to daily life through habituation. In line with what was expressed by Nuryati (2008) revealed that behavior is the result of learning processes carried out by individuals from their environment through habituation, observing and imitating

3.2 *Eco-literacy learning through real applications in the surrounding environment*

Eco-literacy planting in children cannot only consist of the lecture method, but it requires concrete actions that can really be seen and where benefits are felt. Reynolds and Lowman (2013) explained that the approach that can be taken to improve ecological intelligence is one of them by means of learning research, which is an approach where students not only learn about concepts, but learn through real applications through the surrounding environment. Based on this, it can be analyzed that planting eco-literacy in students can be done in various ways, namely by planting concepts and applying them in solving problems in everyday life in the surrounding environment. In other words, eco-literacy does not only require mastery of subject matter, but of creating meaningful relationships between thoughts, hands and also the heart where there must be a concern and it must be continued to be developed (Orr, 1992).

There are many ways to teach children to love the environment or eco-literacy, Freuder (2006) in his research in the United States revealed that introducing eco-literacy in early childhood can be done in an outdoor playing environment. As the teacher uses the school yard to introduce the absorption of sunlight as a process of photosynthesis and also introduces the function of the tree as one of the obstacles to heat from the sun. In harmony with the research of Carr & Luken (2014) in the United States that offers playscapes as an alternative to environmentally friendly playgrounds to engage in physical activity, investigate scientific principles and improve development in all domains through natural play.

4 CONCLUSION

Early childhood education should include environmental literacy so that the children are aware of the importance of loving the universe by instilling eco-literacy. Eco-literacy for early childhood is to introduce children's awareness to maintain the surrounding environment by using developmentally appropriate approaches. The above-mentioned eco-literacy learning practices are a reference in the learning process in fostering an eco-litercy attitude in early childhood. With the eco-literacy attitude that arises in the child so that the child is able to care for and protect from damage in order to aid in the creation of a sustainable life.

REFERENCES

Capra, F. (2002). *Jarring-jaring kehidupan visi baru epistimologi dan kehidupan.* Translated by Saud Pasaribu. Yogyakarta: Fajar Pustaka Baru.

Carr, V. & Luken, E. (2014). Playscapes: a pedagogical paradigm for play and learning. *International Journal of Play* 3(1): 69-83. http://dx.doi.org/10.1080/21594937.2013.871965.

Cresswell, J.W. (2014). *Research design: Qualitative, quantitative, and mixed methods approaches (Fourth).* Los Angeles, London, New Delhi, Singapure: Sage Publications.

Duhn, I. (2012). Making 'place' for ecological sustainability in early childhood education. *Environmental Education Research, 18*(1), 19-29.

Davis, J. M. (2008). "What might education for sustainability look like in early childhood? A case for participatory, whole-of-settings approaches. In I. P. Samuelsson & Y. Kaga (Eds.). *The contribution of early childhopuod education to a sustainable society* (pp. 18–24). Paris: UNESCO.

Elliott, S, & Julie, D. (2009). Exploring the resistance: An Australian perspective on educating for sustainability in early childhood. *International Journal of Early Childhood,* 41(2), 65. https://doi.org/10.1007/BF03168879.

Freuder, T. G. (2006). *Designing for the future: Promoting ecoliteracy in the design of children's outdoor play environments.* Dissertation. Virginia Tech.

Gaziulusoy, A. I., & Carol, B. (2013). Proposing a heuristic reflective tool for reviewing literature in transdisciplinary research for sustainability. *Journal of Cleaner Production,* 48, 139-147.

Hattingh, J. P. (2001). *Conceptualizing ecological sustainability and ecologically sustainable development in ethical terms: Issues and challenges.* Stellenbosch: Stellenbosch University.

Hyun, E. (2000). Ecological human brain and young children's "naturalist intelligence" from the perspective of developmentally and culturally appropriate practice (DCAP). Presented at the Annual Conference of the American Educational Research Association. New Orleans, LA: American Educational Research Association. https://eric.ed.gov/?id=ED440749.

Irwan, Z. D. (2010). *Prinsip-prinsip ekologi: Ekosistem, lingkungan dan pelestariannya.* Jakarta: Bumi Aksara.

Jalal, F. (2002). The golden age: masa efektif merancang kualitas anak. Jakarta: Direktur Jenderal Pendidikan Luar Sekolah dan Pemuda Departemen Pendidikan Nasional.

Johnson, K. (2014). Creative connecting: Early childhood nature journaling sparks wonder and develops ecological literacy. *International Journal of Early Childhood Environmental Education, 2*(1), 126-139.

Juhriati, I. (2016). Analisis pembelajaran dalam menumbuhkan ecoliteracy Anak usia dini di taman kanak-kanak firdaus. Master thesis, Universitas Pendidikan Indonesia.

Jucker, R. (2012). Sustainability? Never heard of it!" Some basics we shouldn't ignore when engaging in education for sustainability. *International Journal of Sustainability in Higher Education, 3*(1), 8–18.

Kementerian Lingkungan Hidup. (2010). *Pedoman penggunaan kriteria dan standar untuk aplikasi daya dukung dan daya tampung lingkungan hidup dalam pengendalian perkembangan kawasan.* Jakarta: KLH.

Kuang, C. H., & Maya K. D. (2015). Basic And Advanced Skills They Don't Have: The Case Of Postgraduates And Literature Review Writing. *Malaysian Journal of Learning and Instruction, 12*: 131-150.

Kim, S. (2016). *A comparative study of early childhood curriculum documents focused on education for sustainability in South* Korea *and* Australia. Dissertation. Queensland University of Technology.

Landriany, E. (2014). Implementasi kebijakan adiwiyata dalam upaya mewujudkan pendidikan lingkungan hidup di SMA Kota Malang. *Jurnal kebijakan dan pengembangan pendidikan, 2*(1), 82-88.

Lickona, T. (2013). *Educating for character, Mendidik untuk membentuk karakter: Bagaimana sekolah dapat memberikan pendidikan tentang sikap hormat dan tanggung jawab.* Translated by Juma Abdu. Jakarta: Bumi Aksara.

Manteaw, O. O. (2012). Education for sustainable development in Africa: The search for pedagogical logic. *International Journal of Educational Development* 32(3): 376-383.
Mamu, A. T. T. S., (2017). Peningkatan ecoliteracy siswa dalam pemanfaatan sampah pelastik melalui pembelajaran project based learning pada pembelajaran ilmu pengetahuan sosial. Tesis, Universitas Pendidikan Indonesia.
Nenggala, A. K. (2007). *Pendidikan jasmani olahraga dan kesehatan*. Bandung: Grafindo Media Pratama.
Nuryati, L. (2008). *Psikologi Anak*. Jakarta: PT. Indeks.
Orr, D.W. (1992). *Ecological literacy: Education and the transition to a postmodern world*. New York: SUNY Press.
Pangeti, P.R.R. (2013). *Towards sustainable futures: exploring ecological learning in early childhood development*. Dissertation. Stellenbosch: Stellenbosch University
Prasasti, R. (2017). *Peningkatan ecoliteracy siswa dalam mengkonsumsi makan sehat di sekolah melalui model pembelajaran project based learning pada pembelajaran IPS*. Master thesis. Universitas Pendidikan Indonesia.
Reynolds, J.A., & Lowman, M.D., (2013). Promoting ecoliteracy through research service-learning and citizen science. *Frontiers in Ecology and the Environment, 11*(10), 565-566.
Rusmana, N. E., & Akbar, A. (2017). Pembelajaran ekoliterasi berbasis proyek di sekolah dasar. *Jurnal Edukasi Sebelas April, 1*(1), 1-12.
Setyowati, T. (2013). Peran keluarga dalam membentuk karakter go green untuk mencegah global warming pada anak usia dini. *Jurnal Penelitian Ilmu-Ilmu Eksakta: Agri-tek, 14*(1), 100-108.
Supardi, I. (1994). *Lingkungan hidup dan kelestariannya*. Bandung: Penerbit Alumni.
Sujiono, Y. N. (2004). *Konsep dasar pendidikan anak usia dini*. Jakarta: PT. Indeks.
Suwasono, H., Suemitro, S.B., & Soekartomo, S. (1986). *Pengantar Ekologi*. Jakarta: Rajawali.
Sterling, S. (2008). Sustainable education-towards a deep learning response to unsustainability. *Policy & Practice-A Development Education Review* 6.
Swilling, M. & Eve, A. (2006). Building sustainable neighbourhoods in South Africa: learning from the Lynedoch case. *Environment and Urbanization, 18*(2), 315-332.
Widiani, D. (2017). Peningkatan *ecoliteracy* siswa dalam bertanam melalui project based learning pada pembelajaran ilmu pengetahuan sosial. Master thesis, Universitas Pendidikan Indonesia.

Early Childhood Education in the 21st Century – Yulindrasari et al. (eds)
© 2020 Taylor & Francis Group, London, ISBN 978-1-138-35203-2

Criticizing the phenomenon of shark exploitations using a critical literacy approach in primary schools

I. Rengganis, Teguh Ibrahim, Mela Darmayanti & Winda Marlina Juwita
Universitas Pendidikan Indonesia, Bandung, Indonesia

ABSTRACT: This research aims is to describe the practice of critical literacy-based learning in primary schools using problem-posing education. This education concept seeks to deal with people with problematic phenomena that disrupt the balance of human life to be addressed critically. The phenomenon that will be criticized is "shark fins, massive exploitation by humans". The critical literacy products produced in this study are illustrated images that are inserted in the framework of critical arguments. Critical literacy learning aims to make people aware of their roles as agencies that always strive for the emancipation of life in its various existential dimensions. The method of this research is a collaborative action research conducted with primary school teachers. The subject of this study is students of the 5th grade at SD X which consists of 20 students. The results of the study show that critical literacy learning using problem-posing education consisted of three stages, namely (1) pre-reading (problematization), (2) reading (critical discourse discussions), (3) post-reading (social transformative actions). Illustrated images and essays produced by students show some positive indications, namely: (1) students are able to name and break down the core of the problem of nature exploitation and its causal relationship with greedy and vile human nature; (2) students are able to illustrate the ideal situation that should occur; (3) students are able to write their hopes and expectations to the fishermen in an argumentative manner. This research has enriched the scientific significance of pedagogic multi-literacy, particularly critical literacy-based learning in primary schools.

1 INTRODUCTION

The phenomenon of shark fins exploitation is a problem in human life that is related to the balance of marine biota ecosystems. In news from Mongabay (2018), Head of Fisheries and Marine Resources and Human Resources Research and Resources Ministry of Maritime Affairs and Fisheries M Zulficar Mochtar in Jakarta on Wednesday (28/3/2018) said that:

> ...the practice of exploiting shark fins is increasingly threatening shark populations in a rapid process. Yet to restore the population, it will take a long time. Furthermore Zulficat Mochtar said that the phenomenon of massive shark exploitations is caused by a fantastic export value that is worth Rp1.4 trillion. China is the main export destination country, with a total export value in 2017 reaching IDR 626 billion, followed by Thailand with IDR 356 billion. While more living sharks are exported to Hong Kong, which is 1,098. The main demand of these countries is the fin to be used as food and traditional medicine...

Behind the fantastic selling value of shark fins, of course, it leaves a sad story and it will trigger an imbalance of marine biota ecosystem in the future. The sad story that arises is that sometimes after the fins are taken, the sharks are dumped into the sea and then die due to difficulty in moving. Without the ability to swim, the shark will experience asphyxia (wikipedia.com).

In the preliminary study, researchers examined the 2013 primary school curriculum thematic textbook documents. The findings of the researchers concluded that the phenomenon of shark exploitation was not discussed in the 2013 curriculum thematic teaching material in primary schools. This phenomenon needs to be bridged by learning models framed with critical dimensions. In this case the researcher proposes critical literacy based defenders by using an approach to the problem and the technique of revealing works with illustrated images and essays of critical argumentation.

2 THEORETICAL FRAMEWORK

2.1 *Historical and theoretical foundations of education against critical problems and literacy*

Historically and theoretically, education with a problem of critical literacy is initiated by Paulo Freire's thinking. Freire is an educational figure came from Brazil. Education with problems is when education makes the problem of human life as a phenomenon that must be sued and fought in school, so the life become more balanced (Freire, 2008; Kesuma & Ibrahim, 2016).

Education with a problem would carry out the concept of education in critical perspective. The purpose is making literacy a force capable of raising critical awareness of the people in Brazil of the reality of oppression that has bound them as human beings. "The development of critical awareness makes people question the nature of their historical and social situation, read their world with the aim of acting as autonomous subjects capable of bringing change towards a more democratic and humanist society" (Kesuma & Ibrahim, 2016).

Therefore the concept of literacy education initiated by Freire is aimed at allowing students to be able to read all forms of social and cultural reality that were around them. The concept of Freire literacy explained that learning is not only reading a text, but also reading the meaning behind the text, finding the link between the text and the context of human life. The concept of *"Read the Word and the World"* from Paulo Freire is the main foundation of learning critical literacy-based languages, such as those suggested by Freire and Macedo (2005):

> The act of learning to read and write has a very comprehensive understanding of the reading world, something that human beings do before reading the words. Even historically, human beings first changed the world, secondly proclaimed the world and then wrote the words. Tlese are moments of history. Human beings did not start naming A! F! N! They started by freeing the hand, grasping the world.

It can be deduced that the critical literacy education initiated by Freire that focus on learners' awareness to their world. Then student teach to be more transformative and participate in life. The awareness process uses a Praxis approach, namely through critical reflection on their social and historical situation, reading their "the world", which then provokes the birth of actions that can bring social change (Hendriani & Ibrahim, 2017).

2.2 *Urgency of critical literacy*

Literacy in a new perspective has a broad meaning, not limited to just the ability to read and write words. Now literacy is an ability to read, understand, make meaning and interpretation and solve the problems of human life (Kist, 2005).While the word 'critical' is generally defined as an attitude to questione and skeptical something that generally accepted (Cooper & White, 2008). Then critical literacy can be interprete as an ability to question or doubt all phenomena of life that are not working properly. Furthermore, Tilaar, Paat and Paat (2011) expressed that critical literacy can be briefly understood as the ability to read texts actively and reflective with the aim of obtaining a better understanding of the abuse of power, inequality or inequality, and injustice in human relations. Whereas Johnson and Freedman (in Priyatni, 2012) revealed that critical literacy is combination of critical thinking skills, awareness to social justice, to political issues, and power relation in the text.

In the early years of primary school, literacy-based learning is considered critically important for student learning development. Learning critical language-based literacy explores the relationship of language and strength and focuses on the need to create critical speakers, readers, and writers who can deconstruct the surrounding text and interpret them, both as products and processes of certain social practices. Furthermore, Lee (2017) claims that critical literacy practices are good and analytical procedures can broaden students' understanding and perspectives on issues of power and social inequality. Critical literacy encourages the reader to examine the point of view from which the text is composed, and to think of other perspectives that might be included in the text, helping the reader to diversify their understanding, beliefs and perspectives about the text and its social context.

Furthermore, critical literacy-based learning also helps students to give voice to experiences in oppressive social systems (Lankshear & McLaren, 1993; Rosenblatt, 2004). More than that, literacy is defined as "the act of consciousness and resistance" which is very contextual (Giroux, 1993). Referring to the opinions of several experts, it can be interpreted that critical literacy is an ability to empower literacy skills and critical thinking in revealing hidden phenomena or facts that represent inequalities in various dimensions of human life.

2.3 *Text based on critical literacy*

According to Abidin (2012), being critical on a text, due to situation that the text has been influenced by the author's perspective so that the text is no longer neutral. For this kind of text, students must be able to carry out critical thinking activities so that students not only gain insight into the content of the discourse in depth but are also able to find things that need to be criticized for matters that need to be criticized can be either problematic provocative persuasive or hidden messages made by the author contained in the discourse. Rosenblatt (2004) emphasized that critical literacy allows the reader not only to play the role of a "code breaker", but also the role of textual criticism.

Furthermore, other experts add that:

> Some of the objectives of critical literacy are to recognize aspects of language that are not neutral, examine power relations in the text, identify many voices in the text and handle their own belief systems in response to a text.
>
> (Gee, 2004; Lesley, 2004)

In addition, critical literacy skills can also be interpreted as the ability to criticize a text based on different perspectives. The aim is to oppose the status quo to question the partiality of an authority. Problems that are criticized in a text can be in the form of social problems such as various community diseases such as corruption, collusion, suppression of nepotism of sexual harassment, gender issues and others. As well as scientific issues such as technological problems, environmental pollution, and problems of a nature (Abidin, 2015).

Viewed from the opinions of Abidin, this study presents critical literacy texts that contain the phenomenon of the marine environmental crisis caused by greed and savagery of a handful of men. The phenomenon raised in this study is shark finning.

3 METHOD

3.1 *Research design*

This study uses the CAR (Class Action Research) model developed by Kemmis and McTaggart (2000). This model consists of four stages: planning, implementation, observation, and reflection. These four stages flow like a continuous cycle to overcome various problems. This is in accordance with the functions and goals of CAR, according to McNiff (2013):

CAR is a reflective cycle-shaped study conducted by the teacher himself, whose results can be used as a tool for curriculum development, school development, development of teaching skills, etc.

In this section, the researcher raises the description of the implementation of critical literacy-based learning and raises research artefacts in the form of illustrations and arguments.

3.2 *Subjects and research sites*

This research was carried out on class V students in SD X, a total of 20 students, that is, 10 male students and 10 female students. Researchers were interested in conducting research in this school, because the researchers was finding that literacy learning still focuses on literacy in reading, writing, mathematics and science. The implementation critical literacy on a text is still rarely found.

3.3 *Data collection and data analysis*

The process of data collection was carried out by implementing education towards critical literacy-based problems in several action. To collect these data, researchers used several instruments including the observation format, the format of assessment of students' critical literacy skills. The three data will be described in the results and discussion section.

4 RESULTS AND DISCUSSION

4.1 *Description of learning steps for critical literacy with problem-posing approach*

The Design of Critical Literacy base on Problem-Posing Approach, which is a modification model of Paulo Freire (in Kesuma and Ibrahim, 2016) and the Critical Schematic Method (Abidin, 2012). The stages are described in the following paragraphs.

Problematization stage (pre-reading). This stage consists of two steps, namely codification and modification. This stage is the stage of literacy education in concrete contexts and theoretical contexts (through images, folklore, etc.). The stage of *codification (marking)* is a process where educators give marks in the form of illustrations (pictures, folklore, etc.) of problematic themes that are to be built based on the reality experienced by students. Codification is a knowledge object that bridges between educators and students revealing the veil of life.

The codification (parsing). Teacher and students analyzes their own lives, in long discussions they release their sharpness of vision towards themselves which is involved with the world (objective reality).

In this study, the problematization stage was carried out by presenting a problematic story video that represented the "shark finning" phenomenon. At this stage the researcher also acts as a teacher, the researcher begins learning by displaying two contradictory videos. The first video shows the lives of sharks in the ocean that are beautiful and they do not hurt each other.

Figure 1. Video of sharks' life, from https://www.youtube.com/watch?v=BDxwT8wqUzU.

In the video, researchers conducted a question-and-answer session with students about the beautiful marine life and also about human relationships with sharks. In the next video, the researcher presents a video that represents the corpse of a shark without fins.

Figure 2. The second video about the shark massacre, from https://www.youtube.com/watch?v=zep7BlesW-M.

After the second video, the researchers began asking questions related to what problems students could find in the video. After the video was shown, there were some expressions from students who were sad and a little scared. After the question-and-answer session, the researcher presents the story behind the massacre.

The researchers told the students a story about a greedy and vile fish entrepreneur. The story contains the phenomenon of exploitation of shark fins which are associated with greedy, vile, and bribed man. After the storytelling, researchers and students build some dialogue and brainstorm about the characterization of the characters in the story, the storyline, and student experiences related to the story. The researchers help students mark (codification) and decipher (decodification) the social inequalities that exist in the story and its relevance to the problems they experience in everyday life. At this stage the researcher gives pre-reading paper sheet for each student to explore the students' schema in the story presented. The student sheet Pre-reading can be seen in Figure 3.

Referring to Figure 3, it can be concluded that the researcher questions the character on the first sheet. The aim for students is to understand the characteristics on the story. After that, students are faced with causality questions related to their dislike of the character. Students' reasons also vary and have represented a negative response to crime, cunning and cruelty. In the second student sheet image, the researchers begin to confront students with the core of the problem being discussed, students are asked to give their responses related to the problem of what they see/find in the text or story that has been delivered. Then they are asked to express their feelings by first drawing the emoticon. There are students who expressed sadness and

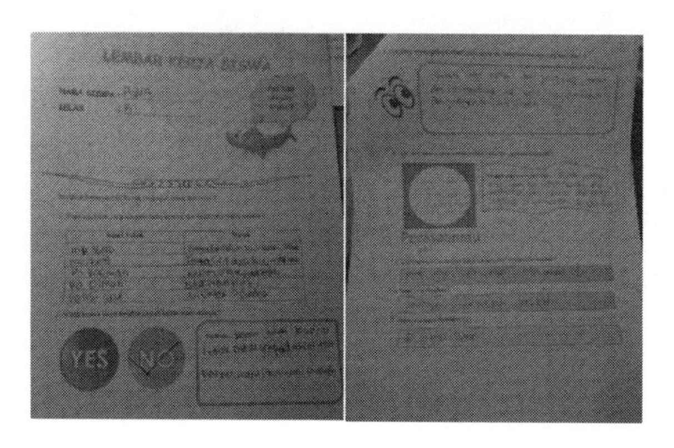

Figure 3. Student sheet pre-reading.

anger, even the arguments that some remember are responding to the pain of sharks with some violence.

Regarding the type of questions posed by the student sheet, critical literacy learning uses probe questions that "encourage thinking, ask for clarification or translation" (MacKnight, 2000). Examples of probe questions adopted by MacKnight (2000) from Socratic Questioning compilation Richard Paul Recommendations cover the following aspects: clarification, assumptions, reasons and evidence, origin or source, implications and consequences and finally points of view or perspective.

Cultural Discussion Stage (Reading). In this stage, students form units of small work groups that try to discover the context of the text that is being discussed using key words (Kesuma & Ibrahim, 2016). In this stage, students read words and find the relationship between text and context.

At this stage the teacher divides students into several groups. Students read stories that were previously read by the teacher, then they discuss with their friends to find social issues. In this section, students are asked to draw pictures that illustrate the problematic phenomena they have marked. The images that have been made by students can be seen in Figure 4.

This step is conducted with the goal of developing critical reading skills in describing an inequal reality. Because critical literacy is basically about how to position the writer as a worker with the agenda of using language to describe reality. This indicates to be going beyond the nominal value of the text and questioning the representation of reality in the text (Koh, 2002). Fairclough (1992) recommends that students be given the opportunity to practice writing in different positions and guided to realize the effect of their language choices on others. They also must be responsible for the risks they take when they oppose social inequality that occurs.

Furthermore, according to Rosenblatt (2004), the critical reading process allows readers not only to play a code-breaking role, in a sense of writers and text readers, but also to act as critics of the text. Next, Anstey and Bull (2006) emphasize the dangers faced by students if they are not taught how to read critically because "they can be marginalized, discriminated against, or they cannot act actively in life relations; in short, students will not control their social life in the future". It can be concluded that critical literacy skills involve an activity such as reading text involving analytic and critical thinking to find social inequalities problems.

Social Action Stage (Post-reading). This stage is the real "praxis" stage in which the actions of each person or group become a direct part of reality. This step's goal is to realize actions that have been reflected previously in the codification, decodification and cultural discussion (Kesuma & Ibrahim, 2016). Cultural actions are started by investigating facts, opinions and social phenomena within the text and then reflecting it and critically evaluating themselves. After that, students take a real action, one example is writing a paper critically.

Figure 4. The illustrations based on critical literacy.

In response to this, at this stage, the researchers assign students to describe the ideal situation about harmonious marine life. An argumentation essay is written with the aim of convincing a reader. According to Finoza (in Dalman, 2014), "argumentation essays are essays that aim to convince readers to accept or take a certain doctrine, attitude, and behavior, while the main requirement for writing arguments is that the writer must be skilled in reasoning and composing logical ideas".

Social actions to write essays are considered suitable for children in primary school. The arguments written by students are quite good and critical, they contain facts and opinions, they are reflective, they propose changes, and oppose all forms of social inequality for the sake of a better life. Figures 5 and 6 show an argument example written by one of the students.

Figure 5. Student's sheet.

Figure 6. "Hopes" student's sheet.

The themes of the arguments in Figure 5 are hopes, messages and invitations proposed by the students on the phenomenon of shark exploitation by greedy fishermen and fish traders. Referring to the quotation written by students, it can be understood that through critical literacy education students can act as agencies that provide criticism and solutions to social inequalities that occur around their environment. Cooper's (2008) explained that "critical literacy is the capacity to read words and the world, connecting the development of self-efficacy, a search of attitude, and the desire to influence positive social change". In addition, according to Freire (2008), if a critical view has been manifested in action, then an atmosphere of hope and self-confidence will develop and demand humans in trying to overcome situations that restrict their critical space.

It can be understood that the emphasis on critical literacy education is of course the existence of utopian consciousness, namely the desire to obtain a complete, emancipatory, moral and heavenly order of life.

5 CONCLUSION

This research is a classroom action research that seeks to carry out critical literacy education using a problem-posing approach. The technique used to reveal students' critical literacy skills is through illustrative images and essays of critical argumentation. The learning stages consist of: (1) Problematization stage (Pre-Reading), (2) Cultural Discussions stage (Reading), (3) Social Actions stage (Post-Reading). The phenomenon that is criticized in this study is the massive exploitation of shark fins. This study concludes that critical literacy education is a necessity that can be applied at the level of primary school education. Through the application of critical literacy education, students are able to analyze the environmental problems of marine biota and associate them with greed, vile, and accessible bribery. Students are also able to illustrate the ideal situation that should occur, in which hopes and positive messages are received and fishermen no longer destroy shark lives.

REFERENCES

Abidin, Y. (2012). *Character based language learning*. Bandung: PT Refika Aditama.

Abidin, Y. (2015). *Multiliteration learning: an answer to the challenges of 21st Century education in the context of Indonesia*. Bandung: PT Refika Aditama.

Anstey, M., & Bull, G. (2006). *Teaching and learning multiliteracies: changing times, changing literacies*. Newark, DE: International Reading Association.

Cooper, K., & White, R.E. (2008). Critical lteracy for school improvement: an action research project. *Improving Schools*, *11*(2), 101–113.

Dalman. (2014). *Writing skills*. Jakarta: Raja Grafindo Persada.

Fairclough, N. (1992). *Critical language awareness*. London: Longman.

Freire, P. (2008). *Education of the oppressed*. Yogyakarta: LP3S.

Freire, P., & Macedo, D. (2005). *Literacy: Reading the word and the world*. London: Routledge Classics.

Gee, J.P. (2004). *Situated language and learning: A creative of traditional schooling*. New York: Routledge.

Hendriani, A., & Ibrahim, T. (2017). Pedagogik literasi kritis: sejarah, filsafat dan perkembangannya di dunia pendidikan [Pedagogic critical literacy: history, philosophy and its development in the world of education]. *Pedagogia Jurnal Ilmu Pendidikan*, *16*(1). DOI: http://dx.doi.org/10.17509/pdgia.v16i1.10811.

Kesuma, D., & Ibrahim, T. (2016). *Pedagogic fundamental structure (Dissecting Paulo Freire's thought)*. Bandung: Refika Aditama.

Kist, W. (2005). *New literacies in Action: teaching and learning in multiple media*. New York: Teachers College Press.

Koh, A. (2002). Towards a critical pedagogy: creating 'thinking schools' in Singapore. *Journal of Curriculum Studies*, *34*(3), 255–264.

Lankshear, C., & McLaren, P.L. (1993). *Critical literacy: politics, praxis and the postmodern*. Albany: State University of New York Press.

Lee, Y.J. (2017). First steps toward critical literacy: interactions with English narrative text among three foreign language readers in South Korea. *Journal of Early Childhood Literacy*, *17*(1), 26–46.

Lesley, M. 2004. Looking for critical literacy with post-baccalaureate content area literacy students. *Journal of Adolescent & Adult Literacy*, *48*(4), 320–334.
MacKnight, C.B. (2000). Teaching critical thinking through online discussions. *Educause Quarterly*, *4*.
McNiff, J. (2013). *Action research: Principles and practice*. New York: Routledge.
Mongabay. (2018). *Ekosistem biota laut [Marine biota ecosystem]*. Retrieved from http://www.mongabay.co.id/.
Priyatni, T. (2012). *Reading literature with critical literacy*. Jakarta: Bumi Aksara.
Rosenblatt, L.M. (2004). The transactional theory of reading and writing. In Ruddel, R.B., & Unrau, N.J. (Eds.), *Theoretical models and processes of reading, 5th edn*. (pp. 1363–1398). Newark, DE: International Reading Association.
Tilaar, H.A.R., Paat, J.P., & Paat, L. (2011). *Critical pedagogics: development, substance, and development in Indonesia*. Jakarta: Rineka Cipta.
Youtube Video. *Kehidupan ikan Hiu [The life of sharks]*. Retrieved from https://www.youtube.com/watch?v=BDxwT8wqUzU.
Youtube Video. *Pembantaian ikan Hiu [Shark massacre]*. Retrieved from https://www.youtube.com/watch?v=zep7B1esW-M.

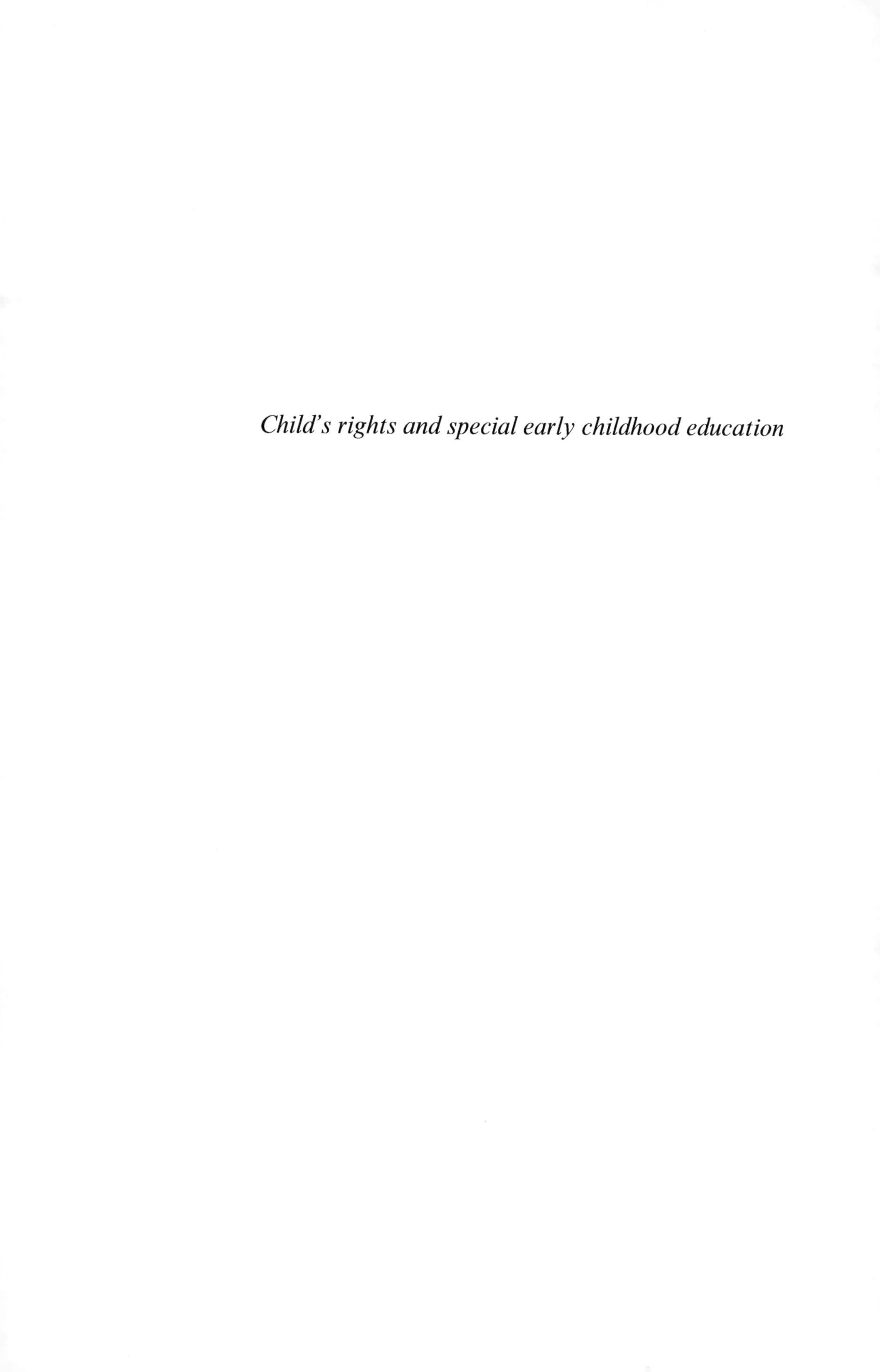

Child's rights and special early childhood education

Early Childhood Education in the 21st Century – Yulindrasari et al. (eds)
© 2020 Taylor & Francis Group, London, ISBN 978-1-138-35203-2

Burnout in special needs teachers in kindergartens

A. Wulandari & H. Djoehaeni
Universitas Pendidikan Indonesia, Jawa Barat, Indonesia

ABSTRACT: The Indonesian Government Regulation on inclusive schools is an effort to prevent various discriminatory actions in schools and to bring up a multiculturalism attitude so that it is more open to various differences. In early childhood education institutions, the concept of inclusive schools provides opportunities for young children with special needs to enter regular classes and produce changes in schools conceptually, organizationally, and structurally. One very important aspect in implementing inclusive schools is teacher competence. Every child has the right to get the same treatment, so the teacher must respect every difference that exists in the child, because the attitude of the teacher is very influential on the child, where when the teacher cares more and shows respect will motivate the child to learn. However, teachers in inclusive schools often experience burnout. Burnout will affect the quality of teacher work, so that it will have an impact on children as students. Therefore, it is necessary to know the burnout that is felt personally by every teacher who teaches children with special needs directly in early childhood education. This chapter aims to explore the burnout felt by early childhood teachers in teaching early childhood who have special needs.

1 INTRODUCTION

Teaching children with special needs is certainly not as easy as teaching children in general, especially in early childhood who have special needs that really need more guidance and preparation in learning. Children with special needs in the regular class are increasing, this shows the beginning of the development of attention and concern for inclusive education in Indonesia. Through inclusive education accompanied by solidarity between generations, it is hoped that it can bridge generations to share common interests, share affection, care for each other, and be able to feel a sense of security (feeling protected) for children or people who have psychological vulnerabilities or differences (Logvinova, 2016; Lukešová & Martincová, 2015, Ezechil, 2013, Kitlinska, 2014).

Inclusive education can accommodate all children to take part in learning in one environment regardless of physical, intellectual, social, emotional, linguistic, family or cultural background, and/or other marginal groups (Ministry of Education, 2009). The development of academic skills and adaptive behavior of children with special needs in inclusive classrooms compared to children with special needs in special schools, is evidence of inclusive benefits for children with special needs (Hastings & Suzanna, 2003). However, the acceptance of children with special needs in general education produces conceptually, organizationally and structurally significant changes in schools (Lee, Yeung, Tracey & Barker, 2015). These changes make a difference in the attitude of the teacher in the field. In some cases, teachers have a more positive attitude than parents of children with special needs (Hastings & Suzanna, 2003), but not all teachers care about children with special needs. Some teachers may even experience burnout because they have to teach children with special needs.

The development of academic skills and adaptive behavior of children with special needs in inclusive classrooms compared to children with special needs in special schools, is evidence of inclusive benefits for children with special needs. The journal found that simple support was provided by teachers to include children with special needs in inclusive classes, and the longer someone was involved with implementing inclusion, the better acceptance and treatment of children with special needs in inclusive classes (Lee, Yeung, Tracey & Barker, 2015).

The fact that education is inclusive in Indonesia means that there are many attitudes of teachers who have not shown concern for children with special needs. There are teachers who even experience burnout, that is, stress, and are unable to manage their feelings of depression, resulting in a situation that leads to a reluctance to teach or participate in his work because they feel intimidated by their colleagues (Leiter & Maslach, 2005). This is supported by the experience of the author as a teacher who teaches early childhood with special needs, and some interesting facts that the author found from the teacher's complaints that teach children with special needs and parents who have children with special needs children.

The existence of complaints from teachers does not reduce the number of teacher roles that must be performed in carrying out their duties. The existence of complaints from parents is proof that the teacher's role is not done optimally. The role of teachers in schools is not only to teach, but many other tasks are as conservators, innovators, transmitters, transformers, and organizers. So, it is difficult for teachers to carry out their role if the teacher experiences burnout (Bastas, 2016).

In this case the author himself had experienced burnout when teaching children with special needs. In the end, the writer resigned from his job because he felt ineffective, a feeling of not being able to handle early childhood with special needs appeared and felt progress in children was minimal so the writer felt that he was a quality teacher (Maslach, Jackson, & Leiter, 1997). Educators must have confidence in carrying out their roles and not have burnout where someone feels hopeless because they feel depressed and lack of high self-achievement, for example when the teacher fails, the teacher is difficult to restore his confidence to meet the demands in teaching, so the learning process does not will be maximal because stress experienced as an educator makes children unable to accept learning according to their needs (Sulsky and Smith, 2005; Wardhani, 2012; Bandura, 1994).

The emergence of complaints from parents of children with special needs shows the lack of maximum role of teachers in schools and lack of achievement of goals and expectations of inclusive education. Many factors cause this to happen, such as the number of interactions that become a pressure for the teacher, both the demands of roles in the job (boss, principal, parents), workload, as well as pressure from coworkers so that the teacher feels stress that cannot be overcome, and causes burnout that reduces the effectiveness and quality of the teacher in carrying out his role as an educator (Maslach, Jackson & Leiter,1997).

A conclusion that can be drawn from this is that every school should implement inclusive education, but from some research and temporary data it can be seen that the application of inclusive education is a burden for teachers. So, it is very interesting to investigate how burnout of teachers in dealing with children with special needs in kindergarten regarding the experience of teachers in teaching children with all the differences and limitations and different backgrounds for each individual child. This research was conducted to help the success of inclusive education, where maximum service can be provided by teachers in teaching children with special needs because of burnout on the teacher, which will harm the teacher, coworkers, parents, and children.

2 METHOD

The method used in this article is the research library (library research), a series of studies relating to the method of data collection library, or research object of research explored through a variety of information literature (books, encyclopedias, journals, newspapers, magazines, and documents). The research literature or review of the literature (literature review, literature research) is a research that examines or critically review the knowledge, ideas, or findings contained in the body of academic-oriented literature (academic-oriented literature).

The data used in this research is secondary data. Secondary data is data obtained not from direct observation, but rather from a systematic review of the existing literatures.

3 DISCUSSION

3.1 *Burnout*

In several studies, one of the variables that most influenced the success of inclusion schemes was the attitude of the teacher. The results show that of 93 assistants who participated in the training had little support regarding their attitudes in handling special needs children (Hastings & Suzanna, 2003). Other journals also found that simple support was provided by teachers to include children with special needs in inclusive classrooms, and the longer someone was involved with the implementation of inclusion, the better their acceptance and treatment of children with special needs in inclusive classes (Lee, Yeung, Tracey & Barker, 2015). In Thailand preschool teachers who have fully implemented inclusion policies have a clear understanding of the process of inclusion in managing classes and teaching disabled children. In contrast, preschool teachers who have not fully implemented inclusion do not yet have knowledge about classroom management and teach children with disabilities (Sukbunpant et al., 2013). What then is the attitude of teachers who experience bornout in dealing with early childhood? How can burnout occur in early childhood education teachers in Indonesia?

Burnout is a global phenomenon (Schaufeli, Michael & Christina, 2008). For researchers and practitioners in America and other countries, it is agreed that burnout is a social problem that needs to be addressed and addressed (Schaufeli, Michael & Christina, 2008). Burnout appears as part of stress consisting of three dimensions, namely emotional fatigue, depression, and decreased personal achievement due to physical and mental fatigue resulting from involvement in intense emotional situations (Pines & Aronson, 1988; Schaufelli et al., 2008; Luthans, 2005).

The journal is related to the attitudes and views of teachers and the readiness of teachers to organize inclusive classes, at TK Aisiyah, Malang. The journal written by Saidi (2015) found results that good awareness and preparedness were already possessed by the teacher from the kindergarten. However, the lack of knowledge of teachers about inclusion and children with special needs, differentiated curriculum, facilities and infrastructure, the refusal of acceptance and learning together with special need by students is an obstacle found in conducting inclusion classes. So that in this journal can be a supporter that apparently the obstacles that arise can be the cause of burnout in the future. So the questions related to the cause of the emergence of burnout in preschool teachers in Indonesia.

3.2 *Factors that cause burnout*

Maslach, Jackson and Leiter (1997) mention the factors that cause burnout, namely: work overload, which can reduce individual quality. Lack of work control, making it difficult to innovate, take choices and decisions and take full responsibility for the work. Rewarded for, because individuals feel the work they do is worthless so they don't get a good affection from their superiors. A breakdown in community causes a lack of positive individual bonds in the work community. The feeling of being treated unfairly can also lead to the emergence of burning out. Individuals can reduce their performance because they do things that are contrary to their values, beliefs, integrity and self-respect because of pressure or rules from other individuals so that they must deal with their conflict values.

In addition, there are also causes of burnout resulting from self-perception. Maslach, Jackson and Leiter (1997) states that burnout is shown by the emergence of three dimensions, namely exhaustion, cyniscism, and ineffectiveness which will cause a decrease in self-performance for those who experience burnout. Individuals who experience prolonged physical, mental, and emotional exhaustion that cannot be controlled, so that fatigue will not heal just by resting. Cynicism is also one of the causes of burnout which is shown by a cynical attitude and tends to withdraw from the work environment, this is intended to reduce disappointment in oneself, but it actually affects the effectiveness of one's work. When someone feels ineffectiveness in doing their duties, then everything that is done is a burden for him and things that are in vain, because this belief in oneself and the beliefs of others will decrease.

The competence of teachers in dealing with children in class greatly contributes significantly to depresionalization and personal achievement. In addition, information from students regarding teacher reports provides information regarding several aspects of teacher mental health and class processes (Evers, Tomic & Brouwers, 2004). From the results of this study it can be seen that teachers who have good competence are able to provide good contributions and self-achievement, but conversely if the teacher does not have good competence in handling children in class, this will worsen the teacher's mental health as well as the classroom learning process become not good (Evers, Tomic & Brouwers, 2004). Looking at the things that have been explained that this can be a cause of burnout where teachers feel ineffective in teaching, of course this is a big problem for teachers, students, and the existing education system (Evers, Tomic & Brouwers, 2004; Maslach, Jackson & Leiter, 1997; Hughes, 2001).

When teachers experience stress due to situational factors or personal factors, teachers must remain calm, professional, and positive in facing children, (Hussain, 2010; Evers, Tomic & Brouwers, 2004). Personal factors are demographic features, age, gender, years of service, duration of service, ways to deal with stress, focus of control, and motivation factors (Evers, Tomic & Brouwers, 2004). Situational factors are poor behavior observed by students, tension in the school atmosphere, inadequate support and respect for their work, lack of material support to carry out their work, and lack of administrative support (Evers, Tomic & Brouwers, 2004).

The results of research conducted by Lau et al (Evers, Tomic & Brouwers, 2004) show that young age which causes a lack of teacher experience in dealing with early childhood is a factor of teachers experiencing burnout, compared to older teachers with more experience. In addition, burnout is positively related to the number of different conflicts in the workplace, the demands of work, and the monotony of work. Organizations that function well, open communication, employer work, and job control are good characteristics of the worker's community and thus do not cause burnout (Karisalmi, 1999). However, psychological work demands and conflicts have the strongest relationship with burnout, where the more work demands become workload becomes heavier and also conflicts that occur with coworkers will make burnout appear on the teacher and can be interpreted medically or non-medically (Maslach, Jackson & Leiter, 1997; Schaufeli, Maslach & Leiter, 2008; Karisalmi, 1999).

3.3 *Impact of burnout*

Not only the causes of burnout Leiter and Maslach (2016) also describe the impact shown by individuals who have experienced burnout. Feeling prolonged and uncontrolled fatigue causes individuals to feel inadequate so that their work productivity decreases due to burnout is lost energy. In addition, burnout is lost enthusiasm to work, so that individual enthusiasm to be more creative decreases and causes his quality to decrease. Not only is burnout shown with lost confidence in carrying out trust which is his responsibility because of the loss of energy and lack of participation in work.

Professionals who have burnout will experience difficulties in performing the worktasks (Evers, Tomic & Brouwers, 2004). (Evers, Tomic & Brouwers, 2004; Logvinova, 2016; Lukešová & Martincová, 2015, Ezechil, 2013; Kitlinska, 2014; Bastas, 2016).

Evers, Tomic and Brouwers (2004) teachers who experience burnout may experience physical symptoms such as: headache, fatigue, stomach problems (such as stomach ulcers), anxiety, cardiovascular problems, and neurological problems. Psychological symptoms, namely the emergence of anger, indications of depression, excessive tension, low self-esteem, anxiety, confusion, hesitation, chronic anxiety, long-term feelings of weakness, substance addiction, unfocused disappointment, anxiety attacks, filling in others for their duties. There will be indications of behavior that is coded by deteriorating interactions, having a way of taunting other people, the absence of deteriorating quality of services and delaying or avoiding work that has become their responsibility.

Teachers who experience burnout not only harm themselves and students, but significantly affect the education system (Hughes, 2001). Burnout usually affects teacher performance in the learning process, absent teacher teaching, decreased expectations of student development, lack of interest, affection and idealism towards students, administration and parents of students, and develop negative feelings towards students as people served (Evers, Tomic & Brouwers, 2004; Bastas, 2016).

4 CONCLUSION

Achieving goals and expectations of inclusive education requires effectiveness and quality of the teacher in carrying out his role, but if the teacher experiences burnout, the professionalism of the teacher will decrease. The teacher cannot make learning preparation, and the maximum evaluation, because the teacher has difficulty in dealing with the causes of burnout appear. Not only that, children's development will not develop optimally as expected because the teacher does not provide the needs that must be met by the child. In addition, there is a need for awareness and cooperation between teachers and special teachers and other colleagues, principals and parents to achieve the goals of inclusive education. Burnout is a term coined by Freudenberger which is considered a significant danger for teachers, because it has a negative impact, namely the emergence of negative teacher behavior that might be detrimental to students (Bastas, 2016).

REFERENCES

Bandura, A. (1994). Self-efficacy. In *Ramachaudran, V. S. (Ed.). Encyclopedia of Human Behavior*, *4*, 71-81.

Hastings, R. P., & Suzanna O. (2003). Student teachers' attitudes towards the inclusion of children with special needs, educational psychology. *An International Journal of Experimental Educational Psychology*, *23*(1), 87–94. DOI: 10.1080/01443410303223.

Hughes, R. E. (2001). Deciding to leave but staying: teacher burnout, pre-cursors and turnover. *International Journal of Human Resource Management*, *12*(2), 288–298.

Hussain, H. (2010). *A thesis submitted in fulfilment of the requirements of Bournemouth University for the degree of Doctor of Professional Practice*.

Leiter, M. P., Maslach C. (2016). New insights into burnout and health care: strtegies for improving civility and alleviating burnout. *Medical Teacher*, *39*(2), 1–4. http://dx.doi.org/10.1080/0142159X.2016.1248918.

Maslach, C., & Leiter, M. P. (2008). Early predictors of job burnout and engagement. *Journal of applied psychology*, *93*(3), 498.

Leiter, M. P., & Maslach, C. (2005). *Banishing burnout: six strategies for improving your relationship with work*. New Jersey: John Wiley & Sons.

Luthans, F. (2005). *Organizational behavior 10th edition*. Yogyakarta: Andi.

Maslach, C., Jackson, S. E., & Leiter, M. P. (1997). Maslach burnout inventory, third edition. In C. P. Zalaquett & R. J. Wood (Eds.), *evaluating stress: a book of resources* (pp. 191–218). Lanham, MD: Scarecrow Education.

Saidi, N. (2015). Analisis kesiapan guru dalam pengelolaan kelas inklusi [Analysis of teacher readiness in management of inclusion classes]. In *Seminar Psikologi dan Kemanusiaan*. Malang: Pshycology Forum UMM.

Ministry of Education. (2009). Pendidikan inklusif bagi peserta didik yang memiliki kelainan dan memiliki potensi kecerdasan dan/atau bakat istimewa [Inclusive education for students who have miscellaneous and have the potential of intelligent intelligence and/or talent]. Indonesia. Retrieved from http://kelembagaan.ristekdikti.go.id/wp-content/uploads/2016/11/Permen-No.-70–2009-tentang-pendidiian-inklusif-memiliki-kelainan-kecerdasan.pdf.

Pines, A., & Aronson, E. (1988). *Career burnout causes and cures*. New York: New York Press.

Schaufelli, W. B., Leiter, M., & Maslach, C. (2008). Burnout: 35 years of research and practice. *Career Development International* *14*(3), 204–220. DOI 10.1108/13620430910966406.

Sukbunpant, S., Arthur-Kelly, M., & Dempsey, I. (2013). Thai preschool teachers' views about inclusive education for young children with disabilities. *International Journal of Inclusive Education*, *17*(10), 1106–1118. DOI: 10.1080/13603116.2012.741146.

Sulsky, L., Smith, C. (2005). *Work stress*. New York: Thomson Wadsworth.

Karisalmi, P. T. S. (1999). Impact of working life quality on burnout. *Experimental Aging Research*, *25* (4), 441–449.

Wardhani, D. T. (2012). Burnout di kalangan Guru Pendidikan Luar Biasa di Kota Bandung [Burnout among Special Education Teachers in the City of Bandung]. *Jurnal Psikologi Undip*, *11*(1).

Early Childhood Education in the 21st Century– Yulindrasari et al. (eds)
© 2020 Taylor & Francis Group, London, ISBN 978-1-138-35203-2

Beyond the 'limited desk' to the better world for children: An autoethnography approach of a Korean ECE consultant in Indonesia

E.J. Won
Korea National Open University
Soongeui women's college
Universitas Pendidikan Indonesia, Bandung, Indonesia

ABSTRACT: This chapter adopts an "auto ethnography" method to unpack experience of a Korean ECE (Early Childhood Education) consultant who has been living in Indonesia and working with the local Indonesian government in the field of ECE for two years. She had been dispatched by KOICA (Korea International Cooperation Agency) in order to help develop the ECE field in a regency of Indonesia working with around 200 kindergartens. Despite the fact there had been 72 Korean ECE specialists dispatched to 38 institutions all over Indonesia, it is hard to find a record of their voices about ECE in Indonesia. This chapter attempts to explore the meaning of 'limitation' in the field of ECE in Indonesia through of a common symbolic icon, which can be shown by the medium of 'the desk' according to the view of a foreign consultant's own experience and narrative.

1 INTRODUCTION

The Early Childhood Education (ECE) had developed rapidly under the motto of "Education for All" during the 1990s. (World Bank 1999, re-quote) As such, in Indonesia the ECE priority project had been kicked off by using the support of the World Bank in 1998 in four states; West Java, Banten, Bali, and South Sulawesi. In 2001 the ECE department under the Ministry of Non formal Education had been established in the government, called PAUD. Moreover, a forum for the national policy for the ECE was held in 2002. As a result, Indonesia's ECE has experienced a rapid growth for the past 20 years. At around the same time, the government of the Republic of Korea had started dispatching the specialists in ECE field to Indonesia. Since 2001 to date, around 72 specialists have been dispatched to 32 institutions, and some have supported foreign aid funds to develop the ECE institutions in the regions under the management of the [1]*KOICA* (Korea International Cooperation Agency (footnote description).

Among the 79 countries, the KOICA had dispatched the highest number of ECE specialists to Indonesia. (KOICA statistics 2018, Jeong 2014, 65). This strongly indicates that the government of Indonesia has moved with great concern for the development of the ECE since 2001. The local government had requested the number of specialists and the area to which they would like the specialists to be dispatched.

1. The Korea International Cooperation Agency (KOICA) was established as a governmental agency dedicated to providing grant aid programs of the Korean government in April 1991. KOICA endeavor to combat poverty and support the sustainable socioeconomic growth of partner countries. By doing so, KOICA established and strengthens friendly ties with developing countries. Since its establishment of KOICA in 1991, it has dispatched 22,444 Korean Volunteers to 79 countries for the last 27 years. (Source: Korea International Cooperation Agency website)

In spite of the Korean government sending a considerable number of specialists to the ECE fields in Indonesia for the last 18 years in order for the ECE to grow, it is difficult to find records of their voices and experiences concerning ECE in Indonesia. Foreign volunteers' multi-cultural experiences have ample values, and I believe it is very important to investigate it through the auto ethnography method.

In this chapter, the researcher uses her own experiences while helping the local government in Indonesia in order to explain personal and social meaning of 'restriction' during her working journey. This chapter can provide a guidance to ECE specialists who are preparing to be dispatched from KOICA to Indonesia. And also outlines the benefits in providing an objective viewpoint when dealing with Indonesian ECE workers.

2 USING AN AUTO ETHNOGRAPHY

Auto ethnography is an emerging qualitative research method that allows the writer to unpack a highly personalized story by drawing on his or her own experience to extend understanding about a societal phenomenon. It is grounded in postmodern philosophy and is linked to a growing debate about reflexivity and voice in social researches (Sarah, 2006, p. 146). This qualitative research method is still rare in Indonesia, however internationally this narrative method is has seen a rapid increase recently in the fields of international cultural studies, pedagogy and curriculum studies for teachers and researches of minorities or multicultural people.

This chapter unpacks several meaningful stories from my experience in the ECE field in Indonesia by looking back and using self-reflective data. Currently I am working on my Master's degree, majoring in ECE at UPI (Universitas Pendidikan Indonesia). Prior to this I had been working as an ECE consultant in an Indonesia regional education office from 2003 to 2005 under the program of KOICA. My main objectives were to train ECE teachers and parents and to support and establish new ECE institutes in that area.

It is not easy to recall all of my past memories and analyze them because a lot of time has passed since I was first dispatched from Korea. However, I have collected as much data as possible that was related to my two years of experience; such as photos, videos, diaries, working letters and reports, and also attempted to understand and extend my own personal experiences to link with socio-cultural meaning. I wrote more than 20 pages of self-reflection and self-observation stories based on private memories and collected data. After writing, I revised it many times and extracted several main points of personal and social issues. Then I coded it as three kinds of 'desk' and used Won (2008)'s [2] *Ten kinds of analysis strategies* to analyze the writings.

While proceeding the analysis and interpretation, I repeatedly became an insider and outsider through continuous self-reflection and revision and supplemented the analysis. I used several auto ethnography Writing skills designed that has been modified by Chang (Chang 2008). These were descriptive-realistic writing, confessional-emotive writing, analytical - interpretive writing and imaginative-Creative writing. Four kinds of writing skills were reconstructed by mixing. In order to secure objectivity and truth from data, the previous writings were read and interpreted several times, and I also visited the working region where I interviewed people who had worked with me 15 years ago and had memories integrated with my own. There have been many debates about auto ethnography; whether it can be a scientific method or not, because it has been focused on the researcher's self-narrative. However, it directly brings the voice of insider without misinterpretation by others (Adriany et al., 2017).

2. 10 Kinds of analysis strategies: (1) Search for Recurring Topics (2) Look for Cultural Themes (3) Identify Exceptional Occurrences (4) Analyze Inclusion and Omission (5) Connect the present with Past (6) Analyze Relationships Between Self and Others (7) Compare cases (8) Contextualize Broadly (9) Compare with Social Science Constructs (10) Frame with Theories (Chang, 2008, pp. 131–137)

3 MY STORY

> *The desk makes me sit down and focus on things on the desk only.*
> *When a desk is given to me, I couldn't learn a lot of things beyond the desk*
> (September 2018 researcher's self-reflection record)

3.1 *Going beyond my limited desk in Korea*

I went to Indonesia, which I had never known before. By going to Indonesia, it was possible to look beyond my desk where I was supposed to sit and work, which was a mental set up by me and social customs. Korea had experienced a high economic growth in a very short period. Because of that life goals of most young Koreans had changed. Their goals are to earn a lot of money, get an expensive early childhood development education, get into a good school, obtain a good job and earn a high salary so they can enjoy a comfortable life. Likewise, my plan was also to obtain a good education, and after graduating from college to prepare for study abroad after saving up enough money. However, I decided to go beyond my 'limited desk' space and join an overseas volunteering service in Indonesia. At the time it was developing countries. Indonesia was a 'beyond my desk' country for me. To others this is like going to a stagnant place rather than going for successful growth. For this reason, my family and friends strongly advised me not to join an overseas volunteering service.

3.2 *Facing empty desks of public officials.*

On the first day at work at one of the local education offices there was an empty desk and a chair placed on one side of the office for me. It was the only place for me to work as an ECE consultant at the time. The local staff welcomed me, a single female, and stranger from Korea. No one including myself knew what I could do here. Not only me, but it was also the first time for them to work with a KOICA ECE specialist. I sat on my chair in front of an empty desk and turned on my laptop and started to write what I could do here. First, I had to figure out the present status of ECE in the region in order to find out what I could do to support them.

> A Staff of Education Office: Anisa (Researcher's local Name), how can you help us? Do you have any grants from KOICA?
>
> Researcher: Yes, we have some. But first, I have to understand the local status of ECE. I can't provide direct support without information.
>
> (Conversation with local staff of education office, self-recall data, August 2003)

When I was sitting at my chair, the staff of the education office were regrettably sitting at their desk drinking tea and talking to colleagues, and not working humbly. It seemed clear that if there is no order from the boss, nothing was happening at the desk. This was so different from the office scene in Korea because they are busy every minute of the day. I am a Korean who likes 'fast and busy' so I was always in a hurry. Since I started to work at the office, I constantly requested the present status of ECE institutes and visiting schedule for ECE fields but it was delayed continuously and they kept asking me if I had a budget.

> I came all the way from Korea on flight and I came to the office everyday by car taking one hour, but I did not get anything from them. Not even the field data. Why are they not working? Actually supporting ECE institutes is not only my job but also their job, but why don't they work actively? Do they only want the grant of KOICA, and not my cooperation? What can I do here? Is it going to change at this region in two years?
>
> (self-reflection data, September 2003)

As I came to Indonesia by going beyond my 'limited desk', I also wanted to jump over the desk of the government officials. Unfortunately, the first impression of Indonesian local government officers was quite disappointing. In Korea, the government officials are often criticized as 'desk' officials. What it means is that they only push papers from their desk. The country would not change by only pushing papers from the desk, but rather needs to walk on their two feet and resolve the practical needs of the country. When I came to Indonesia, I felt the same from the Indonesian government officials. If I had been sitting at the desk like them, I would not have given them realistic help. For instance, the KOICA's subsidies established to support the early childhood development programs could not have been used in its entirety to help ECE in Indonesia. It is sorry to say, but most of the portion may have been used privately in the name of administrative expenses rather than for the actual needs of the local kindergartens and the children.

I wanted to visit the ECE fields with local education officers and find out ECE field's needs, but they did not want to take any actions unless there was a budget for it. They knew that the Koreans earned a lot more money than Indonesians did, and their mindset was that I should use the budget given from the Korean government if I came to help them out. However, my thoughts were different from them. I had to turn their thoughts upside down by seeking out what makes them not cooperate with me. The ECE development of Indonesia should be the goal of this country and not of the other country. They should understand that they are not helping my work, but instead I was helping their work. I told them that I came to this country as a volunteer after resigning from a high salary job. I also told them that I worked not just regular work hours, but also visited the ECE fields during my personal free time. Of course, there was no extra pay for this. I did my best to try to attend all of the teacher's activities. I went to the public market with them and bought the things just like them and used *Ojek* and *Angkot* as they did. After a while, the local officials started to slowly open their hearts after seeing the lifestyle of this foreign consultant and recognizing that it was not any different from theirs. I started to visit the kindergartens and began to express their difficulties and hard feelings during open seminars for people at the Education Department. It is true that the teachers from Indonesia received less salary than Koreans did. However, if they do not work hard simply because their salary was lower than other countries, then it would also be very difficult to see any kind of changes. Likewise, In Korea the salary of the teachers in kindergartens was much less than other jobs. However, their working hours are ten hours long from 8 a.m. to 6 p.m. Also, if there are any kindergarten activities during the new semester, they normally work late till after 9 p.m., and they do not receive any additional pay for it. In spite of their poor working conditions, they worked hard with pride in their job. As a result, the recognition and treatment toward the teachers of the kindergartens changed a lot during the past ten years. Few of the employees of the education department started to cheer me up and opened their hearts after seeing my passion toward my work and moving around on foot.

I could have done the work much faster if I had used a part of the budget set by KOICA for labor and use native government officials to do some of the work, but I did not do that because that was not why I came here. If I asked them to do the work by giving them money, it would have become my project and my work. Instead of acting as an agent by giving the country money, I chose a method that would make them find out their needs and solve the problems together. By depending on ownership, the attitudes of Indonesian ECE's toward work would change drastically. It their responsibility to work on the issues in educational fields, and not for me to stay and work in this country. That is why I chose the hard and slow way rather than choosing an easy way. When there was a request from the teachers, I visited ECE fields using *Angkot* and rental cars, and provided training. Often these trainings would be on Saturdays, which is a non-working day here in Indonesia. After a few months of the process, I finally met up with a government official who wanted to visit the ECE fields.

I even visited all the kindergartens and all the teacher's meetings in the region with him on Saturdays, and we delivered the support money in the form of not cash but in the form of the training material and equipment instead. I did not use a single penny for administrative expenses such as gas and food. For this reason,100% of the KOICA's project funding got delivered to the local ECE institutions. Using this money that was enough to build one kindergarten, we were able to provide supports for the ten ECE facilities. I still feel grateful and cannot express enough thanks to the officer who had been sharing and executing my thought of jumping over the desk to visit the sites on foot with me.

3.3 *Again facing empty 'desks and chairs' limited children's space!*

In order to get information about the regional ECE status I started to work at a national kindergarten, which is located near the education office. Instead of being in front of an empty desk in education office where nothing could be done, working at kindergarten where I could see children's activities every day, was more satisfying. I felt alive and realized why I came here in the first place.

The scenery of the education office and the scenery of kindergarten were very similar. There were only children's desks and teacher's blackboards in the classrooms, and no interesting play areas and various educational tools. On the first day at the education office, I faced an empty desk with no working data. Staring out empty window, I felt depressed and though what can I do. It's the same situation for children who sit in front of an empty class. If children sit on the chair in front of an empty desk which is lined toward the front they cannot do anything without receiving instructions from the teacher. Even if the teacher teaches singing, reading, counting numbers in front of blackboard, the size of children's imagination cannot exceed the size of the board if that is the only equipment used. Young children have a great influence on the physical environment because they cannot choose or control their own environment. (Child Policy Institute 2009c). Physical environment of an institution affects infant development and behavior, and a good physical environment attracts infant interest and manifests various types of play behavior. The infants were born to navigate, play more and also appear to have more active interaction with peers and teachers in Korea's superior physical environment in several studies. The higher the level of physical environment, the higher the creativity and cognitive abilities of young children. I filled up the vacant physical facilities and educational environment of the National Kindergarten here with my teachers, beginning my first year with me.

When we tried to change the location of teacher-centered classroom environments such as blackboards, desks and chairs, some teachers expressed difficulty about what their parents' perceptions would be as they had not yet changed. Parents might not be sent to kindergarten unless their kids learn how to read, write, and calculate in kindergarten. Similar anxieties were in the 1980s and 1990s in Korea, so I could understand their position well enough and realize that not only teachers but also parents should understand first. Because a better world for children should be achieved by collaborating with not only teachers, but also parents, teachers, kindergartens and local communities.

> Don't limit your child's thoughts to the desk size. Your children can learn in a wider playing area using grass, stones, and wood as paper to spread their imagination instead of sitting on a chair to learn writing and counting. The size of an idea between a child who receives only a piece of paper, and a child who receives a large piece of paper, an entire classroom play area, an outdoor playground is incomparable. The later's thought process always out weight the prior's thought process. Children learn more while enjoying the environment and facilities of a kindergarten which meets the needs of the later child, and it helps to increase their thinking, creativity, problem-solving, and imagination. Play is learning for children.
>
> (Lecture contents, self-memory data during parenting seminar, September 2004)

As I wanted to jump over the officer's limited 'desk' I also wanted to jump over the 'desk' in front of the children. The first time I saw the children in kindergarten, their appearance was just sitting in front of the desk and only looking at the blackboard and the teacher aimlessly. I did not want to limit the children's infinite possibilities in their desk. I immediately moved the desks in the middle of the classroom by dividing into several areas and have placed a mat on the floor so that the children could take off their shoes and play in the classroom comfortably. This process would not have been possible without the cooperation of the local teachers and the understanding of parents. They were afraid of the new changes at first, but they accepted it positively through constant dialogue and education. The children began to expand their possibilities in every corner of playing areas in the classrooms and outdoor playing area, and not in front of the desk only.

Finally, I found three desks I wanted to jump over; The desk that is in front of me; the desks of the public officials; and the desk of the children. When I had jumped over my restrictions and limitations, I found another desk in front of the others, and I also hoped to jump over the desk of 'boundaries and limitations' in front of them. I wished to cross over it with them, and I had spent two years crying and laughing with them. I sincerely hope that my research will give a strong motivation and challenge to the readers to cross over the 'desk' involving the meaning of restrictions and limitations.

4 CONCLUSION

I was working at local education office in Indonesia as a specialist for ECE development for two years. While I was continuously reviewing and analyzing my narrative writings I found a common symbolic icon, which can be shown by the medium of 'the desk'. All that drew me to the commonness of its limitation and its restriction. The desk makes me sit down and focus on things on the desk only. When a desk is given to me, I couldn't learn a lot of things beyond the desk. However, when I go beyond 'The Desks', there was a huge change to them and to me. Therefore, I recommend the following to improve the results that are beyond the limitations and restrictions that are blocked-off by the thoughts common to the understanding of desk works.

First, I recommend that government officials' work is carried out using the field-centered method rather than the top-down method of the office environment. I am certain that there will be a rapid change in quality if the officials leave their desks; go to the field and listen to the voices of the teachers; and work closely with them, rather than top-down method of daily business by signing the documents.

Second, I recommend that teacher training locations and times should be decided by local teacher's work schedule and not for the convenience of government officials. I often have seen teachers skip a day of teaching in order to attend the teacher's training program run by the ministry of education. It is because the teachers training program takes place in the morning the same time as the children go to the kindergarten. In Korea, they plan for an intensive training programs usually during the vacation so that teachers do not skip the classes due to their training requirements. If the training event is only to distribute the materials, then the use of online service can be the most effective alternative.

Third, I recommend to utilize the activity and local situations reports from the KOICA specialists when the future specialists are dispatched. The specialist agents can reduce the initial adaptation time if the statistics for the corresponding field and district of their interest that they have requested while the KOICA dispatches their agents. To date, many specialist agents been dispatched to Indonesia. However, the sad situation is that their reports and data is not properly used. If the valuable work experiences of the current workers are recorded and distributed to the future volunteers dispatched to Indonesia, then they can obtain a great amount of help for their adaptations..

Fourth, I recommend a child-centered desk setting rather than a teacher-centered. It is often seen that children act passively on teacher's directions when the desks in the kindergarten are faced to the blackboard in front of the classroom. When there is no direction from the teacher, the children seem forced to be quiet. However, children like to work together and talk to each other, and when the desks are facing each other and not facing the front they are enabled to do this. Children can also choose the activities that interest them the most in all areas of activities. Children can improve responsibility, self-directing ability, problem solving, self-achievement, autonomy, and so on within this environment (Kim, 2017; Lee, 2014).

This study suggests recommendations of the researcher by transcribing the narrative related to the local culture and the current educational status. It includes her early experiences and emotions while she was dispatched to the ECE department of the government and ECE institutions in Indonesia. For the future studies a collaborative auto ethnography approach with other specialists in the ECE field in Indonesia could help expand, develop and integrate meaning to the ECE status in Indonesia.

REFERENCES

Adriany, V., Pirmasari, D. A., and Satiti, N. L. U. (2017). Being an Indonesian feminist in the North. *Tijdschrift Voor Genderstudies, (20)*3, 287–297.

Kim, M. K. (2017). *Effects of operating styles using free choice activities on autonomous behaviors of 5-Year-olds.* Thesis, Korean Foreign Language University of Graduate.

Lee, J.H. (2014). *The Young Children's Experience on Discussion of Free Choice Activities Photography.* Thesis Korean Education University of Graduate.

Sarah, W. (2006). An Autoethnography on learning about Autoethnography. *International Journal of Qualitative Methods*, *5*(2), 146–160.

Won, J. H. (2008). *Autoethnography as a method*, Walnut Creek, CA: Left Coast Press.

World Bank. (1999). *Education sector strategy*. Washington, DC: The International Bank for Reconstruction and Development/World Bank.

Early Childhood Education in the 21st Century – Yulindrasari et al. (eds)
© 2020 Taylor & Francis Group, London, ISBN 978-1-138-35203-2

Analysis of a scientific approach in child creativity development with autistic spectrum disorders at ECE Akramunas Institution, Pekanbaru

Y. Novitasari & S. Fadillah
Lancang Kuning University, Pekanbaru, Indonesia

Hidjanah
Muhammadiyah Institute of Teacher Training and Education, Bogor, Indonesia

ABSTRACT: The purpose of this research to describe the implementation of a scientific approach in child creativity development with autistic spectrum disorders at ECE Akramunas Institution. This research used a qualitative descriptive approach. Through the scientific approach, children learn by seeing, holding, listening, gathering information to find their own knowledge to stimulate their creativity. The results indicated that the scientific approach has been implemented in learning activities and has been followed earnestly by early childhood children with autistic spectrum disorders in the classroom together with teachers and peers. Furthermore, the results explained that learning activities using scientific approaches could stimulate children's creativity with autistic spectrum disorders at ECE Akramunas Institution.

1 INTRODUCTION

Children of 0-6 year-old have their own uniqueness. Early childhood period (0-6 year-old) is known as the golden age, because children experience growth and development rapidly; it is not repeated and cannot be replaced. Therefore, it is very important to make sure that the surrounding environment could enrich the development of the children (Novitasari, 2017). Keep in mind that not all children are born with a strong physical condition and healthy mentality. Children with different physical and mental conditions are usually called children with special needs. Therefore, in some cases, we need to adjust the learning environment to the needs of the child.

Children with special needs are children with special characteristics. Special characteristics are not only at physical differences but also mental and social differences, as described by the National Council for Special Education (2014). Special characteristics include a restriction in the capacity of the person to participate in and benefit from education on account of an enduring physical, sensory, mental health or learning disability, or any other condition which results in a person learning differently from a person without that condition. Children with special needs need to have adjustments in various fields. This adjustment is done so that children with special needs continue to get the same rights as other children, one of which is the right to education. These adjustments should enable all children with special needs to optimize growth and development.

Early childhood with autistic spectrum disorders is included in the criteria of children with special needs. Hallahan and Kauffman (Mangunsong, 2009) explain that these there are three areas in which autistic spectrum disorders are seen: communication skills, social interaction, and repetitive and stereotyped behavioral patterns (Mangunsong, 2009). The results of observations conducted by researchers in the field at Early Childhood Education Akramunas indicated that children had difficulty in interacting with their friends and they rarely communicate. However, children continue to do their activities according to their preferences.

Strock explained that today scientists have not yet definitely found out what causes a child to have autistic, but certainly the cause is more neurobiological than interpersonal (Hallahan & Kauffman in Mangunsong, 2009). It also has been found that children on the autistic spectrum come from family backgrounds with various levels of socioeconomic, intelligence, geographics, ethnic, and race (Widyawati in Mangunsong, 2009). Autism is not a certain economic class issue. Children from any socio-economic background can have autism. We conducted reseach at Akramunas, an upper middle class ECE centre. Thus our research participants came from affluent families.

In addition to experiencing interaction, communication, and behavioral disorders, children on the autistic spectrum also have additional characteristics, including disorders in cognition, sensory perception, motor, affection or mood, aggressive and dangerous behavior, and sleep and eating disorders (Hallahan & Kauffman in Mangunsong, 2009). The existence of these disorders influences children's attitudes and certainly inhibits other developments including creativity development (Ward, Smith, & Vaid, 1992). This explains that creativity is the result of children's cognitive processes. Therefore, stimulation is needed in the form of adjustments in early childhood education, especially for children with special needs. The hope is maximizing child growth and development in order to minimize the possibility of a bad life in the future of children.

Through early childhood education, children should be able to develop their potency in four core competencies: spiritual attitudes, social attitudes, knowledge, and skills. However, in the field there are a lot of teachers who experience difficulties in providing stimulation in the form of adjustments to children with special needs with various characteristics. The adjustments in question are environmental adjustments that can accommodate the needs of all children. The adjustments are in the form of the ability, skills, and knowledge of educators, adjusting learning facilities and infrastructure, and adjusting peers and adjusting learning activities to models, strategies, and approaches that are carried out with the aim of stimulating child growth and development. In this study, the adjustments analyzed were in the form of a scientific approach to early childhood learning activities with autistic spectrum disorders.

The scientific approach is a learning approach that provides opportunities for children to gain learning experience through observing, asking questions, gathering information, reasoning, and communicating (Permendikbud, 2013). All learning processes using a scientific approach support children in building their own knowledge through various kinds of creativity so that children are able to update information that has been obtained both in the form of attitude and creativity. Clearly, Soetardjo (Sunarto, 2015) states that a scientific approach is an approach that leads to the development of basic mental, physical and social abilities as a driving force for higher abilities in students' individual. Arkamunas uses a scientific approach in its teaching and learning activities, especially for children with autism. The approach enabled autistic children explore their creativity.

Basically, every child was born with creative potential. It is just that as time goes by the children's creative power decreases. This is due to the unnecessary restrictive rules at home and at school. Simonton (2000) explained that family environment plays a significant role in creativity development. These factors include birth order, early parental loss, marginality, and the availability of mentors and role models (Simonton, 2000). Parents and teachers have a large influence on increasing or decreasing children's creative power. All of this depends on stimulation offered to children. Therefore, stimulation or adjustment made by the teacher in the form of learning programs must be able to maintain the creative potency of children, including early childhood with autistic spectrum disorders.

2 METHOD

This research used qualitative research approach. This study aims at assessing the implementation of a scientific approach in the creativity learning process for children with autistic spectrum disorders. Through a scientific approach, children learn by seeing, touching, listening, and gathering information to construct their own knowledge so that the development of children's creativity can be enhanced.

The data collection technique in this study was by using observation. Observations were made in two sessions, where the first observation was made during learning activities using a scientific approach with five stages taking place in the classroom. Furthermore, the second observation was carried out to obtain further data regarding the development of children's creativity through five college activities with teachers and peers. The results of data collection were then followed up by analysis. The analysis process began with assembling raw materials and taking an in-depth overview or total picture of the whole process. This was then interpreted in the form of discussions and conclusions.

3 RESULTS AND DISCUSSION

We found that early childhood children with autistic spectrum disorders were five years old and had participated in joint learning activities with other children. Learning activities carried out at ECE Akramunas Institution using a scientific approach, this is adapted to the demands of the 2013 curriculum. The scientific approach includes observing, questioning, gathering information, reasoning, and communicating.

Table 1. Child observation results learning activities using the scientific approach.

Theme		: Animals
Name of Activity		: Make friends with fish
The Activities		: Get to know the fish in the tube
Num	Aspects	The Results
1	Observing	Children watch the fish in the tube with friends. The child also objected when asked to take turns with his friend. Sometime later the child rubbed the tube with his palm. The child also brings his ear to the tube. Next, the child engages himself without caring for other friends.
2	Asking	The child makes a sound, facing the teacher. However, it is not clear what children say. So the teacher gives stimulation by pointing towards the fish in the tube. The child makes a lot of noise that cannot be interpreted concretely.
3	Collecting Information	The child shows an inquisitive attitude by willingness to lift the fish tube. It's just stopped and continued by putting a hand into a fish jar. The child touches the water but does not touch the fish. Children have preoccupied themselves, but still under the supervision of the teacher.
4	Reasoning	Children point towards fish in turn.
5	Communicating	The child communicates well through body language that is trying to talk to the teacher while pointing towards the fish in the tube. The child makes a sound and feels happy when the teacher responds.

Stimulation in the form of adjustments using a scientific approach is indeed not easy for children with special needs like early childhood with autistic spectrum disorders. Odom et al. (2005) stated that due to its complexity of special educational research is one of the hardest to do. One feature of special educational research that makes it more complex is the variability of the participants. But looking at the observations clearly the scientific approach with five steps can provide the widest opportunity for children to explore. Early childhood with autistic spectrum disorders responded well to a fish in the tube.

The potency of creativity is a gift which God gives to every child, including early childhood with autistic spectrum disorders. Therefore, the child has the ability to learn something according to his own way (Rachmawati & Kurniati, 2010). The activities with a scientific approach enabled autistic children to explore, and to express their aspiration, because science has unlimited possibilities for experiment and discoveries (Nichols & Stephens, 2013). This is proven by the results of observation, early childhood children with autistic spectrum disorders

Table 2 . Observation notes of scientific approach for developing children's creativity.

Theme		: Animals
Name of Activity		: Make friends with fish
The Activities		: The collection of fish pictures from the paper
Num	Activities	The Results
1	Thicks the line to form a fish picture	Children begin to follow the teacher's instructions by holding a pencil. Children thicked the dash line around the picture in a hurry. As a result, the children failed to follow the dashed line neatly. Next, the children completes the fish picture with the help of the teacher.
2	Choose Free Color Paper	The children chooses the same colored paper as the fish in the tube which is orange.
3	Tear the paper into small sizes	The children only hold the paper and have not ripped it. Children see friends ripping each paper; slowly after that the children start tearing paper. However, the children then follow other children tearing of paper, but failed to rip it into small pieces. We saw some children also helping other children tearing the paper.
4	Pasting paper with glue	The children pasting the paper-fish to orange papers slowly. However, the children put too much glue on the paper-fish fish. So the fish collage became too moist and slightly blackened.
5	Sticking the fish collages on the board	Children follow their friends to stick fish collages they have made on the board. The child stuck it in the lower left corner.

begin to participate in activities with good response. Children choose a paper with the same color of the fish that they observed before. With this activity, the child can produce a work even though the results are not perfect. But the development of children's creativity has emerged.

4 CONCLUSION

Based on the results of the research and discussion, the following conclusions can be drawn: (a) the scientific approach has been implemented in learning activities and has been followed earnestly by early childhood children with autistic spectrum disorders in the hospital together with teachers and peers; and (b) learning activities using a scientific approach that has been done are able to stimulate the creativity of early childhood children with autistic spectrum disorders at Early Childhood Education Arkamunas.

REFERENCES

Permendikbud. (2013). *Peraturan menteri pendidikan dan kebudayaan nomor 81 a tahun 2013 lampiran iv tentang implementasi kurikulum*. Jakarta: Departemen Pendidikan Nasional.

Mangunsong, F. (2009). *Psikologi dan pendidikan anak berkebutuhan khusus*. Jilid Kesatu. Jakarta: LPSP3-Fakultas Psikologi Universitas Indonesia.

National Council for Special Education. (2014). *Children with special education needs*. Trim: National Council for Special Education.

Nichols, A. J., & Stephens, A. H. (2013). The scientific method and the creative process: implications for the k-6 classroom. *Journal for Learning Through the Arts*, *9*(1), 1–13.

Novitasari, Y. (2017). Development of child activity sheet by using the scientific approach at ethnic subtheme to introduce Indonesian cultural variety. In *Proceeding the International Conference on Education Innovation 1*(1), 116–120.

Odom, S. L., Brantlinger, E., Gersten, R., Horner, R. H., Thompson, B., & Harris, K. R. (2005). Research in special education: Scientific methods and evidence-based practices. *Exceptional Children*, *71*(2), 137–148.

Rachmawati, Y. & Kurniati, E. (2011). *Strategi pengembangan kreativitas pada anak usia dini taman kanak-* kanak. Jakarta: Kencana.

Simonton, D. K. (2000). Creativity: Cognitive, personal, developmental, and social aspects. *American Psychologist*, *55*(1), 151.

Sunarto. (2015). *Penerapan Pendekatan Saintifik untuk Meningkatkan Hasil Belajar Subtema Sikap Kepahlawanan Siswa Kelas IVA SDN Perak 1 Jombang*. Magister Thesis, Universitas Negeri Surabaya.

Ward, T. B., Smith, S. M., & Vaid, J. E. (1997). *Creative thought: An investigation of conceptual structures and processes*. Washington, DC: American Psychological Association.

Early Childhood Education in the 21st Century – Yulindrasari et al. (eds)
 ISBN 978-1-138-35203-2

Learning difficulties in children with ambidexterity

A.R. Pudyaningtyas & M.S. Wulandari
Sebelas Maret University, Surakarta, Indonesia

ABSTRACT: Children adopt stable hand use preference at the age of three years old or earlier, settling on their preference when they reach the age of five years. The dominance of a child's hand is very important not only they will become independent, but also for conducting learning activities when they enter school. In general, the world population is 90% right-handed, followed by 9% left-handed, and then a rare 1% being mix-handed or ambidextrous. Children who performed with no dominant hands face many difficulties, especially in activities involving the fine motor skills of physical ability. The research conducted in the form of qualitative descriptive research with an observation. The data collecting techniques were observation, test, and documentation analysis involving the photos of learning activities. The subject of this study was 4–5 years old children. The child with ambidexterity gave rise to poor performance in learning outcomes that were thought to be caused by weak cognitive perceptions manifested in poor perceptions of movement (fine motor skills), which was characterized by misdirections in writing (left to right), reversed letter errors, and confusions on transcribing the letters and numbers using the different hands. Verbal skill was developed slower and resulted in longer task completion time. This was identified by involving three activities of cutting, drawing and writing.

1 INTRODUCTION

Hand use preference is very important for children because children will develop more specialized abilities such as writing, cutting, holding crayons, and other special abilities that involve controlling their hands. These abilities will be the basis for developing a skill needed for learning. When a child is growing through the preschool years, hand control is very important not only because it does not just help the child to become more independent but also it is also connected with problem solving and learning (Woolfson, 2006).

Motor physical ability is a predictable ability, as its growth and development can be planned. Motoric development follows the law of the direction of development, in which motoric development can be predicted as indicated by evidence of changes in rough/general activities into special/precise activities (Hurlock, 2013). Early age is an ideal opportunity for children to learn to develop control of their muscles and movement (Beaty, 2013).

Children actually have a choice regarding their hand use preference. In early childhood children begin to show a tendency to prefer to use either their left or right hand, usually starting to become clear at the age of three years and reaching stability in hand use preference at the age of five years (McManus et al., 1988; Paula et al., 2003; Fisher, 2006; Woolfson, 2006; Papalia, Olds, & Feldman, 2009).

In the population of the world, handedness generally referred to the dominance of the hand-use in humans. Handedness was the most studied aspects of the asymmetry of the human brain (Ocklenburga, Besteb, & Güntürkün, 2013). Handedness could be classified as right hand, left hand, and mix-hand/ambidextrous. Right-hand and left-hand domination were the most common. However, there are several groups of people who could use both their hands, right and left, which were called mix-hand or ambidextrous. The right hand was for humans with the ability and skills to use their right hand with the dominance of the left hemisphere, the left hand was for humans with the ability and skills to use their left hand with right hemisphere domination, while mix-hand was the ability and skill to use both, right and left hands without the dominance of any of the two hemispheres.

Previous researchers found that children with ambidextrous hand use had twice the difficulty with learning disabilities during their school years (Brown, 2017; Bryner, 2010; Rodriguez et al., 2010). Hand use preference and hemispheric dominance were the determinants of children's success in learning. Children without hand use preference and domination of both hemispheres would experience many obstacles on their learning tasks. There was the assumption that ambidextrous children had many advantages with its effectiveness, but ambidextrous actually could inhibit the brain specialization (Brown, 2017). This was because the two hemispheres of the brain would give the same command signal, so when transmitted to an activity it would cause confusion. There was no command specification on the hemisphere of the brain so there would be no command settings. The study showed that ambidexterity (mixed hands) were "warning signs" of the potential damage and also as the tendency of the brain in developing learning disabilities and organizational problems caused by weakness in the "middle line crossing" of the hemisphere. There was a need for specialization of the cerebral hemisphere to organize tasks that had been programmed to be done (Sender, 2015; Brown, 2017). These children would experience twice as much difficulties twice as children with hand use preferences and dominance of one hemisphere of the brain.

2 METHOD

This study used a qualitative approach by conducting observations, involving very smart children age 4–5 years old, who had no hand use preference. The data collection methods used were: the first stage, observation, conducted to know the condition in the field related to ambidexterity learning difficulties; and second stage is the tests, using the McManus Test (1988) which was also used by Sender (2015) in identifying the preference of hand use in children. The test included 10 activity items, including: (a) drawing face; (b) writing name or color square; (c) throwing ball; (d) threading beads; (e) turning three cards; (f) using spoon; (g) cleaning teeth; (h) using comb; (i) blowing the nose; and (j) taking sweets. Tests were carried out on the subject to find out whether the child was right in being capable of using two hands or being ambidextrous; and the third stage is analysis of documents, to identify the learning difficulties in the activities of cutting, drawing, and writing, involving photo results of learning activities.

3 RESULT AND DISCUSSION

Learning outcomes became a reference for learning success and referred to children's cognitive abilities development. Hand use preference was an indicator of learning success in children. One important determinant of children's cognitive development was the hand-use of either the left or the right (Johnston, Nicholls, Shah, & Shields, 2009). Weak and non-existent hand use preferences were associated with developmental disorders of cognitive function in certain populations (Fagard, Chapelain, & Bonnet, 2015).

It was assumed that ambidextrous children had many advantages. But the usage of both hands could actually inhibit brain specialization (Brown, 2017). Previous studies showed that ambidexterity (mixed hands) was a "warning sign" for the potential damage and tendency of the brain for learning disabilities and organizational problems (Sender, 2015). The results of previous studies and theories of finding poor learning outcomes were associated with weak brain specifications in organizing commands. The results found that children with ambidexterity gave rise to poor performance in learning outcomes that were thought to be caused by weak cognitive perceptions that were manifested in poor perceptions of movement (fine motor skills). Learning difficulties in children's with ambidexterity were mainly seen in the activities of writing, cutting, and drawing. The impact of "middle line crossing" weakness on ambidexterity were difficulties in fine motor skills influencing writing, drawing, and cutting (Sender, 2015).

3.1 *Writing*

Writing is the one skill that required preparation. Writing was the result of cognitive perception that had occurred in perception motion involving smooth muscles. Writing was a combination of manual skill, hand-eye coordination, muscle, and ripeness of highly complex nerve and visual perception skills (Woolfson, 2006).

At the Age 4 to 5 years old children on the phase hand-eye coordination development. So the children could write their first name without needing to imitate (Woolfson, 2006). Children aged 4 and 5 years must be agile in writing letters and numbers so they were illegible, as the ability of perceptual and memory helped children form, store, and take the concept of form (Beaty, 2013). In fact, children with ambidextrous manner experienced difficulty in the name of and illegible letters (words); when subjects were asked to write their names, it was found that all letters were reversed in the directions the letters were facing (especially in letter a, e, h, o, s) with the wrong writing direction (from right to left). In addition, there were many errors in writing in letters (especially letters b, p, d) or numbers (almost all numbers) on children using different hands. Children with ambidexterity did not have brain dominance, so this caused a weak cognitive perception which resulted in difficulty for the children in comprehending letters and numbers.

The natural development of writing and reading took place simultaneously in children (Beaty, 2013). It became reasonable when the ambidextrous children experienced difficulty in develop skill writing, as there were delays in verbal skill of pre-reading. Writing and reading for children was developed together, and nothing was developed first. The results found that verbal abilities that were developed more slowly were intended abilities that involved letter recognition (subjects had difficulty distinguishing, especially letters b, d, p), known as literacy skills, especially in pre-reading activities, developed more slowly. The capability related to the ability and skill of language (such as memorizing letters, hadiths, and communicating) were developed similarly to those of their age.

Children with ambidexterity actually had weakness in "middle line crossing" in the brain, so the brain hemisphere could not specify the commands (Sender, 2015). This difficulty was hypothesized to develop because the brain hemispheres compete with one another, as both sides of the brain were given same sensory information (Brown, 2017). This resulted in weak cognitive perception on forms of words and numbers that caused the study results to be bad, the children's early pre-reading ability included. Ambidextrous children produced worse performance in all activities because the brain could not give a signal to one side of the body to do the desired activity, which caused the development of the skills and learning skills that must be developed to be learned more slowly.

3.2 *Drawing*

Drawing and coloring images was one of the activities that were endeared and attractive to preschool children. Almost all children liked to draw with the fingers of the hand, even the ones who did not showing interest drawing and coloring (Woolfson, 2006). Ambidextrous children showed the same interest in drawing and coloring activities. The findings showed that images of ambidextrous children were no different from children of their age. The result of the picture was a combination of lines and circles by including text to indicate the meaning of the image.

From the age 2 - 4 years old, some put more streaks full of lines and circles (Beaty, 2013). The difference was the preference of the use of hands who settled in finding how to make scrawl fine line with the direction and circle definitely (only have one way) while in an ambidextrous child, direction make scrawl become more random. Scrawl direction could perceive as developing initial writing skills related to the direction of writing and developing printed letters. Supposedly, there was no standard rule that making scribbles had to start from the right, or left, from top or bottom. However, a definite direction helped children to perceive form images in relation to their hand control to develop skills and knowledge of printed letters. Writing and drawing both represented symbolic objects (Beaty, 2013).

3.3 *Cutting*

When the child reached the age of 3 years the child began to use scissors, in part because the size of his fingers and hands increased but also because his grip became more mature (Woolfson, 2006). Ambidextrous children had a higher likelihood of learning difficulties and poor development of fine motor physical skills. Cutting involved skills in operating the tool. The form of fine motor rounds was one of the ways children complete learning tasks. Fine motor skills techniques were involved in cutting activities. The results showed that children with ambidextrous were more likely to finish the task longer when performing this activity. This group of children needed to observe and conduct experiments via trial and error to find a way to complete the "cutting" learning task.

The results found that when an ambidextrous child was asked to cut a circle's shape, the direction of cutting was the opposite. It was different from cutting a circle that involves the operation of cross between hands (right hand to the left, left hand to the right), cutting straight lines (only involving the directions of left and right). In this case the right and left for the right hand and the left hand were the same so it did not apply cross-hand operation. This made it easier for them and, as a result, they spent less time working.

4 CONCLUSION

Hand use preference was very important for children, not only to help children to become more independent but it would be developed into learning skills. Children without hand-preference or also known as ambidexterity had learning difficulties involving fine motor skills especially in the activities of writing, drawing, and cutting. It was important for teachers and parents to pay more attention to the difficulties showed by children, so the detection could happen earlier, in order to identify the specific needs of children, provide intervention more quickly, reduce healing time, and increase the percentage of results.

REFERENCES

Beaty, J. J. (2013). *Observasi perkembangan anak usia dini*. Jakarta: Kencana Prenadamedia.

Brown, J. (2017). Ambidextrous brains: How handedness affects the brain. *Live Science Managing Editor*.

Bryner, J. (2010). Ambidextrous children may have more problems in school. *Live Science Managing Editor*.

Fagard, J., Chapelain, A., & Bonnet, P. (2015). How should "ambidexterity" be estimated? *Laterality, 20* (5), 543–570.

Fisher, J. R. S. (2006). *Psychosocial differences between left handed and right handed children* (Unpublished thesis). Department of Counseling, Educational and School Psychology and the Faculty of the Graduate School of Wichita State University, Wichita, Kansas, United States of America.

Hurlock, E. B. (2013). *Perkembangan anak jilid 1*. Jakarta: Erlangga.

Johnston, D. W., Nicholls, M. E. R., Shah, M., & Shields, M. A. (2009). Nature's experiment? Handedness and early childhood development. *Demography, 46*(2), 281–301.

McManus, I. C., Sik, G., Cole, D. R., Mellon, A. F., Wong, J., & Kloss, J. (1988). The development of handedness in children. *British Journal of Developmental Psychology, 6*, 257–273.

Ocklenburga, S., Besteb, C., & Güntürkün, O. (2013). Handedness: A neurogenetic shift of perspective. *Neuroscience and Biobehavioral Reviews, 37*, 2788–2793.

Papalia. D. E., Olds, A. W., & Feldman, R. D. (2009). *Human development: Perkembangan manusia*. Jakarta: Salemba Humanika.

Paula, M. L., Doril, H. A., Firas, E., John, T., & David, T. N. (2003). Measuring normal hand dexterity values in normal 3-, 4-, and 5-year-old children and their relationship with grip and pinch strength. *Journal of Hand Therapy, 16*(1), 22–28.

Rodriguez, A., Kaakinen, M., Moilanen, I., Taanila, A., & McGough, J. J. (2010). Mix-handedness is linked to mental health problem in children and adolescents. *Pediatrics, 125*(2), e340–e348.

Sender, Y. (2015). An intervention program for improving writing and information retrieval among students with ambidexterity. *Social and Behavioral Sciences, 209*, 565–571.

Woolfson, R. C. (2006). *Anak yang cerdas: memahami dan merangsang perkembangan anak anda*. London: Karisma Publishing Group.

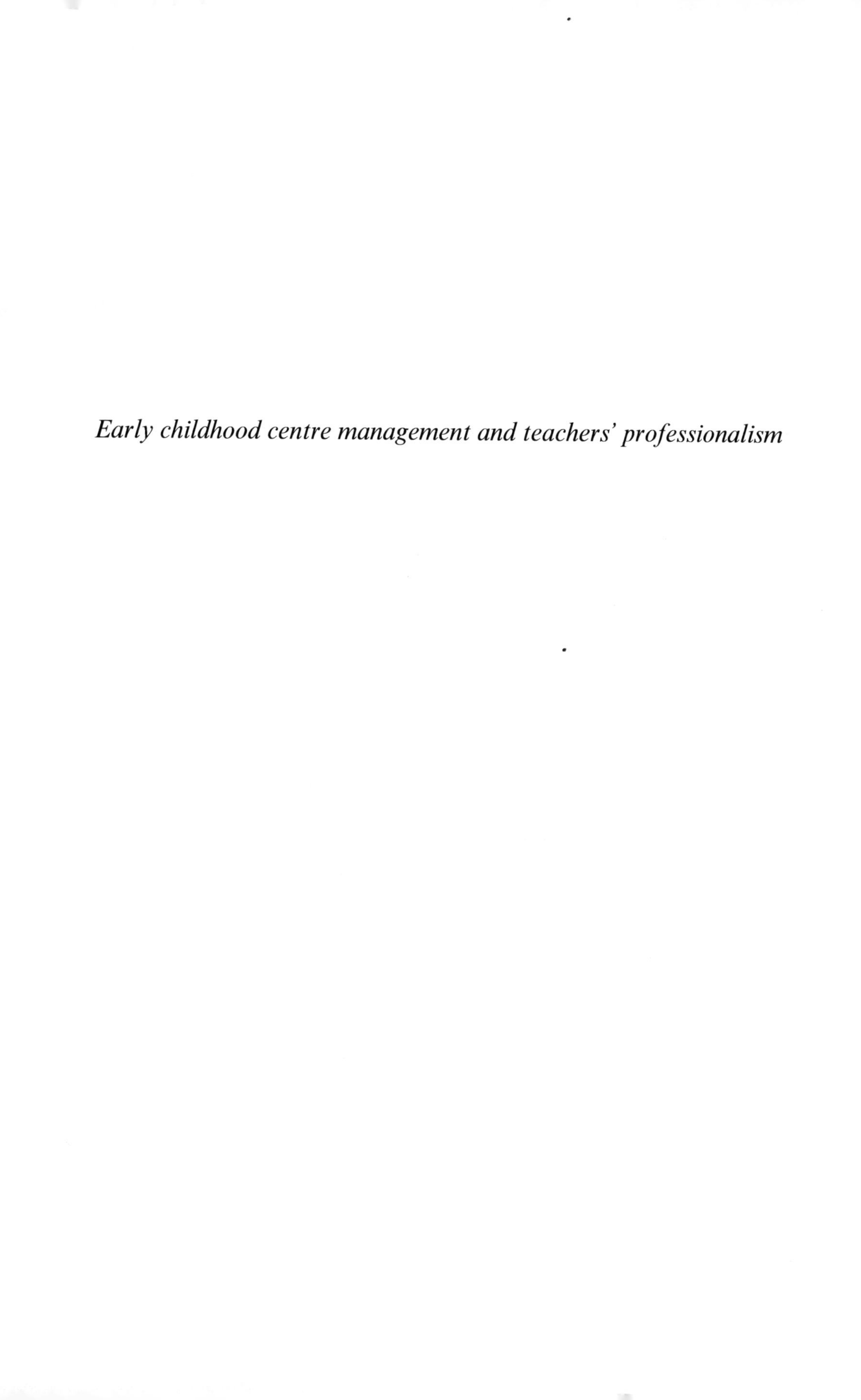

Early childhood centre management and teachers' professionalism

Early Childhood Education in the 21st Century – Yulindrasari et al. (eds)
© 2020 Taylor & Francis Group, London, ISBN 978-1-138-35203-2

The perspectives of kindergarten teachers on the demands of teacher professionalism in rural areas

M. Nur & Y. Rachmawati
Universitas Pendidikan Indonesia, Jawa Barat, Indonesia

ABSTRACT: This research was motivated by certain unjust and unsubstantiated criticism toward teachers in Indonesia stating that many teachers are allegedly less qualified and professional all without seeing actual condition of teachers while holding those educational demands. Motivated by these occurrences, the researchers were interested in exploring teachers' perspective on professionalism demands in a certain rural and remote area.

1 INTRODUCTION

Preschool education is considered to be the early foundation in developing all aspects of children development including their physical, cognitive, social, and emotional and language abilities. Therefore, regarding its significant nature, it is fair that the government and society require kindergarten teachers to be qualified and professional. The Indonesian government has specifically regulated national standards for educators under No. 19 (2015) article 19 paragraph 1, stating that educators are required to (1) acquire a minimum education qualification of diploma four (D-IV) or a Bachelor's degree (S1); (2) possess higher education background of early childhood education, other fields of education, or psychology; and (3) obtain teachers' certification for early childhood education. Moreover, they are also obligated to meet another four standards: (1) personal competence, (2) professional competence, (3) pedagogical competence, and (4) social competence. The Ministry of Education and Culture of Indonesia regulation No. 58 (2009) regarding educational standards of early childhood education, chapter III, on the standards for teachers and educational staff states that early childhood educators are professionals who are responsible for planning, preparing and implementing the learning process, assessing learning outcome, providing guideline and advice, and ensuring the safety and protection of students. Early childhood educators in Indonesia serve in various types of educational services comprising of both formal and non-formal education such as kindergartens, *Raudhtaul Athfal* (Islamic kindergarten), *Kelompok Bermain* (children's playgroups), and other services.

In addition to teaching, teachers are also required to fill in educational basic data forms, participate in teacher forums such as the Association of Kindergarten Teachers Indonesia, to take a test for pedagogical and professional competence, to actively participate in online training programs of continued professional development, and also strive to meet the expectations of the superintendent and the foundation. Handling those demands has never been an easy task especially for those who are in rural and remote areas which are basically limited in Internet access, public transportation and facilities. Consequently, teachers have to strive to fulfill these demands within such limitation. Within their limited salary around Rp. 500.000 or equivalent US$35.5, to Rp. 1000.000 or equivalent of US$71, teachers have to spend Rp. 60.000- or equivalent to US$ 4.26, daily for using a public transportation. Unfortunately, there are certain unjust and unsubstantiated criticisms found in studies and newspapers toward teachers in Indonesia, stating that many teachers are allegedly less qualified and professional (Kartowagiran, 2011), without taking into account actual condition of teachers.

Much has been written regarding professionalism, either on a domestic or international scale . Motivated by the fact that there is unjustified criticism regarding professionalism in rural areas, specifically in the tea plantation area of Pasirjambu, researchers are inspired to conduct research related these issues to fill in the gap in the literature and help teachers cope with frustration and stress. As Sato (2014) said that the majority of teachers undergo stressful situations due to the nature of the teaching profession which is extremely challenging. In addition to that, teacher's frustration and stress may be caused by the lack of teacher support, disappointments, excessive workloads, and inadequate classroom management (Breaux & Wong, 2003). Next, this research aims to answer the following questions: (1) What are teachers' perspectives on teacher's professionalism?; (2) How do teachers handle those demands within limitations while living in such premises; and (3) What are teachers' expectations toward the government and other stakeholders?

Six different kindergarten teachers from a remote area, situated in the tea plantation Patuha, participated in the study the teachers are employed as kindergarten teachers by the plantation company as well, and as a result, during holidays, they have to work as tea farmers.

2 METHODS

This study adopted a qualitative approach using the paradigm of phenomenology as it offers greater flexibility in exploring ways in which kindergarten teachers perceive professionalism in their profession and it also restricts preconceived assumptions about their experience, feelings as well as responses toward their profession. Regarding interview protocol, researchers used content mining and content mapping to allow respondents to broaden their perspectives, to stimulate their minds, clarify their experience and their holistic understanding (Legard, Keegan, & Ward, 2003). Probing questions were posed to gain detailed description of respondents' experience when dealing with professional demands (Legard et al. 2003).

As for data analysis, emerging themes and categorizes were obtained using constant comparative to code and analyze data to develop concepts and to make comparisons among respondents (Kim, Park, Malonebeach, & Heo, 2016; Legard et al. 2003). Within data analysis, researchers followed the following steps. Firstly, raw data were analyzed. Next, comparisons among preliminary categories were conducted to determine consistent classification. Following that, researchers developed clusters of ideas.

3 RESULT AND DISCUSSION

After analyzing the data, it was revealed that it has one major theme that is a gap between awareness and actual conditions. It falls under three categories: (1) salary-related issues; (2) location- and facilities-related issues; and (3) organization-related issues. As per expectation, it falls into a major theme which gives more attention toward teachers in rural and remote areas.

3.1 *Disparity of awareness and real condition*

Although the terminology of professionalism in the teaching profession varies among countries, most of definitions are in line that professionalism demonstrates behavior that portrays the knowledge and skills of the profession (Creasy, 2015). Furthermore, Creasy (2015) explained that professional teachers' responsibilities would include demonstration of responsibilities toward profession, students, and the school district and of course the community. When professionalism within educational institutions is valued, respect among all educational aspects will increase, and it will result in the achievement of what has been expected (Sockett, 1993). In relation to that matter, all respondents were completely aware of the importance of professionalism in their profession as kindergarten teachers and they actually strive to fulfill those demands, but challenges and barriers were too much to be handled.

3.2 *Salary-related issues*

Low salary level is the first barrier in the implementation of professionalism. As shown in research conducted by Gonzalez, Brown, and Slate (2008) low salary is one of the most predominant reasons for leaving the teaching profession, aside from the lack of administrative support and difficulties with student disciplines (Burgess & Ratto, 2003). The majority of the respondents were positive about this issue as it somehow affects many life aspects of respective respondents including their households and live expenses (Firestone, 2014). For example, Ida (32) stated:

> Ida is aware of that being teacher must be professional and being professional teacher must acquire certain academic qualification which is bachelor degree. Nonetheless, I'm honestly burdened if I have to go through academic process to gain such degree, not that i did not want to, to improve my knowledge by gaining degree, but the distance and cost would become my greatest challenge. I might be able to deal with distance, but when it comes to the tuitions and my salary, I could not do anything; moreover I have family to take care of. What important for me in this condition is that I will do everything what I can do to ensure all children in this tea plantation area undergo early childhood education as it is really important.

Reflecting on that, it can be seen that certain professionalism demands do burden kindergarten teachers serving in the area. With regard to the significant aspects of professionalism, they are quite aware of that, yet such challenges definitely hinder their expectation to meet those qualifications specifically if related to salary as without a doubt that with certain amount of money or salary, teachers might be able to fulfill basic needs (survival). Further, people will definitely attempt to fulfill minimum resources necessary for their life before moving to subsequent level (Pope-Davis, 2003). Aside from that, according to Westling and Whitten (1996) professionals or in this case, kindergarten teachers on those premises might make excellent teachers.

3.3 *Location- and facilities-related issues*

Another sub-theme that emerges is related to location and facilities issues. Lack of accessibility toward public transportation, higher education, sufficient facilities, knowledge sources, and teaching aids all contribute to the emerging challenge in meeting the required level of professionalism (Westling & Whitten, 1996). It is very common and natural that early childhood education should be equipped with particular learning instruments, various toys and so on, yet because rural areas have less access, such beneficial instruments were scarcely found. Hence, teachers here mostly use traditional games and devices to conduct the education process; they believe that despite the fact that they have no sufficient learning instruments or facilities due to living in these areas with certain limitations, they will strive to provide a proper education for their students. One of the ways is through using traditional games.

For example, Sri (34) said:

> I realize that educating early children should be careful, additionally, we have to be patient as today children are quite smart. Since we live in rural area, children are completely active especially their physical and motoric abilities. Unfortunately, our classroom is not that spacious so we have to be really creative.

While conducting observation in those two schools, researchers discovered that the teaching process was considerably good. During which time, the children learnt how to plant flowers then learnt to draw. Second observation was conducted in another school located in the tea plantation; researchers were quite interested that teachers were quite creative as they can creatively set such small learning spaces into comfortable classrooms using plywood and so on.

Regarding the online training program of continued professional development for teachers, they have trouble in answering the questions as they have no previous knowledge because they don't have sufficient Internet access or college education. Other factors such as limited access of Internet and information are related to teachers' inability to properly and accurately make daily, weekly, semester and annual classes tend to imitate from the school nearby. A subsequent effect is related to filling in basic data online because it leads to another expenditure in order to hire teachers or computer experts to do such administration tasks because they are less computer-savvy.

3.4 *Organization-related issues*

Organizational problems are seen to be another contributor for teachers' problem in handling professional demands (Burgess & Ratto, 2003; Westling & Whitten, 1996). As schools situated in rural areas and owned by private factors or certain foundations ruling these areas, kindergarten teachers in that premises often receive challenges from the principal or the school owners, who often neglect teachers' needs and demands. Teacher participation in forums conducted by work groups or Indonesia Kindergarten Associations is basically helping the teachers to improve their professionalism. Nevertheless, such participation is quite scarce as no financial sources are present to support them in active participation. For instance, Marni (35) states:

> IGTKI and KKG are definitely beneficial but unfortunately we have no proper access for transportation as we have to go to the meeting which is basically 3 to 4 hours travel. Sometimes, we have to use our limited money around Rp.120.000 for going there. For me it was expensive. We hope toward the government and all stakeholders to support teachers like us by giving much effort to build good facilities and proper public transportation, giving us continually monthly subsidy like we used to earn around Rp.300.000 within three months.

4 CONCLUSION

It is apparent that kindergarten teachers in the rural area of the Pasirjambu sub-district are aware of the significance of professionalism in their profession but unfortunately while striving to meet those qualifications, from their perspective, there is one major theme: the discrepancy between their awareness and actual condition falling into three sub-themes. As for expectations, they hope that all stakeholders could fund salary increases to support them in handling the professional demands in this era of globalization.

REFERENCES

Breaux, A. L. & Wong, H. K. (2003). *New teacher induction: How to train, support, and retain new teachers*. Mountain View, CA: Harry K. Wong Publications.

Burgess, S. & Ratto, M. (2003). The role of incentives in the public sector: Issues and evidence. *Oxford review of economic policy*, *19*(2), 285–300.

Creasy, K. L. (2015). Defining professionalism in teacher education programs. *Online Submission*, *2*(2), 23–25.

Firestone, W. A. (2014). Teacher evaluation policy and conflicting theories of motivation. *Educational researcher*, *43*(2), 100–107.

Gonzalez, L. E., Brown, M. S., & Slate, J. R. (2008). Teachers Who Left the Teaching Profession: A Qualitative Understanding. *Qualitative Report*, *13*(1), 1–11.

Kim, J., Park, S. H., Malonebeach, E., & Heo, J. (2016). Migrating to the East: A qualitative investigation of acculturation and leisure activities. *Leisure Studies*, *35*(4), 421–437.

Kartowagiran, B. (2011). Kinerja guru profesional (Guru pasca sertifikasi). *Cakrawala Pendidikan*, (3), 463-473

Legard, R., Keegan, J., & Ward, K. (2003). In-depth interviews. *Qualitative Research Practice: A Guide For Social Science Students And Researchers*, *6*(1), 138–169.

Pope-Davis, D. B., Heesacker, M., Coleman, H. L., Liu, W. M., & Toporek, R. L. (2003). *Handbook of multicultural competencies in counseling and psychology*. California: Sage.

Sockett, H. (1993). *The moral base for teacher professionalism.* New York: Teachers College Press.
Sato, M. (2014). Mereformasi Sekolah: Konsep dan praktek komunitas Belajar. *Tokyo: JICA Publication, Translation.*
Westling, D. L., & Whitten, T. M. (1996). Rural special education teachers' plans to continue or leave their teaching positions. *Exceptional Children, 62*(4), 319–335.

Early Childhood Education in the 21st Century – Yulindrasari et al. (eds)
© 2020 Taylor & Francis Group, London, ISBN 978-1-138-35203-2

Teachers' knowledge about the assessment of child development in the Jabodetabek region

S. Hartati & N. Pratiwi
Department of Early Childhood Education, Faculty of Education, Universitas Negeri Jakarta, Jakarta, Indonesia

ABSTRACT: This study describes Early Childhood Education (ECE) teachers' knowledge of child development assessment in Jabodetabek. Ninety teachers of Early Childhood Education (ECE) who worked in the Jabodetabek region participated in the study. Data collection of teachers' knowledge of child development assessment used multiple choice test. The data was analyzed using a quantitative approach. The teachers in this study had diverse educational background and teaching experiences. The results show that 60% of teachers were able to answer more than 50% of questions regarding preparation for early childhood assessment. The data also showed that 31% of teachers were able to answer more than 50% of questions regarding implementation of early childhood assessment and 21% of teachers who were able to answer more than 50% of questions regarding reporting on early childhood assessment.

1 INTRODUCTION

Assessment has a pervasive influence in schooling as it affects on how children learn and how teachers teach. Assessment as important as learning in measure child development (Pang & Lang, 2011). Assessment of the learning development is very important for teachers to provide fundamental data on the knowledge, understanding, skills, interests, and dispositions of children. The data can then be used as a fundamental information of curriculum development to strengthen competencies and provide appropriate experiences to support children's learning and development (Schweinhart, Debruin-Parecki & Robin, 2004; Nah & Kwak, 2011; Huber and Skedsmo, 2016; Stringher, 2016).

In fact, many assessments carried out in educational institutions still focus on the assessment of children's learning outcomes, not the collection of information about children's learning processes and child development. Therefore, the research focusing on assessment of child development. This shows that ECE educators in Jakarta have known the importance of conducting developmental assessments for early childhood. This study was carried out in the Jabodetabek area to examine the knowledge and abilities of teachers about child development assessment.

Teachers play an important roles in students' achievements (Harris & Sass, 2007; Canales & Maldonado, 2018). It is urgently required for early childhood educators to understand child development. Knowledge about child development becomes the fundamental knowledge for ECE educators in developing learning activities that are appropriate to the age, interests, and needs of children (Massyrova et al., 2015; Pauker et al., 2018). Educators need to take into account the developmental needs of the children in designing educational activities. Developmental assessment used as a strategy for teachers to gather information about the children's developmental needs, therefore teachers should have knowledge about early childhood assessment (Goldstein & Flake, 2016).

Assessment is the activities that teachers and students undertake to get information that can be used to alter the process teaching and learning (Wang, 2014; Amua-Sekyi, 2016) required carefully design. Assessment serves as a form of formative assessment (Nicol & MacFarlane-Dick, 2006; Kearney, Perkins & Perkins, 2011; Agathangelou, Charalambous & Koutselini,

2016) plays important roles in the teaching and learning process; therefore, it is central to classroom practice and is sensitive to the learning process (Mulliner & Tucker, 2017).

Assessment has pervasive influence in schooling as it affects on how children learn and how teachers teach. Assessment as important as learning in measure child development (Pang & Lang, 2011). Assessment of the learning development is very important for teachers to provide fundamental data on the knowledge, understanding, skills, interests, and dispositions of children. The data can then be used when working on curriculum development to strengthen competencies and provide appropriate experiences to support children's learning and development (Schweinhart, Debruin-Parecki & Robin, 2004; Nah & Kwak, 2011; Huber & Skedsmo, 2016; Stringher, 2016).

In fact, many assessments carried out in educational institutions still focus on children's learning outcomes, not the collection of information about children's learning processes and child development. Therefore this research aims to assess teachers' knowledge about children development assessment.

2 METHOD

This study used a paper-test on teachers' knowledge of assessment for early childhood. The study was conducted from May to September 2017 in the Jabodetabek area and involved 90 teachers. Jabodetabek consists of DKI Jakarta, Depok, Bogor, Tangerang, and Bekasi. Data from this study are described to obtain an overview of ECE teachers' knowledge of early childhood assessment. Data analysis was conducted based on a test that was completed by teachers in ECE institutions. The teacher completed several multiple choice tests to see the teacher's knowledge about assessment. The distributions of score given are listed below (Table 1).

3 RESULT AND DISCUSSION

3.1 *Result*

All the teachers hold a Bachelor's degree in education with various study programs, such as early childhood teacher education programs, Islamic primary education teacher programs, Islamic studies, economics, and counseling. The teachers' teaching experiences ranged between 2–20 years.

The data analysis shows that ECE teachers' knowledge of children's development assessment is diverse. This is influenced by educational background and teaching experience. The teachers know and understand the general development of the child. Teachers understand that the child development is important in creating learning activities and designing developmentally appropriatc. In conducting assessments, the teacher is also familiar with various techniques, but assessment techniques have not been widely used when observing child behavior. The detail of the finding can be seen in Table 2.

Table 1. Research aspect and indicator.

No	Aspect	Indicator	Item	Score
1	Preparation for early childhood assessment	Understand the early childhood development	1,12,21,25	10
		Design an assessment model for early childhood development	3,4,20,23	15
2	Implementation of early childhood assessment	Observe the behavior of early childhood	2,6,10,14,24	20
		Use unique assessment model	7,15,16,17	25
3	Reporting on early childhood assessment	Document child development	9, 11,13,19	15
		Report the assessment results	5, 8,18,22	15
Total			100	

Table 2. Result 1.

Statistics		
	Valid	90
N	Missing	0
Mean		60,8444
Median		60,0000
Mode		68,00
Std. Deviation		14,19599
Variance		201,526
Range		52,00
Minimum		36,00
Maximum		88,00

Based on the data analysis of the results of the study on 90 respondents regarding the assessment knowledge of the results of the workmanship and with a range of theoretical values from 0 to 100 it is known that the lowest value of knowledge about assessments is 36 while the highest score is 88. Data was calculated to have mean 60.84, median 60, mode 68, standard deviation 14.20, and variance 201.53. Frequency distribution and the assessment value of knowledge from 90 respondents are shown in Table 3.

Table 3. Result 2.

Interval					
		Frequency	Percent	Valid Percent	Cumulative Percent
Valid	36–42	10	11,1	11,1	11,1
	43–49	15	16,7	16,7	27,8
	50–56	16	17,8	17,8	45,6
	57–63	7	7,8	7,8	53,3
	64–70	19	21,1	21,1	74,4
	71–77	10	11,1	11,1	85,6
	78–84	8	8,9	8,9	94,4
	85–91	5	5,6	5,6	100,0
	Total	90	100,0	100,0	

Table 4 shows the distribution of knowledge values of assessments that have been divided into eight groups. Based on the results of Table 3, the histogram graph is obtained as shown in Image 1.

Based on the data in Image 1, it can be seen that the distribution of the highest values is between 64–70, or 21.1% of the total respondents. Based on the histogram, it can be seen that 53% teachers are below the average value. The histogram pattern appears to follow the normal curve, even though there are some data that appear outlier, but in general the data distribution follows the normal curve, so it can be concluded that the data is normally distributed.

Based on the ANOVA test conducted on the respondent's test scores it appears that three aspects are significant with F 20,494. The Scheffe test results (Table 5) shows that the results of the teacher's knowledge test preparation for early childhood assessment were significant to the teacher's knowledge of the Implementation of early childhood assessment and reporting on early childhood assessment. However, the results of the teacher's knowledge of implementation and reporting on early childhood assessment are not significant. This shows that

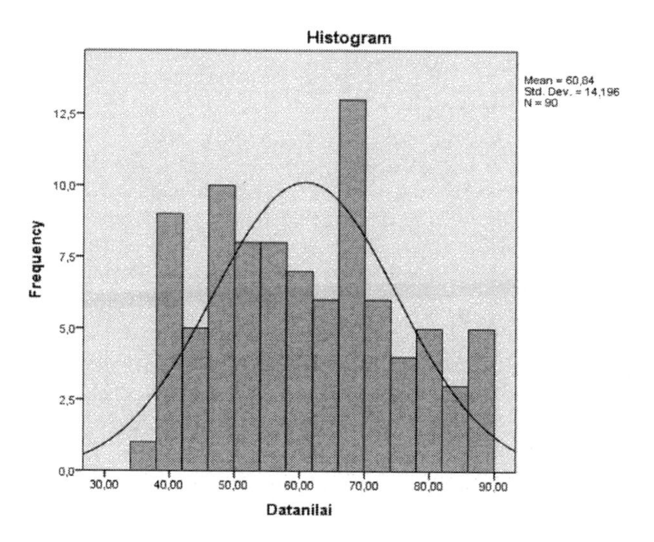

Image 1. Research histogram.

Table 4. ANOVA result 1.

ANOVA					
skor					
	Sum of Squares	df	Mean Square	F	Sig.
Between Groups	89,652	2	44,826	20,494	,000
Within Groups	584,011	267	2,187		
Total	673,663	269			

Table 5. ANOVA result 2.

Multiple Comparisons						
Dependent Variable: skor Scheffe						
					95% Confidence Interval	
(I) klpkm	(J) klpkm	Mean Difference (I-J)	Std. Error	Sig.	Lower Bound	Upper Bound
Preparation	implementation	,98889*	,22047	,000	,4462	1,5316
	Reporting	1,36667*	,22047	,000	,8240	1,9094
Implementation	preparation	-,98889*	,22047	,000	-1,5316	-,4462
	Reporting	,37778	,22047	,232	-,1649	,9205
Reporting	Preparation	-1,36667*	,22047	,000	-1,9094	-,8240
	implementation	-,37778	,22047	,232	-,9205	,1649

*. The mean difference is significant at the 0.05 level.

teachers are more understanding about preparation for early childhood assessment than implementation and reporting on early childhood who are at the same level of knowledge.

The average teacher is able to answer 6 out of 8, or 75%, of questions given regarding preparation for early childhood assessment which includes understanding early childhood development and designing an assessment model for early childhood development, whereas the average teacher can answer 5 out of 9 questions, or 56%, of the total questions regarding the implementation of early childhood assessment, and only answer 4 of the 8 questions, or 50%, of the total questions about reporting on early childhood assessment.

As many as 60% of teachers were able to answer more than 50% of questions regarding preparation for early childhood assessment. The data also showed that 31% of teachers were able to answer more than 50% of questions regarding implementation of early childhood assessment and 21% of teachers who were able to answer more than 50% of questions regarding reporting on early childhood assessment.

3.2 *Discussions*

A research shows that teachers master the basic concepts of assessments but are weak in implementation and scoring, which causes the teacher to only be able to implement the assessment simply. This is consistent with research conducted by Odaba, Cimer and Cak (2010) in Turkey, where teacher has lack of knowledge about the performance assessment. Some studies show that the assessment training that teachers receive as undergraduate students does not prepare them to feel comfortable with the decisions they have to make on a regular basis (Mertler & Mertler, 2009; Furtak et al., 2016). Some studies show that teachers were not adequately prepared to assess student learnin and do not have sufficent knowledge about children assessment (Craig, 2003; Mertler & Mertler, 2009). The provision of additional knowledge about the implementation of assessments was needed by teachers to support the course of learning in the classroom.

According to Brookhart and Llc (2011), the teacher should be able to understand the basic assessment especially implementation and scoring. Teachers must be able to build assessment schemes that measure student performance in class assessment about students, classrooms, schools, and districts that lead to increased student learning, growth, or development. In addition, teachers must also be able to manage external assessments and interpret their results for decisions about students, classrooms, schools, and districts. Teachers must be able to articulate their interpretations of the results of the assessment and their reasons for educational decisions based on the results of the assessment on the education population they serve.

The teacher's understanding of authentic assessment will influence how the teacher obtains information about a child's behavior without the assumption of the teacher. The teacher's skills in conducting authentic assessments are important in conducting an accountable assessments. Authentic assessment provides comprehensive picture of the child's development. Goldstein and Flake (2016) argue that teacher must understand the multidimensional nature of assessments and the restrictions that they must administer. Kostelnik et al. (2018) also argues that teacher involves in an ongoing, strategic, and purposeful assessment and evaluation. Daily, they are active in documenting what students' process in acquire knowledge, the progress being made toward learning and developmental goals, and whether various aspects of their programs support each child's growth.

Furthermore teacher can report the results of the assessment to parents so that parents are being informed about their children's development. The assessment results should benefit children's learning in school by connecting the experiences and knowledge that children have. The assessment must be generate clear and meaningful information for each child. It also should include all assessment strategies to provide a comprehensive picture of the progress and children's needs. The teacher chooses a relevant assessment method to get the information needed and to use the result in planning the curriculum and learning. Teachers should also pay attention to individual differences and provide appropriate learning plan for each child.

The teachers are able to carry out observation activities to assess the children; however, the assessment was often limited to the anecdotal record, rating scale, and portfolio. This is different from teachers in several other countries. Teachers in Italy carry out complex assessments ranging from cognitive, metacognitive and socio-affective motivational assets of the students (Stringher, 2016). Teachers in South Korea conduct a comprehensive assessment of the attitudes, skills, and knowledge of children rather than a focus on specific achievements, affective domains and daily experiences (Kitano, 2011; Nah & Kwak, 2011). Teachers in South Korea utilize a variety of systematic methods such as observation, examination of work samples, interviewing, and gathering information from various sources (Nah & Kwak, 2011). Assessment must align with the main principles embedded in educational goals and student learning objectives (OECD, 2013).

4 CONCLUSION

The results of the study shows that most ECE teachers in the Jabodetabek region had knowledge of child development assessment. that teachers master the basic concepts of assessments, but that they are weak in implementation and scoring, this causes the teachers are only be able to implement the simply assessment. In conducting assessments, most teachers use anecdotal records and checklists. However, they have not used running records, event sampling, time sampling and other technological tools. In the future, teachers should be given training on other authentic assessment methods comprehensively.

REFERENCES

Agathangelou, S. A., Charalambous, C. Y., & Koutselini, M. (2016). Reconsidering the contribution of teacher knowledge to student learning: Linear or curvilinear effects?. *Teaching and Teacher Education*, 57, 125–138. doi: 10.1016/j.tate.2016.03.007.

Amua-Sekyi, E. T. (2016). Assessment, Student Learning and Classroom Practice: A Review. *Journal of Education and Practice*, *7*(21), 1–6. Available at: http://search.ebscohost.com/login.aspx?direct=true&db=eric&AN=EJ1109385&site=ehost-live.

Brookhart, S. M., & Llc, B. E. (2011). Educational Assessment Knowledge and Skills for Teachers, *30*(1), 3–12.

Buddin, R., & Zamarro, G. (2009). Teacher qualifications and student achievement in urban elementary schools. *Journal of Urban Economics*. Elsevier Inc., *66*(2), 103–115. doi: 10.1016/j.jue.2009.05.001.

Canales, A., & Maldonado, L. (2018) 'Teacher quality and student achievement in Chile: Linking teachers' contribution and observable characteristics', *International Journal of Educational Development*. Elsevier, 60(April 2017), 33–50. doi: 10.1016/j.ijedudev.2017.09.009.

Craig, A. (2003). Preservice versus inservice teachers â€™ assessment literacy : does classroom experience make a.

Goldstein, J., & Flake, J. K. (2016). Towards a framework for the validation of early childhood assessment systems. *Educational Assessment, Evaluation and Accountability*. Educational, *28*(3), 273–293. doi: 10.1007/s11092–015–9231–8.

Harris, D. N., & Sass, T. R. (2007). Teacher Training, Teacher Quality, and Student Achievement', *National Center for Analysis of Longitudinal Data in Education Research*, 3 (march),1–63. Available at: papers3://publication/uuid/ADEB13F1-B084–4F44-A7F1–596FE0B6A2CE.

Huber, S. G., & Skedsmo, G. (2016). Assessment in education—from early childhood to higher education. *Educational Assessment, Evaluation and Accountability*, *28*(3), 201–203. doi: 10.1007/s11092–016–9245-x.

Kearney, S. P., Perkins, T. and Perkins, T. (2011). ResearchOnline @ ND Improving Engagement : The Use of "Authentic Self and Peer Assessment for Learning" to Enhance the Student Learning Experience, (January).

Kitano, S. (2011). Current issues in assessment in early childhood care and education in Japan. *Early Child Development and Care*, *181*(2), 181–187. doi: 10.1080/03004430.2011.536639.

Furtak, E. M., Kiemer, K., Circi, R. K., Swanson, R., de León, V., Morrison, D., & Heredia, S. C. (2016). Teachers' formative assessment abilities and their relationship to student learning: findings from a four-year intervention study. *Instructional Science*, *44*(3), 267–291.

Massyrova, R., Bainazarova, T., Meterbayeva, K., Smanov, I., & Smanova, G. (2015). Future Teacher Training For The Children Under School Age Reative Abilities Development. *Procedia-Social and Behavioral Sciences*, *190*, 164–168.

Mertler, C. A. & Mertler, C. A. (2009). Improving Schools assessment professional development assessment professional development. doi: 10.1177/1365480209105575.

Mulliner, E., & Tucker, M. (2017). Feedback on feedback practice: perceptions of students and academics. *Assessment and Evaluation in Higher Education*, *42*(2), 266–288. doi: 10.1080/02602938.2015.1103365.

Nah, K. O., & Kwak, J. I. (2011). Child assessment in early childhood education and care settings in South Korea. *Asian Social Science*, 7(6), 66–78. doi: 10.5539/ass.v7n6p66.

Nicol, D., & MacFarlane-Dick, D. (2006). Formative assessment and selfregulated learning: A model and seven principles of good feedback practice. *Studies in Higher Education*, *31*(2), pp. 199–218. doi: 10.1080/03075070600572090.
Odaba, S., Cimer, Õ., & Cak, I. (2010). Teachers ' knowledge and practices of performance assessment, *2*, 2661–2666. doi: 10.1016/j.sbspro.2010.03.391.
OECD. (2013). Synergies for Better Learning. doi: 10.1787/9789264190658-en.
Pang, N. S.-K., & Lang., Z. L.-M. (2011). Teachers' competency in assessment for learning in early childhood education in Hong Kong. *Educational Research Journal*, *26*(2), 199–222.
Pauker, S., Perlman, M., Prime, H., & Jenkins, J. (2018). Caregiver cognitive sensitivity: Measure development and validation in Early Childhood Education and Care (ECEC) settings. *Early Childhood Research Quarterly*, *45*, 45–57.
Schweinhart, L. J., Debruin-Parecki, A., & Robin, K. B. (2004). Preschool assessment : policy recommendations. *Educational Research*, (734), 1–12.
Stringher, C. (2016). Assessment of learning to learn in early childhood : an italian framework assessment of learning to learn in early childhood : An Italian framework', *8*, 102–128. doi: 10.14658/pupj-ijse-2016-1-6.
Wang, T. H. (2014). Developing an assessment-centered e-Learning system for improving student learning effectiveness. *Computers & Education*, *73*, 189–203.

Early Childhood Education in the 21st Century– Yulindrasari et al. (eds)
© 2020 Taylor & Francis Group, London, ISBN 978-1-138-35203-2

Character-based school culture and climate management in kindergarten

A.N. Hidayat
Universitas Islam Nusantara, Bandung, West Java, Indonesia

O. Setiasah
Universitas Pendidikan Indonesia, Bandung, West Java, Indonesia

V. L. Ayundhari
Universitas Islam Nusantara, Bandung, West Java, Indonesia

ABSTRACT: The purpose of this study is to understand the application of Character-based management of kindergarten in West Bandung-Indonesia. This research is a qualitative case study. The data was obtained through observation and interviews addressed to teachers and principals. Field notes of teachers and principals in managing the character-based school culture and climate were also compiled. The findings indicate that teachers and principals have designed, implemented, and evaluated character-based management through faith, *piety*, and nationalism approach. The three of them are applied separately and integrated in learning. Teachers and principals also pointed out that the school culture and climate management can develop students' discipline, independence, responsibility, and cooperation.

1 INTRODUCTION

Climate is behaviour, while culture comprises the values and norms of the school (Heck & Marcoulides, 1996). Schein (1996) also stated that norms, values, rituals and climate are all manifestations of culture. Character-based management in kindergarten is supposed to be implemented by all kindergartens as they are the main institutions that help to develope children's character. Character-based management should be implemented by principals and teachers in preschool activities primarily integrated in learning. Character-based management is one of the alternative managements that supports the achievement of the education objectives in a school. For that, then character-based management should be combined with school activities programs.

Character-based management is the responsibility owned by principal and teachers. When implementing, they certainly should pay attention to the development and potential of each child. If character-based management is not implemented to the maximum effort, then the response of parents will be negative against the school and the community. They will give 'negative' label the kindergarten. Therefore, the principal and teachers must carry out character-based management as it will give changes to the children and parents. Before organizing character-based management, the kindergarten should uncover and examine data from parents and the community to make planning. School principals and teachers should have maximum effort to carrying out planning and evaluation. Besides, they should always work together with parents to carry out character-based management, so that the atmosphere at school has no difference to home.

In doing character-based management, principals and teachers try to use humans, facilities, and other resources around the centre. The utilization of existing sources will be very useful for the implementation of character-based management in kindergarten especially when managed correctly, including the children's development and potential management, as well as carried out preparation before drawing up of plans, planning with cooperative making, implementation, monitoring and evaluation, doing construction, using inputs, processes, and outcomes as indicator of success (Mulya & Jayadiputra, 2012:2). Research conducted by Mulya and Hidayat 2013:2) indicates that character-based management should be carried out continuously in school. The school community is directed to become a fully discipline worker, thorough and tenacious. Research done by Mulya and Karwati (2014:2) indicates that character-based management should be implemented with integrated learning. The implementation of character-based management would not have reached a maximum result if done frivolously and not working to the school community. Research conducted by Karwati and Efendi (2015:2) shows that the implementation of character-based management should be integrated to the learning and tasks and consider the characteristics and potential of the school. The implementation will not reach the highest result without cooperation among the stakeholders including the community. Karwati and Mulya (2016:1) the research results showed that character-based management implementation should exploit the human resources, determine the character values, develop character, determine the indicators of success, compiling character-based management design which is the success-oriented and carrying out a proper evaluation. Muttaqien and Efendi (2017:2) state the research results in the implementation of character-based management should be preceded with a sincere intention. Integrate it in a variety of activities and conduct first by the teacher before children.

From these research results, it can be concluded that when executing character-based management in kindergarten: do with proper, continuous, directional management as well as integrated learning, determine its potential, utilize existing human resources, determine the character values that will be developed, draw up the design, lead to a goal, and do the evaluation. The proper culture and climate management will support the achievement of goals against education in kindergarten. When established continuously, the children will be affected positively and will overcome negative influences from the outside world. The management is directed in school program, integrated with learning to create conducive climate. The activities will be seriously executed by them who have certain potential in accordance to the carried activities.

Based on the conclusions outlined earlier, character-based management can be implemented well if done with precision. Proper implementation should be supported by the doers that have in common to the views, feelings and actions. This is because character-based management should always be supervised and evaluated by the leaders as if the monitoring and evaluation not fully succeeded. Therefore, the character-based management management should be done as a whole started with planning, implementation, supervision, and evaluation.

The character-based management implemented at the kindergarten should supported by problem analysis results among others: graduates of kindergarten are still not convinced to have character aspects to be developed in elementary school, not having readiness over the knowledge, attitudes and skills, not having the spirit and required learning habits in elementary school, as well as having yet to find any specific guidelines to implement character-based management in kindergarten.

Based on this description, the researchers conducted research in kinderganten with the problem "*How is* character-based management implemented *in a West Bandung kindergarten?*"

2 METHOD

This study is a qualitative or naturalistic case study with issues related to character-based management in one of West Bandung kindergartens. This research is carried out in the natural setting, to understand and to interpret behaviors, views and interpretations of both school principal and teachers about children as well as the relationship among them in school. Researchers sought to examine how character-based management is performed by the principal and teachers in West Bandung kindergarten. The study will describe the subject in a comprehensive manner, the behavior, and the factors that influence data, look for associated cases examined, get to know more about the issues, solve the problems that is experienced by the subjects, and understand their behavior. The research results can be applied but the scope is limited. Data-collecting techniques used in this research are observation, interview, and documentation. The purpose of the data search intends to describe condition in one of West Bandung kindergartens.

The stages in this study are: (1) preparatory stage is drawing up the research design, arranging the research draft consisting of revising design, review, and revise the instrument, devise and reproduce it which is oriented to the subject and related parties; (2) implementation is a stage that maintains contact with the purpose to trust each other and to obtain the data and information according to existing condition; (3) check the accuracy the collected data or information, the researchers also did member checking with the aim to assess the research results. Researchers conduct a validation which relies heavily on credibility, dependability, transferbility, and confirmability. The activities realization in this research was carried out through a completely recapitulation of raw data collected from the field. The analysis results are among others: did the data selection, made a summary in the form of a picture and a systematic synthesis, devised by customizing themes, objectives and conclusions; and (4) compiling reports of research as a whole.

3 RESULTS AND DISCUSSION

Research results of interviews with principals and teachers showed that they have always tried to prepare various things needed for drafting plans to character-based management including building a good relationship with the children's parents. Both principal and teachers did deal-making and the leeway to parents to educate children since the beginning school, as well as to educate them reciting the Qur'an and especially the religious values application. They also arranged character-based management plan, implemented it and carried out an evaluation.

Before drawing up the plan, the principal and kindergarten teachers did preparation process by exposing ideas, opinion, and results related to character-based management. It was done through the excavation of teacher's aspiration with regard to character-based management. The results were discussed jointly and then finally stacked the activity plan which contains the backgrounds why activities are carried out, the purpose, the tasks structure and division, and schedule of activities. It is all arranged to do the activities that have been mutually agreed. The principal motivated the teachers in order to carry out activities earnestly. In addition, the principal also conveyed the carried out activities to the children's parents. Cooperation with parents would always be instilled from the beginning, especially to the freedom of getting familiar first to the environment when their children entered the school. Other things are in regard to the condition of learning in and outside of classroom, the facilities and infrastructures school grounds, school programs, the execution and implementation of the evaluation. It is intended to let parents and children to get to know the condition of the school, so they will not feel forced in choosing the school that mostly resulted school transfer.

At the time of drawing up character-based management plan, principal and teachers will be involved as the committee and they will determine the character values that will be developed as well as determine funding, strategy, and a resource person who could support the plan, discuss it with teachers to support character-based management implementation. Afterwards, they get to know the needs of teachers and children as well as the factors that support the directness implementation of character-based management. The needs of children and parents were expressed, discussed and concluded jointly with teachers, and of course it should be in accordance with the situation and condition of the school. The principal should always try to devise activities that can develop the children's character and potential. Therefore, the principal should always be ready to implement ideas that have been discussed together and jointly agreed upon.

When drawing up the planning, there are things that need to be taken care of by the principal about the drafting plan: (1) create a comfortable school for developing characters; (2) discuss the values development at each meeting in the school; (3) be responsible for the education success in school; (4) do character-based management thoroughly; (5) craft a program that can be done by the teacher; (6) develop the school's character and culture; (7) pay attention to the rules set forth by the Department of Education; (8) seek ways to become for teachers to follow in managing character-based management.

After the plan was drawn up, it should be socialized to the teachers and children's parents as well as preparing teachers to implement character-based management, doing activities that can support the goals and purpose of character-based management implementation, creating a comfortable working atmosphere to implement character-based management. One more thing is assured the teachers about the school's culture, described activities undertaken related to character-based management and oriented to the success of it.

In the character-based management implementation, the principal should perform the blending of ideas, feelings and actions of teachers against Character Based School Culture and Climate Management, utilizing existing human resources, creating the rules for teachers, determining workable character-based management activities, integrating the activities of character-based management to learning—all activities leading to the success of Character Based School Culture and Climate Management, and carrying out the steps of character-based management directed to developing the character values.

Activities in the character-based management implementation should support the success of school, improve the quality of graduates, ensure the quality by using the optimisation of effectiveness and efficiency, productivity, and the suitability of implementing character-based management by leading to the success of national education. Principal and teachers should put efforts to integrate character-based management in learning and school activities. It is established so that the character values from religious teachings combined with learning material will be adapted to the children's condition, and the teachers themselves. Learning activities constitute important areas and as the main requirement of the existence of the education organization in the school. Therefore, if the implementation of learning quality is lacking, then the quality will decrease.

In carrying out the evaluation, principals and teachers worked with parents. Together they make an observation towards the children. The principal uncovered the changes that happen to children at home. So did about children's success in school is revealed by the principal and teachers, and then discussed with the parents. So it looks clearly which ones changed permanently and which are incidental. The implementation results of evaluation should be analyzed and discussed by the principal and teachers to serve as material when making a plan of character-based management on next time.

From these two opinions, it can be concluded that the management of character-based management has a very important role in developing good character values for teachers and children. Therefore, it should be done by the school especially for the kindergarten as it is the basic foundation to form the character.

4 CONCLUSIONS

Before drawing up character-based management plan, the principal and teachers should prepare facilities and infrastructure, material activities, implement activities, determine the character values that would be developed, and supporting factors to implement education activities including deals and discretion to educating children.

In the implementation of drafting character-based management plan, the principal and teachers tried to explore human resources in school, developed the character and values to manage character-based management, created a climate conducive and condition so that the school community integrate the management in school activities.

In the implementation of character-based management, principal gives priority to the application of the character values taken from the teachings of the Islamic religion, realizes the teachers and children potential as well as realizes supporting factors for the success of character-based management implementation.

In carrying out the character-based management evaluation, the principal and teachers worked with parents in regards to the changes that occur in children at school and home. The evaluation results are discussed and addressed so clearly and served as a basis for the plan next year.

REFERENCES

Heck, R. H., & Marcoulides, G. A. (1996). School culture and performance: testing the invariance of an organizational model. *School Effectiveness and School Improvement*, *7*(1), 76–96.

Hidayat, A. S. (2011). *Manajemen sekolah berbasis karakter. Disertasi*. Bandung. Sekolah Pascasarjana UPI. Unpublished.

Mulya, Dj. B., & Jayadiputra, E. (2012). *Pengelolaan Lingkungan Sosial Budaya Sekolah Berbasis Karakter* (PLSBSBK). Uninus Bandung.

Mulya, Dj. B., & Hidayat, A. N. (2013). *Pengelolaan Lingkungan Sosial Budaya Sekolah Berbasis Karakter* (PLSBSBK). Uninus Bandung.

Mulya Dj. B., & Karwati, E. (2014). *Pengelolaan Lingkungan Sosial Budaya Sekolah Berbasis Karakter* (PLSBSBK). Uninus Bandung.

Karwati, E., & Efendi, G.Y. (2015). *Pengembangan Karakter Siswa Melalui Pengelolaan Iklim Sosial Budaya Sekolah* (PKSMPISBS). Uninus Bandung.

Karwati, E., & Mulya, Dj.B. (2016). *Pengembangan Karakter Siswa Melalui Pengelolaan Iklim Sosial Budaya Sekolah* (PKSMPISBS). Uninus Bandung.

Muttaqien, K., & Efendi, G.Y. (2017). *Implementasi manajemen iklim budaya sekolah berbasis karakter* (ICharacter Based School Culture and Climate Management). Uninus Bandung.

Schein, E. H. (1996). Culture: the missing concept in organization studies. *Administrative Science Quarterly*, *41*, 229–240.

Early Childhood Education in the 21st Century – Yulindrasari et al. (eds)
© 2020 Taylor & Francis Group, London, ISBN 978-1-138-35203-2

Competency-based selection for caregivers in daycare

W. Cahyaningtyas & M. Agustin
Graduate School of Early Childhood Education, Universitas Pendidikan Indonesia, Bandung, Indonesia

ABSTRACT: The number of working mothers has increased in recent years. This means that parents leave their children in daycare, so the task of parenting is switched to caregivers. The role of caregivers becomes very important because it can affect children's development. However, maltreatment of children carried out by caregivers in daycare is still common. For this reason, a selection system for caregivers in daycare is needed, so that these maltreatment cases can be minimized. This chapter will discuss about competency-based selection system for caregivers in daycare using in-depth interview methods such as Behavioral Event Interviews.

1 INTRODUCTION

About 90 percent of the child's brain development process occurs in the first five years of life. At this age the function of the child's brain is like a sponge that absorbs all information quickly. Children are in a period that is sensitive to experiences that support their development (Britto et al., 2016).

A baby's brain develops as a consequence of their relationship with their caregivers (Gerhardt, 2004). Establishment and strengthening of the relationship between brain cells and nerves is partly determined by genetic code, but it is supported and influenced by relationships with parents and adults in the vicinity, as well as the environment (Doherty & Hughes, 2009). In brain development, once a key pathway is formed, tthis will survive until adulthood (Zeedyk, 2006). The intervention must be carried out to establish a pathway that allows children to learn and reach their full potential (Gerhardt, 2004). Therefore, it can be said that early intervention for childcare at early age is important.

Childcare is didactic, prepares cognitive development of children, and helps them participate in the learning process that starts from birth. Quality childcare doesn't neglect children's educational needs but combines learning activities as a part of the curriculum. Furthermore, childcare staff, especially in daycare, should pull together with parents to help them learn how to support children's learning at home. The main purpose of parenting is to enable optimal development of children overall and supporting efforts to achieve this goal (Morisson, 2008).

These days, many mothers choose to work; as long as the mother works, the child is entrusted to caregivers either at home or at daycare. There are more than 70% of working mothers in Washington DC with children less than three years of age, in fact it is very common for mothers to return to work after six weeks of giving birth (Morisson, 2008). Caregivers then replace parenting tasks to their children. This has led to a shift in management services for caregivers, from traditional to innovative which require work capabilities that include aspects of knowledge, skills and work attitudes that are relevant to the implementation of tasks and job requirements according to the development of science and technology (Kepmenakertrans, 2014).

The Indonesian National Standard of Work Competencies defines caregiver as someone who helps parents to keep, take care of, nurture and educate children when the parents are not in the house. taking care of children includes provide treatment, nurture and stimulate the children. Take care of children is the act of treat and keeping up the health of children both in preventive and preservation. Nurturing is knowledge, experience, expertise in performing care, protection, giving love and direction to children.

The role of caregivers in daycare is very influential to the development of the child. This is because long-term cognitive and social development is at stake. The National Institute of Child and Health and Human Development (NICHD) in the United States found that children who received a high quality childcare had higher academic and cognitive achievements than those who with a lower care quality (Vandell et al., 2010). Interestingly, a study in America showed that the positive effects of high-quality childcare can be influential until adolescence (Leana et al., 2009).

However, unfortunately most of daycare does not meet the needs of children's health, safety, warm relationship, and learning, so they become more susceptible to disease due to unmet basic hygiene conditions for diaper replacement and feeding. In addition, babies are often at risk to be in danger because of safety issues and lack of supportive relationships with adult. There is also the risk that the babies could beleft behind in learning because they lack books and toys needed for their physical and intellectual growth (Roopnarine & Johnson, 2009).

Errors in selecting and hiring a caregiver can be fatal for children. This is supported by the results of a study conducted by Gilbert and his team, which showed that child abuse by parents, teachers, caregivers, and other adults can have long-term effects on mental health, misuse of drugs and alcohol (especially in girls), at risk of experiencing sexual deviations, obesity, and criminal behavior that can last into adulthood (Gilbert et al., 2009). For this reason, the selection process for caregivers in daycare is important to note in order to reduce the risk of maltreatment in children.

Daycare managers are required to choose the right caregivers based on the competencies of the caregivers, whether or not that is in accordance with the competencies needed to become a caregiver. But, in practice, caregivers are often selected and employed only based on their work experience without really taking into account compatibility between the competencies of prospective caregivers and the competencies needed to become a caregiver (Goodstein & Davidson, 1998). Based on literature review, this chapter will argue for the importance of conducting competency-based selection for caregivers in daycare and one of the selection methods that can be used so that the goal of childcare can be achieved.

2 COMPETENCY

Competency is a pattern of knowledge, skills, abilities, behavior and other characteristics that can be measured required by a person to perform a job role or job function properly (Aamodt, 2007). A competency is a basic characteristic of an individual that is causally related to criterion-referenced effective or superior performance in a job or situation (Spencer & Spencer, 1993). Competence is a concept that can be applied to all categories of employees (Berman, 1997). Competency divided into personal competencies and technical competencies. A person can be successful at work only when there is a match in both sets of competencies.

Personal competencies describe personality or emotional character, habits where someone deals with their worldly life, especially with work and other people. These characteristics equal or even more important to be considered in determining whether someone will succeed or not at work.

Technical competence is relatively easy to understand and easier to assess than personal competence. However, personal competence is important in determining the success at work. As illustrated in Figure 1, technical competencies regarding knowledge and skills tend to be seen and appear on the surface, on the characteristics of an individual. Personal competencies regarding self-concept, traits, and motivation tend to be more more complex to be mastered by the caregivers.

Competencies that appear on the surface related to knowledge and skills competencies are relatively easy to develop, training is the most cost-effective way to secure these employee abilities. Core competencies that are at the base of personality icebergs such as motives and trait competencies are more difficult to assess and develop, will be most cost-effective to select for these core competencies such as motives and characters, then provide training related to

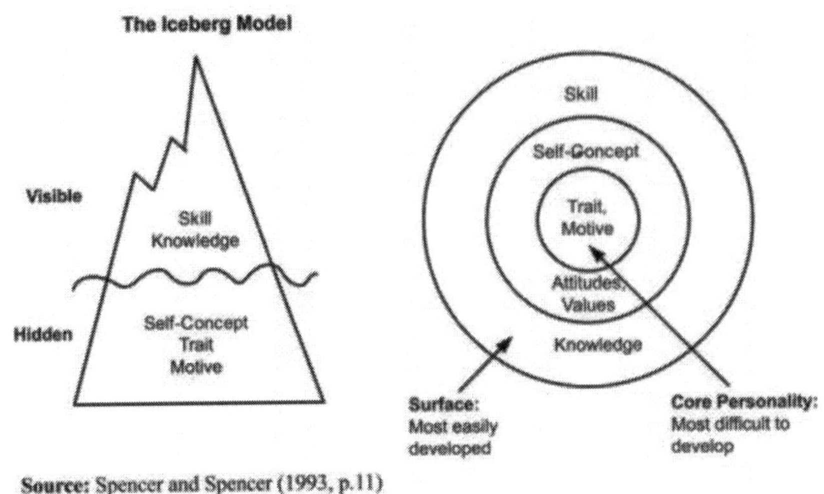

Figure 1. Central and surface competencies.

knowledge and skills needed to do certain jobs. As one illustration of the importance of this personal competence, more than 80 percent of all cases of termination of employment in the United States are a result of personal or interpersonal problems (Goodstein & Davidson, 1998).

2.1 *Competency of caregivers in daycare*

The National Association for the Education of Young Children (NAEYC) states that caregivers need to pay attention and do several things, such as: understand the growth and development of infants and toddlers, observe and record the progress and development of the children (Santrock, 2002). In Indonesia, based on research that conducted by Rizkita (2017) on caregivers in daycare showed that the competencies needed by caregivers in daycare are responsible, love the world of children, know the characteristics of children, have experience in caring for children, have a good personality, and meet the educational qualifications.

The Indonesian National Standard of Work Competencies provides guidance on the list og competencies that need to be possessed by babysitters (Kepmenaker, 2014). The list competencies standard is the formulation of ability that must be owned by someone to be able to do a job based on knowledge, skills, and work attitudes which is in accordance with the performance requirements.

This paper uses the II and III levels of the babysitter profession at The Indonesian National Standard of Work Competencies. The competencies are shown in Table 1.

Table 1. Competency standards KKNI level II and III.

No.	Unit Code	Title of Competency Unit
1	T. 970000.001.01	Cleaning the baby's bedroom environment
2	T. 970000.002.01	Keep the baby's body clean
3	T. 970000.003.01	Take care of baby clothes
4	T. 970000.004.01	Create a food menu for babies
5	T. 970000.005.01	Keep/Storing Milk for the babies
6	T. 970000.006.01	Preparing bottle milk for the babies
7	T. 970000.007.01	Provide food/drink to babies
8	T. 970000.008.01	Babysitting

(Continued)

Table 1. (*Continued*)

No.	Unit Code	Title of Competency Unit
9	T. 970000.009.01	Stimulate baby's growth and development with educational game tools
10	T. 970000.010.01	Implement stimulation in infants
11	T. 970000.011.01	Help mother after childbirth
12	T. 970000.012.01	Applying first aid to infants
13	T. 970000.013.01	Preventing accidents in infants
14	T. 970000.014.01	Attempt to prevent babies from contracting the disease
15	T. 970000.015.01	Keeping the babies with malnutrition
16	T. 970000.016.01	Taking care of sick babies
17	T. 970000.017.01	Handling problematic babies
18	T. 970000.018.01	Handling babies after basic and advanced immunization
19	T. 970000.019.01	Implement occupational safety and health procedures in the workplace
20	T. 970000.020.01	Improve clean and healthy living behavior in the household
21	T. 970000.021.01	Develop emotional maturity and work motivation
22	T. 970000.022.01	Establish working relationships with service users
23	T. 970000.023.01	Communicate with service users
24	T. 970000.024.01	Adapting to the socio-cultural conditions of the destination country
25	T. 970000.025.01	Mastering knowledge about the rules and laws of the destination country

3 COMPETENCY-BASED SELECTION

Competency can predict behavior and performance (Spencer & Spencer, 1993). These behaviors and performance can be predicted based on motives, character, and self-concept. Someone's thoughts include behavioral actions, where through his/her thought we can predict the behavior first. Competence can also predict who does something good or bad, measured by certain criteria or standards (Spencer & Spencer, 1993).

In competency-based human resource management systems, selection and placement decisions based on "fit" or "match" between job competency requirements with individual competencies. The underlying premise of this is: "The better the fit between the job requirements and the competency of a person, the higher job performance and job satisfaction of a person" (Spencer & Spencer, 1993).

Competency-based selection process is based on simple assumption that the best predictor of the future behavior is the past behavior in similar situations. Goodstein and Davidson (1998) said that having competent employees can reduce problems related to supervision, as well as interpersonal problems and conflicts between groups. That way, it can be said that choosing competent caregivers can make parents more calmly leave their child with caregivers because caregivers could work well without full supervision. In addition, it also reduces conflicts between caregivers and children, caregivers with parents, and caregivers with other caregivers.

Goodstein and Davidson (1998) also said that hiring a competent person to fill a job is something that is mutually beneficial for both employment providers as well as for the workers themselves. Competent workers are far more likely to succeed, feel good about what they are doing, and are seen positively by their coworkers than incompetent workers. When they begin to feel secure in their work, their job satisfaction will increase and they bring a higher focus and greater energy to their work. This will eventually motivate, increase productivity, higher self-esteem, and reduce job dissatisfaction.

In this case, selecting a competent caregiver can be beneficial for parents, children, and caregivers themselves. Bakker, Hakanen, Demerouti, and Xanthopoulou (2007) said that when caregivers in daycare are involved in their work, they are able to show high levels of energy and mental endurance while working. Competent caregivers can show satisfying work results for parents and children. In addition, it will make the caregivers comfortable or feel happy with their job, which can make them more excited, easy to get along with the surrounding environment, and feel satisfied with their job.

Identifying certain technical competencies needed for success in work, along with the level of behavior indicators is the first important step in developing competency-based templates. Effective recruitment, selection, training, and performance assessment require competency studies to determine the threshold and differentiation level for each competency. Otherwise, users will face the risk of selection or training for characteristics that are not be able to predict performance.

One of the ways to be able to identify the competencies needed for caregivers is to do a job analysis. A job analysis is a systematic process that explains and collects information about work behavior, activities, and job specifications (Aamodt, 2007). The purpose of job analysis is to identify the task and the key factors of the job, as well as recognize the characteristics needed by the job. Job analysis for the profession of caregivers consists of several stages, including identification of caregiver tasks, make a list of caregiver duty statements, assessment of the list of caregiver duty statements, and determining the necessary competency of caregivers.

After determining the necessary competencies for the caregiver profession, then we could start designing a system to assess the competency of caregiver applicants. There are many sources to assess applicants' competencies, including the curriculum vitae, application forms, background checks, work samples, portfolio analysis, ability and skill tests, and interviews (Jackson et al, 2009).

An in depth interview would be conducted in order to assess the level of competencies. Spencer and Spencer (1993) explain that there is one in-depth interview method that can be used, a Behavioral Event Interview (BEI). The BEI uses techniques by exploring behaviors that have been carried out in real terms, what has been done before can predict what will be done in the future.

3.1 *Behavioral Event Interview (BEI)*

The BEI method produces data about personality and cognitive types of interviewees (what they think, feel, and want to achieve in dealing with the situation). This allows interviewers to assess competencies such as achievement motivation or logical ways of thinking, and problem solving (Spencer & Spencer, 1993). BEI method asks interviewees to focus on the most critical situations they have faced so that it can generate data about the most important skills/ability and competencies. Spencer and Spencer (1993) state BEI is the heart of the Job Competency Assessment process. BEI data are the richest source of hypotheses about competencies that predict superior or effective job performance.

Spencer & Spencer (1993) also said that the basic principle of the competency approach is that what people think or say about their motives or skills is not credible. The purpose of the BEI method is to get behind what people say they have done to find out what they really do. This is accomplished by asking people to describe how they really behaved in a number of specific incidents.

Furthermore, Spencer and Spencer (1993) stated that through the BEI method the interviewees told their stories briefly and clearly about how they handled the most difficult parts, the most important things of their job. Thus, it can demonstrate their competency to do the job. Questions on standard interviews for selection such as "Tell me about your background", "What are your strengths and weaknesses?", "What kind of job do you like and dislike?" are ineffective because most people don't know what their competencies, strengths and weaknesses, or even like the question of whether they really like their job or not..

4 CONCLUSIONS

Children are our most important resource. We must invest in childcare, encourage professional commitment, and raise their psychological well-being. It is important for providers of childcare services to start building a selection system for the caregivers to be employed.

Competency-based selection system could be an option to be considered since it can be a predictor of caregivers' behavior in the future based on their behavior in the past in similar situations.

Competency can predict behavior and job performance based on motives, character, and self-concept. Competency-based selection can be conducted by using the in-depth interview method namely the Behavioral Event Interview, which is the richest source of hypotheses regarding competencies that can predict performance. That way, we hope that maltreatment cases of children that involve caregivers in daycare can be minimized and the goals of child-care can be achieved.

REFERENCES

Aamodt, M. G. (2007). *Industrial/organizational psychology an applied approach*. Belmont CA: Thomson Wadsworth.

Bakker, A. B., Hakanen, J. J., Demerouti, E., & Xanthopoulou, D. (2007). Job resources boost worker engagement, particularly when job demands are high. *Journal of Educational Psychology, 99*, 274–284.

Berman, J. A. (1997). *Competency-Based Employment Interviewing*. London: Quorum Books.

Britto, P. R., Lye, S. J., Proulx, K., Yousafzai, A. K., Matthews, S. G., Vaivada, T., Perez-Escamilla, R., ... Bhutta, Z. A. (2016). Nurturing care: Promoting early childhood development. *Advancing Early Childhood Development: from Science to Scale, 2*, 1–13.

Doherty, J. and Hughes, M. (2009). *Child development: Theory and practice 0–11*. London: Pearson Education.

Gerhardt, S. (2004). *Why love matters*. London: Routledge.

Gilbert, R., Widom, C. S., Browne, K., Fergusson, D., Webb, E., Janson, S. (2009). Burden and consequences of child maltreatment in high-income countries. *Child Maltreatment*, 373, 68–81.

Goodstein, L. D., & Davidson, A. D. 1998. Hiring the "right stuff": Using competency-based selection. *Compensation & Benefits Management, 14*(3), 1–10.

Jackson, S. E., Schuler, R. S., & Werner, S. (2009). *Managing human resource*. Singapura: Cengage Learning.

Kementerian Tenaga Kerja dan Transmigrasi. (2014). *Standar Kompetensi Kerja Nasional Indonesia Kategori Jasa Perorangan Yang Melayani Rumah Tangga; Kegiatan Yang Menghasilkan Barang Dan Jasa Oleh Rumah Tangga Yang Digunakan Sendiri Untuk Memenuhi Kebutuhan Golongan Pokok Jasa Perorangan Yang Melayani Rumah Tangga Bidang Pengasuh Bayi (Babysitter)*. Jakarta: Kepmenakertrans.

Leana, C., Appelbaum, E., & Shevchuk, I. (2009). Work process and quality of care in early childhood education: The role of job crafting. *Academy of Management Journal, 52*, 1169–1192.

Morisson, G. S. (2008). *Fundamentals of early childhood education*. New Jersey: Pearson Prentice Hall

Rizkita, D. (2017). Pengaruh standar kualitas Taman Penitian Anak (TPA) terhadap motivasi dan kepuasaan orangtua (pengguna) untuk memilih pelayanan TPA yang tepat. *Early Childhood: Jurnal Pendidikan, 1*(1), 1–16.

Roopnarine, J. L., & Johnson, J. E. (2009). *Approaches to early childhood education*. London: Pearson Education.

Santrock, J. W. (2002). *Life span development—Perkembangan masa hidup*. Jakarta: Erlangga.

Spencer, L. M., & Spencer, S. M. (1993). *Competence at work models for superior performance*. Canada: Wiley.

Vandell, D. L., Belsky, J., Burchinal, M., Steinberg, L., & Vandergrift, N. (2010). Do effects of early child care extend to age 15 years? Results from the NICHD study of early child care and youth development. *Child development, 81*(3), 737–756.

Zeedyk, S. M. (2006). From intersubjectivity to subjectivity: The transformative roles of emotional intimacy and imitation. *Infant and Child Development, 15*, 321–344.

Early Childhood Education in the 21st Century– Yulindrasari et al. (eds)
© 2020 Taylor & Francis Group, London, ISBN 978-1-138-35203-2

Teacher pedagogical competence to support 21st-century skills

Akbari & Rudiyanto
Universitas Pendidikan Indonesia, Indonesia

ABSTRACT: The purpose of this chapter is to explain 21st-century skills. Explanations of these skills are subsequently supported by explanations of teacher pedagogical competence. Teacher pedagogical competence includes, among others, helping students develop participation, adjusting personalized learning, emphasizing project/problem-based learning, encouraging cooperation and communication, increasing student involvement and motivation, cultivating creativity and innovation in learning, using appropriate learning tools, designing learning activity that is relevant to the real world, developing student-centred learning, developing unlimited learning, and conducting in-depth assessment.

1 INTRODUCTION

Trilling and Fadel's study (2009) shows that secondary school, diploma education, and higher education graduates are still less competent in terms of: (1) oral and written communication, (2) critical thinking and problem solving, (3) work ethics and professionalism, (4) team work and collaboration, (5) working in different groups, (6) technology use, and (7) project and leadership management. Meanwhile, the last century has witnessed a significant shift from manufacturing services to services that emphasize information and knowledge (Scott, 2015a). Knowledge itself grows and expands exponentially. Schools are thus challenged to find ways to enable students to succeed in work and life through the mastery of creative thinking skills, flexible problem solving, collaborating and innovating. Studies such as those by Trilling and Fadel (2009), Ledward and Hirata (2011), Pacific Policy Research Center (2010) and others demonstrate the importance of 21st-century skills for realizing the necessary transformation.

2 21ST-CENTURY SKILLS

Various organizations have attempted to formulate a variety of competencies and skills needed in the face of the 21st century. However, one important thing to note is that educating the younger generations in the 21st century cannot only be done through one approach. Wagner (2010) and Change Leadership Group from Harvard University have identified competencies and survival skills needed by students in facing life, employment, and citizenship in the 21st century, emphasizing seven skills as follows: (1) critical thinking skills and problem solving, (2) collaboration and leadership, (3) agility and adaptability, (4) initiative and entrepreneurial spirit, (5) ability to communicate effectively both orally and in writing, (6) ability to access and analyze information, and (7) curiosity and imagination.

On the same note, the U.S.-based Apollo Education Group has identified ten (10) skills needed by students to work in the 21st century, namely critical thinking skills, communication, leadership, collaboration, adaptability, productivity and accountability, innovation, global citizenship, entrepreneurial ability and spirit, and the ability to access, analyze and synthesize information (Barry, 2012). Based on the results of research conducted by OECD, three (3) descriptions of learning dimensions in the 21st century were obtained, namely information, communication, and ethics and social influences (Ananiadou & Claro, 2009). Creativity is also one of the important components in order to succeed in facing a complex world (IBM, 2010).

3 TEACHER PEDAGOGICAL COMPETENCE IN THE 21ST CENTURY

The following sections describe some perspectives that support the process of teaching and learning that enhance essential competencies in facing a very complex and uncertain future.

3.1 *Helping develop student participation*

Now, it is no longer time for people to study and work in isolation because they can take part in online communities. The emergence of Instagram, Flickr and Twitter can help reporting of the latest student developments that can be uploaded and open to public comments. McLoughlin and Lee (2007) opine that social media can make students want to participate and connect with others. Social media can also support meaningful learning personally through connection, collaboration and sharing in building knowledge.

3.2 *Personalizing and adjusting learning*

According to Redecker et al. (2011), personalization has implications for what, how and where teachers teach. Personalization can occur through collaboration. Collaboration allows the process of sharing innovations to occur faster and make information about students' talents get known more quickly. For the 21st century, teachers are expected to spark students' curiosity and inspire them to explore various applications that support the knowledge and skills they have learned.

McLoughlin and Lee (2008) argue that effective and innovative learning practices will differ according to subjects; however, the emphasis remains on quite similar aspects, namely: digital competencies that focus on individual creativity and performance; strategies for meta-learning, including designed learning; inductive and creative reasoning models and problem solving; collaborative learning content creation and knowledge formation; horizontal (peer-to -peer) learning, and the like.

3.3 *Emphasizing project and problem-based learning*

Woods (2014) argues that project-based learning and problem-based learning ultimately require changes in the teacher's role from being a 'source of knowledge' to being a trainer and facilitator for students to gain knowledge. Trilling and Fadel (2009) explain that the teaching and learning applying the two models in a certain class for such a long time has resulted in students' learning outcomes and various 21st-century skills that are significantly different from those of the class using traditional methods.

3.4 *Encouraging cooperation and communication*

Students can develop knowledge content and learn 21st-century skills such as the ability to work in teams, solve complex problems, and apply acquired knowledge to other situations (Barron & Darling-Hammond, 2008). Different from lecture-based learning, collaborative learning is a form of developing student interaction in building knowledge in groups.

3.5 *Involving and motivating students*

Saavedra and Opfer (2012) state that learning which is merely a "standard transmission" can weaken skill development because of a lack of relevance that causes a lack of motivation, and ultimately may lower the level of learning. Malone and Smith (cited in Meyer et al., 2008) also suggest that teachers must foster motivation by clarifying and sharing learning goals with students.

3.6 *Cultivating creativity and innovation*

Scott (2015c) states that some schools have taught their students to create knowledge, not only teaching students to "digest" knowledge that is static and complete. McLoughlin and Lee (2008) argue that the ultimate goal of learning is to stimulate students' ability to create and produce ideas, concepts and knowledge.

3.7 *Using the appropriate learning facilities*

Redecker et al. (2009) show that the use of social media in learning supports pedagogical innovation by encouraging learning processes based on personalization, collaboration and changing patterns of interaction between students and students, as well as between students and teachers. New technology makes tasks such as searching, filtering, and processing, evaluating, and managing information faster and more efficient. P21 (2007b) explains that digital communication technology has the potential to change schools as well as curriculum.

3.8 *Designing "real world" and relevant learning activities*

According to P21 (2007b), research results show that if teachers create meaningful learning activities that focus on resources, strategies and context in accordance with the lives of students, the level of absenteeism will decrease, collaboration and communication will develop, and critical thinking skills and academic achievement will increase.

3.9 *Building good relationship in teaching and learning*

Leadbeater (2008) emphasizes that students need relationships that motivate them to learn. A good teacher must have the skills to motivate students. A good relationship will make students feel comfortable and cared for.

3.10 *Focusing on student-centered learning model*

Teachers must be comfortable in managing class dynamics and encouraging independent learning; in addition, teachers must support the exploration and acquisition of new knowledge and skills to prepare students for the 21st century (Trilling & Fadel, 2009).

3.11 *Developing unlimited learning (anytime and anywhere)*

Today's students have a variety of choices in learning, not limited to classrooms. The use of various technologies outside the classroom allows students to have various forms of learning (Furlong & Davies, 2012).

3.12 *Conducting more in-depth assessment on students understanding and competency*

Rubrics and other formative assessment instruments will play an important role in the 21st-century classroom because teachers and students have guidelines for the level of attainment of learning outcomes. Students should also be taught how to evaluate their own learning. This will help them master the content and improve their metacognitive skills, including the ability to learn how to learn and to reflect on what has been learned (Saavedra and Opfer, 2012).

4 CONCLUSION

The role of teachers in the 21st century must shift from being "knowledge planter" to that of a mentor, leader of discussions, and assessor of student learning progress (Hampson et al., 2011). The main purpose of 21st-century learning is to build individual learning abilities and support learners to develop into lifelong, active, and independent learners; therefore, teachers need to be "learning trainers"—a role that is very different from traditional classroom teachers. Teachers as a learning trainer will provide guidance to help students develop skills and offer various supports that will help students achieve their learning goals. As a learning trainer, teachers will also encourage students to interact with knowledge—to understand, criticize, manipulate, design, make, and change it.

REFERENCES

Ananiadou, K., & Claro, M. (2009). 21st Century skills and competences for new millennium learners in OECD countries. *OECD Education Working Papers*, No. 41. Paris: OECD Publishing.

Barron, B., & Darling-Hammond, L. (2008). Teaching for meaningful learning: A review of research on inquiry-based and cooperative learning. George Lucas Educational Foundation. https://files.eric.ed.gov/fulltext/ED539399.pdf

Barry, M. (2012). What skills will you need to succeed in the future? Phoenix Forward (online). Tempe, AZ: University of Phoenix.

Furlong, J., & Davies, C. (2012). Young people, new technologies and learning at home: taking context seriously. *Oxford Review of Education, 38*(1), 45–62.

Hampson, M., Patton, A. and Shanks, L. (2011). *Ten ideas for 21st century education.* London, Innovation Unit.

Leadbeater, C. (2008).What's next? 21 Ideas for 21st century learning. London, The Innovation Unit. Mansilla, V.B. and Jackson, A. 2011. *Global Competence: Preparing Our Youth to Engage the World.* New York: Asia Society.

Ledward, B. C., & Hirata, D. (2011). *An overview of 21st century skills. Summary of 21st century skills for students and teachers, by Pacific Policy Research Center.* Honolulu: Kamehameha Schools–Research & Evaluation.

McLoughlin, C., & Lee, M. J. W. (2008). The three p's of pedagogy for the networked society: personalization, participation, and productivity. *International Journal of Teaching and Learning in Higher Education, 20*(1), 10–27.

Meyer, B., et al.(2008). Independent learning: Literature review. Research Report No. DCSF-RR051. Nottingham, UK: Department for Children, Schools and Families.

P21. (2007*a*). *The intellectual and policy foundations of the 21st Century skills framework.* Washington, DC: Partnership for 21st Century Skills.

P21. (2007*b*). *21st Century curriculum and instruction.* Washington, DC: Partnership for 21st Century Skills.

P21. (2008). *21st Century skills, education & competitiveness.* Washington, DC: Partnership for 21st Century Skills.

P21. (2011). *Framework for 21st Century learning.* Washington, DC: Partnership for 21st Century Skills.

P21. (2013). *Reimagining citizenship for the 21st Century: A call to action for policymakers and educators.* Washington, DC: Partnership for 21st Century Skills.

Pacific Policy Research Center. (2010). *21st Century skills for students and teachers.* Honolulu: Kamehameha Schools, Research & Evaluation Division.

Redecker, C., Ala-Mutka, K., Leis, M., Leendertse, M., Punie, Y., Gijsbers, G., Kirschner, P., Stoyanov, S., & Hoogveld, B. (2011). *The future of learning: Preparing for change.* Luxembourg, Publications Office of the European Union.

Saavedra, A., & Opfer, V. (2012). *Teaching and learning 21st Century skills: Lessons from the learning sciences.* A Global Cities Education Network Report. New York, Asia Society.

Scott, C. L. (2015c). *The futures of learning 3: What kind of pedagogies for the 21st century?* UNESCO Education Research and Foresight, Paris. [ERF Working Papers Series, No. 15].

Scott, C. L. (2015a). *The futures of learning 1: Why must learning content and methods change in the 21st century?* UNESCO Education Research and Foresight, Paris. [ERF Working Papers Series, No. 13].

Scott, C. L. (2015b). *The futures of learning 2: What kind of learning for the 21st century?* UNESCO Education Research and Foresight, Paris. [ERF Working Papers Series, No. 14].
Trilling, B., & Fadel, C. (2009). *21st century skills: Learning for life in our times*. San Francisco, CA, Jossey-Bass/John Wiley & Sons, Inc.
Wagner, T. (2010). *Overcoming the global achievement gap* (online). Cambridge, MA: Harvard University.
Woods, D. (2014). *Problem-based learning (PBL)* (online). McMaster University. http://chemeng.mcmaster.ca/problembased-learning

Early Childhood Education in the 21st Century– Yulindrasari et al. (eds)
© 2020 Taylor & Francis Group, London, ISBN 978-1-138-35203-2

Early childhood teachers' voices for excellence in teaching practice

L. Fridani
State University of Jakarta, Jakarta, Indonesia

N. Gandasari
Al Ishlah Community Learning Center, Jakarta, Indonesia

W. Widiastuti
Association of Raudhatul Athfal Teacher, Jakarta, Indonesia

ABSTRACT: Excellence in teaching practice represents the quality of effective teaching in the classroom. Research highlights the quality of teaching as a key factor of differences that leads to cognitive, physical, social, emotional, and moral developmental gains that carry over into later stages of development (Agbenyega, 2013; Berk, 2006; Fauth & Thompson, 2009). The intention of this research is to investigate teachers' perspectives, values and practices, explore their expectations, and investigate concerns that the teachers around Jakarta have to cope with in their classroom. The researchers believe that these factors provide contribution in creating excellence for young children's achievement. This study revealed that teachers viewed the excellence in teaching practice in relation to their competencies in handling children. Three themes emerged from the teachers' Focus Group Discussion (FGD), namely professional knowledge issues, teaching strategy, and workload concern and salary expectation. Findings from this study indicated the need to give greater emphasis to understand the concerns of early year's teachers when trying to maintain a good quality teaching. It also highlights the importance of training for early childhood teachers to enhance their excellence in teaching practices.

1 INTRODUCTION

Research on teachers' practices and their voices have been carried out by many experts in various countries. Berk (2006) explained that early childhood teachers considered various aspects of children's development such as physical, language, independence, competence, self-esteem, and the ability to interact with others. Some scholars (Hamre, Downer, Jamil & Pianta, 2012) described teachers' practice in implementing appropriate program for children's development and explored teacher's quality of learning experience. They found that teacher who applied an enriched curriculum have children with a better motivation, concentration, independence, social interaction and higher order thinking skills compared to other class that applied traditional methods. Other studies have emphasized the importance of teacher's beliefs and values in the creation of quality teacher–children interaction in the classroom (Fleer 2010).

In Indonesia, research on early childhood teachers' voices are still limited in number. Fridani (2014) investigated the perspective and practices of early childhood teachers in Jakarta in relation to children's school readiness and transition using the bio-ecological model. The study has revealed some traditional perspectives that impacts on teachers' practices in preparing children to enter primary school and the shared responsibility of all stakeholders that determine teachers' practices in the classroom.

It is noted that teachers' practices have a direct impact on children's achievement, engagement and motivation for learning. Therefore, it is crucial to achieve excellence in teaching that represents the quality and effective of teaching. Some experts (Agbenyega, 2011; Fleer, 2010) explain that excellence in the practice of learning in early childhood context is related to the methods, ways and approaches taken by the teacher. Therefore, an excellent teacher will guide and support children to develop their understanding and knowledge by carrying out play based learning. Teachers with a good competency also offer pedagogical strategies to establish a positive and supportive class atmosphere in order to improve children learning and adaptation (Berk, 2006; Boethel, 2004). Other scholars believe that teachers can achieve excellence in their roles when they have passion and a strong commitment to create a positive culture in the learning environment that can elevate their children to be motivated to learn (Dockett & Perry, 2006; Laura & Munsch, 2014).

Other aspects that contribute to teachers' excellence in teaching children is the quality of teacher–child interactions (Hamre, Downer, Jamil & Pianta, 2012). According to Agbenyega (2011), the quality teacher–child relationships are often situated within the dilemmas and different situations in the classroom. There is a complex relationship that form different ideas and expectations of the participants involved in the learning (Rosier & Mc Donald, 2011). It is also argued that the cultural and moral diversity have shaped the values and beliefs which influenced teacher's practice in the classroom.

Some scholars (Darling-Hammond, 2000; Harslett, 2000) found that teachers' attitudes toward children contributed to the level of teachers' responsibilities, the way teachers teach and treat children, as well as the way they perceive their professional development. In relation to responsibility and commitment, Coladarci (2002) points out the effectiveness and commitment of teachers to teaching children that derives from their knowledge and perception of children. Other studies explained the quality of teaching practice from the aspects of geographic location of a country and the relationship between schools, family and community (Graue, 2006; Scott-Little, Kagan & Frelow, 2006). Further, Pence and Bame (2008) argued that social, cultural, economic, policy, and historical factors influence how the stakeholders interact to provide support for children.

It is clear from these studies that there are many aspects that contribute to the excellence of teaching practice. Overall, teachers' quality of teaching is shaped by valuable knowledge for professional practice that includes the understanding of early childhood learning. This study seeks to investigate early childhood teachers' voices for excellence in practice in the area around Jakarta. The researchers believe that teacher competencies, values held, and good rewards for teachers tasks, provide contribution in creating excellence for young children's achievement. The research questions of this study namely; (1) *What conceptions of excellence in teaching practice are held by early childhood teachers around Jakarta?* and (2) *How did teacher implement the learning practice?;* (3) *What are the expectation and concerns of the teachers that they have to cope with?*

2 METHOD

The phases of this research consist of conducting a pilot study followed by a questionnaire survey, and a focus group discussion. The pilot study is necessary to guide the selection of items in the questionnaire which would be used in the framing of final questions. The researchers conducted a pilot study using 30 item questionnaires to 35 early childhood teachers around Jakarta region. The questionnaires were revised in the light of teachers' responses and a final version constructed and distributed to 115 teachers (15 male and 140 female). On the whole, the samples were relatively representing a good range of early childhood institution.

The demographic profile contains questions on gender, age, qualification and teachers' teaching experience. The questionnaire statements consist of 30 items that was divided into four main sections which measures (1) teachers' competency, (2) teaching practice, (3) values which teachers' held, and (4) teachers' concern and expectation. Each statement required an evaluation on a five-point Likert type scale, with scale values ranging from very agree (5) to very disagree (1). The questionnaire is summarized in Table 1.

Table 1. Excellence in Teaching Practice Questionnaire

Demographic Profile	Subscale	Number of items
Gender, age qualification teaching experience	competence	7
	teaching practice	8
	applied values	7
	concerns and expectations	8

The second instrument is a qualitative focus group questions which was developed based on the literature to provide a forum for teachers to discuss their voices on excellence in teaching practice. The FGD is important in generating data through group interactions on generic questions in relation to the aims of the research (Creswell, 2009). The question seeks information about teachers' perspectives and experiences in teaching practice and covers the issues that were thought important to achieve excellence in teaching. The samples of the FGD for teachers included: *what is your opinion about excellence in teaching? What methods do you usually apply in teaching practice? Do you have any concern an expectation with your work so far?*

To answer the research questions, the data from the questionnaire were analyzed using descriptive statistics. The qualitative data were obtained through the focus group discussions. The analysis involved the use of Ritchie and Spencer's (1994) framework analysis which involves five key stages: familiarization, identifying a thematic framework, indexing, charting and mapping interpretation.

3 RESULT AND DISCUSSION

There were about one third of participants have more than 11 years' teaching experience and another one third have three to five years' experiences. Almost half of the respondents have a Bachelor's degree; about a quarter are studying for a Bachelor's degree, and less than a quarter graduated from senior high school. The response level was good, 100% (N = 115), even though this is not strong enough for factor analysis.

3.1 *Result on questionnaire*

A descriptive statistical procedure was used to analyze the data generated for all items. This section of the result presents teachers' responses to the items on excellence in teaching practice. Data in Table 2 shows that a great number of teachers belief that maintaining a good communication with their colleagues (97.6%), collaboration with colleagues to do the task (95.1%), paying attention to children's need and their rights (94.3%) and upholding ethics and courtesy toward teachers in the learning process (94.3%) were essential to achieve an excellence in teaching practice.

The result further showed teachers' perspective to have a good quality in teaching is related to their competency to provide stimulation for various aspects of children's development (89.4%). In conjunction with children's developmental aspects, it is interesting that majority of teachers (87.7%) have prioritized religious and moral aspects of developments as the most significant aspect in creating excellence in teaching.

It is quite a worrying situation that almost half percentages of teachers (46.9%) were not sure whether they have implemented a play-based learning for the children. They admitted that they have some difficulties in implementing a fun learning program for young children as they have other lots of target program to be completed. It is assumed that there were still many early childhood programs focus their teaching on academic skills for children's preparation to primary school. Considering this reality, it is important to work collaboratively with these teachers to find a solution to the issues they encounter in their daily work with children.

Table 2. Teachers 'responses to items questionnaire in percentages.

	Items	Strongly agree/agree	Do not know	Strongly disagree/ Disagree
		%	%	%
1	Western theories are importance as the reference for teaching practice at school	26.3	32	41.7
2	My teaching practice refers to our local cultural values to be applied in the current era of globalization	72.2	17.2	10.6
3	I believe that any popular research in the world cannot be compared to the valuable tradition that has been established in our country	40.2	37.7	22.1
4	I prioritize the religious and moral values as the most important aspect in children's development	87.7	9	3.3
5	Teachers have to pay attention to the needs and children's rights in carrying out the learning program	94.3	4.9	0.8
6	Children's ethics and courtesy towards the teacher must uphold in the learning process	94.3	5.7	0
7	There many local cultural values in this country that cannot be applied in the current era of globalization	34.5	39.2	26.3
8	Teacher needs to control the children and apply educated punishment if needed	80.3	11.5	8.2
9	I provide stimulation for various aspects of children's development	89.4	0	10.6
10	I apply various learning methods for children in the classroom	81.2	0	18.8
11	I prepare a variety of media from surrounding for children's learning	81.2	13.9	4.9
12	Teacher needs to collaborate with their colleagues (teacher, staff and principle) at school in order to have the works done well	95.1	4.9	0
13	I establish a good communication with other teachers at school	97.6	0	2.4
14	I share information with colleagues to improve the quality of learning	91	0	9.0
15	I have a patient in handling various children's behaviors.	91	0	11
16	I have not had the opportunity to attend regular training related to early childhood teacher professional development	24.5	60.7	14.8
17	Teacher deserved a a better salary to live a good life	82.8	12.3	4.9
18	Teacher would try to get an achievement if she/he gets an appropriate reward	80.3	13.9	5.8
19	Teacher may implement a good quality of teaching if their workload reduced	63.1	23	13.9
20	Teacher has limited capacity in handling children in a large group	68	23	9.0
21	An good infrastructure and facilities are needed for smooth learning	88.5	9.8	1.7
22	I hope that I am not burdened with the administrative tasks as I have to handle so far	68	27.9	4.1
23	Teachers with a higher background education generally have a better teaching strategy.	53.2	35.2	11.6
24	Teacher often has some difficulties in implementing a fun learning because they have lots of target program to be completed.	49	46.9	4.1
25	Teachers today has many tasks to do, which is not proportional to the reward they get.	73	18.9	8.1
26	The intensity of teacher training provided by the government are still limited.	61.5	28.6	9.9
27	Teacher really need an improvement of welfare financially.	86.9	9	4.1
28	In general, teacher has difficulty to manage their workloads in a range of time given	81.2	15.6	3.2

(*Continued*)

Table 2. (*Continued*)

	Items	Strongly agree/agree	Do not know	Strongly disagree/ Disagree
29	in general, teacher experiences some difficulties in managing their time in the classroom because of their limited competency	71.3	23.8	4.9
30	I think that teacher's job is still not appreciated by our government and society in general	68.8	27	4.2

It is unexpected that more than one-third of the teacher participants (39.2%) considered that many cultural values in this country cannot be applied in the current learning era of globalization, although they confess (72.2%) to adopt the local culture in their own teaching practices. On the other side, they did not agree either (41.7%) that other theories from Western culture be the reference in teaching practice at school. This might be due to the social globalization that interconnects our society since early age to other different values that t provides many choices. It is undeniable that globalization has brought positive and negative aspects, including to the significance of our local culture.

Teachers' voices on excellence in teaching practice were presented in percentages. Table 2 shows the result of the item questionnaire (items 1–30).

3.2 *Result on the focus group discussion*

There are three key themes that emerged from the teachers FGD were (1) Professional knowledge issues (2) teaching strategy (3) workload concern and salary Expectation. Each theme was further explained and typical comments were noted to corroborate the findings. For example, for the question, *what is your opinion about excellence in teaching?* The teachers' responses were categorized under the professional knowledge issues theme. On the question, *What methods do you usually apply in teaching practice?* the participants' answers were categorized under the teaching strategy theme. In terms of the question, *Do you have any concern an expectation with your work so far?* the participants' responses were categorized under the workload concern and salary expectation theme. Representative quotes from FGD have been reproduced to reflect each of the four themes.

Theme 1, professional knowledge issues

The result indicated that in general teachers have some challenges pertaining to their competency in teaching children:

> I think I am not good enough at teaching. I feel terrible in teaching some subjects such as art and sport for children. My background education is not from early childhood. I hope there will be a continuous training for us . . . for free. Further, we have limited facilities at school. I think we need all of these so we can have a good quality of teaching.

Theme 2, teaching strategy

The teachers' comment showed a common practice of teachers in general that focused on developing children's academic skills and set a learning target. It appears that teachers viewed the task given to children as a foundation for future learning:

> I give some activities and lot of tasks for children in the classroom. I have to teach about 40 children in a quite small place so I have to make them discipline in learning.

> I teach in an Islamic kindergarten. I teach them memorizing the verses of holy books, children forget it very quickly.

> In teaching practice, I try to teach moral and spiritual values and also introduce the local cultures to the children. I learn the strategy by following what educational experts' say.

Theme 3, workload concern and salary expectation

The data demonstrated that teachers experience difficulties in relation with many works to be completed at school and have concern on their salary:

> I feel I have lots of things to do at school. I have to do the instructions from school principle as well. Further, sometimes I found it hard to collaborate with my colleagues because we have different rules in teaching children. These problems sometimes influence my emotion and I feel that I lack of sensitivity to children's needs.
>
> Well, we have lots of hope. You know our salary is quite small. We do not know whether the government notice our condition. Sometimes we have problems in our family because of this situation as well. We need protection at health, legal aspects etc.

3.3 *Discussion*

The findings related to teachers' perspective on the importance of maintaining a good communication is supported by Brooker (2008) who argued about the need to practice an effective communication in order to work collaboratively with children and other colleagues. Teachers' problems in Indonesia that related to limited competency in general is due to their educational background which has been a problem there (MoNE, 2010b). Furthermore, large class size would impact on the way teachers support childrens' learning. Therefore, a policy to provide professional training for them is essential in order to improve their educational performance and enhancing excellence in practice. Next, teachers practiced academic tasks and memorization or rote learning in this study showed their prioritization of children's academic skill that may impact to lower children's enthusiasm for learning (Hyson, 2008). The practice contradicts to scholars' recommendation to focus stimulating all aspect of childrens' development, and provide opportunities for children to experience problem solving, exploring and acquiring knowledge (Jensen, 2008). Teacher will complete the tasks excellently when they are motivated. An improvement of teacher's welfare is believed to be crucial in increasing teacher's motivation to perform well.

The FGD results showed that the professional knowledge issues, teaching strategy, workload concern and salary expectation were key variables that informed the quality of teaching practices in this study. Some teacher groups reported their lack of competency in applying various strategy and interacting with children, but other groups felt quite confidence about their skills in teaching. The findings have demonstrated that it is crucial to rethink teachers' access to get opportunity in professional development training.

4 CONCLUSION

The findings demonstrated that the majority of early childhood teachers in this study considered that excellence in teaching practice is influenced mostly by teacher professionalism. Early childhood teachers in this study in general have some challenges to provide appropriate experiences that may supports children's physical, social, emotional, language and cognitive development due to their workload. Cultural, moral, and spiritual values have shaped teachers' belief which have subsequently impacted their practice in the classroom. Further, teachers have an expectation in relation to a better reward or appraisal for their work. The government needs to provide more comprehensive and continuous training for teachers, as well as building a greater collaboration between stakeholders in order to achieve excellence teachers.

REFERENCES

Agbenyega, J., & Deku, P. (2011). *Building New Identities in Teacher Preparation for Inclusive Education in Ghana*. Current Issues in Education, 14(1). Retrieved from http://cie.asu.edu/.

Agbenyega, J. (2013). Teacher motivation and identity formation: issues affecting professional practice. *MIER Journal System* (3), 1.

Berk, L.E. (2006). *Child Development*. Boston: Allyn & Bacon.

Boethel, M. (2004). *Readiness: School, family, and community connections*. Austin, Texas: Southwest Educational Development Laboratory.

Brooker, L. (2008). *Supporting transitions in the early years*. London: McGraw-Hill Education.

Coladarci, T. (2002). Is it a house or a pile of bricks? Important features of a local assessment system. Phi Delta Kappan, 83, 772–774.

Creswell, J. W. (2009). *Research Design: Qualitative, Quantitative, and Mixed Methods Approaches*. 3rd Edition. Thousand Oaks: Sage Publications, Inc.

Darling - Hammond, L. (2000). *Teacher quality and student achievement: A review of state policy evidence. Education Policy Analysis Archives*, 8(1). Retrieved from http://epaa.asu.edu/epaa/v8nl.

Dockett, S. & Perry, B. (2006). *Starting school: A guide for educators*. Sydney: Pademelon Press.

Fauth, B. & Thompson, M. (2009). *Young children's well being, domains and contexts of development from birth to age 8*. London: National Children's Bureau.

Fisher, J. (2008). *Starting from the child: Teaching and learning in the foundation stage*. Maidenhead: McGraw-Hill.

Fleer, M. & Hedegaard, M. (2010*). Early learning and development- cultural historical concepts in play*. Cambridge: Cambridge University Press.

Fridani, L. (2014). School Readiness and Transition to Primary School: A Study of Teachers, Parents, and Educational Policy Makers' Perspectives and Practices in the Capital City of Indonesia. Unpublished PhD thesis. Monash University, Australia.

Graue, M.E. (2006). The answer is readiness-now what is the question? *Early Education and Development*, 17 (1), 43–56.

Hamre, B. K., Downer, J. T., Jamil, F. M., & Pianta, R. C. (2012). Enhancing teachers' intentional use of effective interactions with children. In R. C. Pianta, W. S. Barnett, L. M. Justice & S. M. Sheridan (Eds.), *Handbook of early childhood education* (pp. 507–532). New York: Guilford Press.

Harslett, M. (2000). Teacher perceptions of the characteristics of effective teachers of Aboriginal Middle School students. *The Australian Journal of Teacher Education*, 25(2). Available at: ajte.education.ecu.edu.au.

Hyson, M. (2008). *Enthusiastic and engaged learners: Approaches to learning in the early childhood classroom*. New York: Teachers College Press and Washington, DC: NAEYC.

Jensen, E. (2008). *Brain-based learning: The new paradigm of teaching*. Thousand Oaks, CA: Corwin Press.

Laura, E.L., & Munsch,J. (2014). *Child development: An active learning approach*. London: Sage Publications, Inc.

Ministry of National Education (MoNE). (2009). *Ministerial Decree No. 58/2009 about Early Childhood Care and Education Standard*. Jakarta: Kementrian Pendidikan Nasional/Kemendiknas.

Pence, A., & Bame, N., (2008). *A Case for Early Childhood Development in Sub-Saharan Africa*. Working Paper 51. Bernard van Leer Foundation, The Hague.

Richie, J., & Spencer, L. (1994). Qualitative data analysis for applied policy research. In Bryman and Burgess, eds. *Analysing Qualitative Data* London: Routledge, 173–194.

Rosier, K. & McDonald, M. (2011). Promoting positive education and care transitions for children. *The Australian Institute of Family Studies CAFCA resource sheet*, November 2011. Retrieved from https://pdfs.semanticscholar.org/8d9b/e3ff4f01bd05d916a337244a5d4b9d4b2b06.pdf

Scott-Little, C., Kagan, S.L., & Frelow, V.S. (2006). Conceptualization of readiness and the content of early learning standards: The intersection of policy and research. *Early Childhood Research Quarterly*, 21, (2), 153–173.

Early Childhood Education in the 21st Century – Yulindrasari et al. (eds)
© 2020 Taylor & Francis Group, London, ISBN 978-1-138-35203-2

The roles of leadership in kindergarten

I. Prajaswari & V. Adriany
Universitas Pendidikan Indonesia, Bandung, Indonesia

ABSTRACT: The purpose of this chapter is to describe the leadership role in kindergarten based on the nomenclature of the Minister of Education and Culture Regulation, No. 15 of 2018, as regards to the Assignment of Teacher as Principal, Article 15, namely the Principal Workload and Regulation of the Minister of Education and Culture No. 16 of 2018, relating to the Fulfillment of Teacher's Workload. In the nomenclature, it was explained that the role of the principal was to carry out managerial main tasks, develop entrepreneurship, and supervise teachers and education staff. With the aim of developing schools and improving school quality based on eight national education standards, however, looking at the results of the analysis in the literature, researchers see contradictions between these roles. Therefore, it is anticipated that this chapter can endorse further research to discuss more about leadership roles.

1 INTRODUCTION

The purpose of this chapter is to explain the role of principals in nomenclature, such as to carry out the main tasks in managerial, entrepreneurship development, and supervision of teachers and education staff. In managerial, the details of the principal's duties are: Planning a School Program; Managing the National Education Standards such as Graduates' Competency Standards; Content Standards; Process Standards; Assessment Standards; and Standards for Educators and Education Personnel. Then in the entrepreneurship development, the tasks of the principal are: Planning an Entrepreneurship Development Program; Implementing an Entrepreneurship Development Program; and Implementing an Entrepreneurship Development Program Evaluation. When supervising teachers and educators, the principal has the responsibilities of Planning a Teacher and Education Staff Supervision Program; Supervising Teachers; Supervising Education Personnel; Following Up on the Results of Teachers Supervision to Increase Professionalism, Implementing Evaluation of Teachers Supervision and Education Personnel; and Planning and Reexamining the Results of Evaluation; and Reporting of Teacher Supervision and Education Staff. (Kemendikbud RI, 2018).

Schools are relatively small organizations, nevertheless the challenges that school leaders encounter are neither small nor simple as Portin et al. (2003) have said "In the great scheme of things, schools may be relatively small organizations. But their leadership challenges are far from small, or simple."

With reference to the task of the principal who is dealing with human resources daily, it can also be mentioned that a school principal is an educator who must also be transformative. The research conducted by Mehdinezhad (2016), shows that the role of the principal is very important, one of which is in influencing teacher performance.

The principal is also the school manager who must be able to perform managerial tasks. Managerial tasks include all the elements in the school, for example: teachers and educational staff, students, parents of students, stakeholders and the community. According to research conducted in Singapore, it is influenced by the ability of a good communication strategy, where principals can share their leadership in building relationships with human resources and building school capacity (Slater & Slater, 2009). At the same time, because the principal is also a teacher, henceforth the principal is automatically also subject to obligations under the Teacher Law, which requires professional teacher competencies. One of the development of professionalism competencies is that a teacher with additional duties as a principal is able to conduct research and produce scientific work (Appleby & Aboo, 2013).

Preparation and development of leadership in schools are fundamental in improving effective school systems (Sumintono et al., 2015). Principals need to develop an effective role in organizing and coordinating institutions whose complexity is in the role of school principals as leaders of educational institutions (Hallinger, 2017). The purpose of this chapter is to try to describe the five roles of principals, that is, Supervisor, Educator, Entrepreneur, Manager, and Researcher.

2 HEAD OF SCHOOL AS A SUPERVISOR

As a leader in an institution, the principal must be able to make an academic supervision program plan in order to improve teachers' professionalism, to carry out academic supervision of the teachers by using appropriate approaches and supervision, as well as to follow up on academic supervision of the teachers and evaluate the supervision of teachers and education staff. (Appendix II of the Republic of Indonesia, Minister of Education and Culture Regulation Number 15 of 2018, concerning Fulfilling of Teachers' Workload, Principals and School Supervisors). This becomes very challenging because in completing their duties and responsibilities as leaders, principals must be effective leaders in making the best use of human resources. While being a competent teacher is not a simple problem, to be able to realize and improve teacher competencies require a serious and comprehensive effort. The Ministry of National Education's aim of academic supervision in the 2007 is to develop professionalism, foster motivation and quality control. Therefore, it is necessary for the principal to supervise teachers and the educational process occurs in the school. The education system in Indonesia has experienced significant changes to face the challenges ahead. A principal should be able to make a change regardless of limited time they might have.

The principal's ongoing professional development program focuses on creating regional competitiveness. This program is given to all School Principals in the education units including Kindergarten, Elementary, Middle School and High School/Vocational School with an academic supervision module conducted by the Ministry of Education and Culture starting in 2016 (Kemendikbud RI, 2018). However, we need to further investigate the effectiveness of this program on how the principal carries out his/her duty as a supervisor.

3 HEAD OF SCHOOL AS AN EDUCATOR

In the context of the learning process, principals demonstrate high commitment and focus on curriculum development and learning activities which are at the core of the education process. The principal has to evaluate the teachers' competencies, then to facilitate and to support the teachers improving their competencies continuously for effective and efficient teaching and learning (Holloway, Nielsen & Saltmarsh, 2018).

As an educator, an effective principal is the principal who conducts class visits or interacts during leisure time with teachers and students. For example, the principal of a school that has eight classes, must have sufficient time to visit each classroom (Wahlstrom, 2011).

This is challenging for school principals to do with limited time. For that reason, a leadership division is made as a professional model to be able to reach the expansion of leadership roles in schools in order to improve students' graduation standards by continuously updating teacher competencies (Harris, 2011). The division of leadership from the principal to the teachers and the class coordinators affects the developmental stages of a school (Macbeath & Macbeath, 2017).

In another study, it was claimed that leadership in schools plays a significant role in the quality of education of an institution and the way students learn. All successful principals whom on average carry out basic leadership practices, are responsible and responding to what happens in school, improvising teaching and learning and having the ability to motivate teachers and staff, creating a good work climate; in addition, the principal's ability to distribute his/her leadership to other teachers has a strong influence on the school and on students (Leithwood, Harris & Hopkins, 2008). Whereas other studies state that principals themselves do not have a good teaching model, but they are required to demand that all elements of the school perform good instruction on a daily basis (Lumby, 2013). The study was conducted to see how effective leadership practices are in the context of division of leadership in schools, since basically leadership is the interaction of school leaders, followers and the school situation (Spillane, Halverson, & Diamond, 2004). The principal as an educator is certainly oriented towards learning outcomes that produce superior graduates. With this task, it will be a challenge for when we talk about educators and students. The principal must also have the ability to carry out personal and clinical approaches to both. Is the principal still an educator? Loder & Spillane (2005) states that the approach to educators and students is more personal than to the organization. Subsequent research is needed to be able to see the extent to which the principal's achievement in carrying out this task does not result in the role of conflict and the leadership role that does not take place consistently.

4 HEAD OF SCHOOL AS AN ENTREPRENEUR

The role of the principal as an entrepreneur brings a challenge with the question whether this school is for educational or commercial bussiness. Principals should also be able to make changes and innovations to increase teachers' professionalism and to improve students' learning, Research conducted by Machin (2014) explains that the implications of neo-liberal globalization are not easy since there is a necessity that principals have the knowledge and understanding of commercial densities, the struggle to survive, accountability, managing resources in schools in a philosophical way to seek profit or profit in the context of the school as a business entity.

The principal's attention always goes to the complex and continuous issues that are present in his/her daily life. Learning how to be able to identify, prioritize, synthesize, and act in the context of overcoming existing problems is the biggest interest in career of a principal. Raihani (2008) said, "A school principal had the ability to develop the vision and mission, develop strategies, build capacity, and open networking to achieve the progress and goals of a school".

In Sweden, research was conducted in which it became clear that there was a need to renew the role of schools in entrepreneurship. The biggest challenge is to be creative between the boundaries set by the school authorities where vision, mission and goals are towards building schools in the business sector, building trust, distributing tasks between teachers and staff and having the courage to be able to "think outside the box" (Hörnqvist & Leffler, 2014). One study explained the relationship between leadership and entrepreneurship is a relationship that is rarely empirically studied (Eyal & Kark, 2004).

The findings in the study show that transformational leaders have more opportunities to develop entrepreneurship in a school context that allows them to try to implement a trial and error business culture in entrepreneurship field.

5 HEAD OF SCHOOL AS A MANAGER

Leadership in education; there are two new conceptual models that are empirically assessed as important leadership models, for instance: Instructional Leadership and Transformational Leadership. Transformational Leadership focuses more on building organizational capacity to continue to innovate in the field of teaching and learning, thereupon it can build a vision and commitment to school changes. While Instructional Leadership is a combination of expertise and charisma (Hallinger & Heck, 2010). Although according to Stanavage (2015), the definition of instructional leadership still cannot be clearly described because of the many roles to be carried out. While what actually becomes the day-to-day task of the principal is a perplexing role.

Research on gender and leadership in several countries explained that the lack of representation of women as leaders in educational institutions, even though the majority of the position of teachers throughout the world were women. The terms "glass ceiling" and "glass wall" are defined as horizontal and vertical boundaries for women in the face of their culture and environment as leaders in the fields of education and management (Cubillo & Brown, n.d.).

In Australia, a study was conducted describing a pilot project created to involve teachers in a community project that carries out the practices of new leadership aspects, (Woodrow & Busch, 2008), where the dimensions of leadership in Early Childhood Education are strong elements that must-have, and many practitioners including in Australia are still not ready in the playgroup context. Which in turn, it will produce a new construction in the professionalism preparation of Early Childhood Education teachers as a project of the action and as an activist.

The results of a study in Hong Kong showed that collaboration and teacher participation in making decisions in kindergarten were emphasized in quality assurance and implementation of education schemes before entering primary school, and were influenced by the leadership style of principals in managing their staff (Chan, 2014).

6 HEAD OF SCHOOL AS A RESEARCHER

In the Regulation of the Minister of Education and Culture of the Republic of Indonesia, Number 137 of 2014, concerning National Standards for Early Childhood Education, Appendix II, Educator Competencies, there are four competencies that must be possessed by educators of PAUD (playgroup) for example: Personal, Social, Professional, and Pedagogic Competencies. In Professional Competence, it is stated that a teacher must be able to develop material, structure, and concepts in scientific fields that support and are in line with the requirements and stages of early childhood development; design various creative development activities in accordance with the stages of early childhood development; develop professionalism on an ongoing basis by taking reflective actions (PAUD, 2014). According to Nancy Fichtman Dana in her book *Leading with Passion and Knowledge*, it was explained that teachers with additional assignments as school principals must also be able to develop and improve professional, administrative and school leaders by conducting action research. The principal is expected to be able to identify and explore what interests him/her in conducting research, in terms of building teachers and staff, curriculum, personal actions of a teacher, developing community education, leadership abilities, school performance or building management.

Doing classroom research is a part of teacher's responsibility, yet it is not a formal research. As for academics, if we incorporate the elements of scientific discipline and structure from the day-to-day attitudes obtained, it also means we have done research. As a result, it creates an understanding of how to be able to develop yourself. This is a little different from the provision of training that is usually obtained by a teacher. In professional competence the teacher is expected to carry out his/her own research to develop a professional knowledge for themselves. Which in the end will support what they already know, build a new knowledge from the results of the research. (Appleby & Aboo, 2013). Principals play a role as facilitators for the teachers' research so that the teachers can do their research successfully (Ponte 2005).

7 CONCLUSION

The performance of a school principal is highlighted, because the principal is a profession that reports to stakeholders. The performance of the principal must certainly be shown in behavior and character, mindset, knowledge and skills to the ability to face challenges and pressures. There have been many criticisms of the school principal in carrying out his/her duties, because they are deemed not to have fulfilled the requirements in accordance with the inherent nomenclature when serving as the principal. Many layers of society doubt the competence of school principals since Indonesian education is not progressing well. Ideally, principals play crucial roles in the success of an educational institution. Ineffective leadership of a principal in Indonesian context is a result of conflicts of roles he/she faces. For example conflict between his/her role as an educator versus an entrepreneur.

Leadership in Early Childhood Education is very complex owing to its diversity, and its scale is so large due to strong advocacy and the role of the community. In this case, it refers to more collaborative processes. Early Childhood Education practitioners are committed to increasing collaboration, partnership and positive community role rather than an entrepreneurial approach in the private sector. This again requires a lot of further research in Early Childhood Education (Muijs et al., 2004).

REFERENCES

Appleby, Y., & Aboo, O. (2013). Action research for professional development: Concise advice for new and experienced action researchers. *Studies in Continuing Education*, *35*(1), 128–29. https://doi.org/10.1080/0158037X.2013.767506.

Chan, C. W. (2014). The leadership styles of Hong Kong Kindergarten principals in a context of managerial change. *Educational Management Administration and Leadership*, *42*(1), 30–39. https://doi.org/10.1177/1741143213499263.

Cubillo, L., & Brown, M. (n.d.). Women into educational leadership and management : International differences ?. https://doi.org/10.1108/09578230310474421.

Eyal, O., & Kark, R. (2004). How do transformational leaders transform organizations? A study of the relationship between leadership and entrepreneurship. *Leadership and Policy in Schools*, *3*(3), 211–35. https://doi.org/10.1080/15700760490503715.

Hallinger, P. (2017). Instructional leadership and the school principal: A passing fancy that refuses to fade away instructional leadership and the school principal, 0763. https://doi.org/10.1080/15700760500244793.

Hallinger, P., & Heck, R.H. (2010). Leadership for learning: Does collaborative leadership make a difference in school improvement?. *Educational Management Administration and Leadership*, *38*(6), 654–678. https://doi.org/10.1177/1741143210379060.

Harris, A. (2011). Distributed leadership: Implications for the role of the principal. *Journal of Management Development*, *31*(1), 7–17. https://doi.org/10.1108/02621711211190961.

Holloway, J., Nielsen, A., & Saltmarsh, S. (2018). Prescribed distributed leadership in the era of accountability: The experiences of mentor teachers. *Educational Management Administration and Leadership*, *46*(4), 538–55. https://doi.org/10.1177/1741143216688469.

Hörnqvist, M. L., & Leffler, E. (2014). Fostering an entrepreneurial attitude—challenging in principal leadership. *Education and Training*, *56*(6), 551–61. https://doi.org/10.1108/ET-05-2013-0064.

Kemendikbud RI. (2018). *Peraturan Menteri Pendidikan dan Kebudayaan Republik Indonesia nomor 6 tahun 2018 tentang penugasan guru sebagai kepala sekolah [Minister of Education and Culture Republic of Indonesia regulation number 6 of 2018 concerning the assignment of teacher as principals]*. Jakarta: Kementerian Pendidikan dan Kebudayaan Republik Indonesia.

Leithwood, K., Harris, A., & Hopkins, D. (2008). Seven strong claims about successful school leadership. *School Leadership and Management*, *28*(1), 27–42. https://doi.org/10.1080/13632430701800060.

Loder, T.L., & Spillane, J. P. (2005). Is a principal still a teacher? U.S. women administrators' accounts of role conflict and role discontinuity. *School Leadership and Management*, *25*(3), 263–79. https://doi.org/10.1080/13634230500116348.

Lumby, J. (2013). Distributed leadership: The uses and abuses of power. *Educational Management Administration and Leadership*, *41*(5), 581–97. https://doi.org/10.1177/1741143213489288.

Macbeath, J., & Macbeath, J. (2017). Leadership as distributed: A matter of practice leadership as distributed : a matter of practice. https://doi.org/10.1080/13634230500197165.

Machin, D. (2014). Professional educator or professional manager? The contested role of the for-profit international school principal. *Journal of Research in International Education*, *13*(1), 19–29. https://doi.org/10.1177/1475240914521347.

Mehdinezhad, V. (2016). School Principals: Leadership Behaviours and Its Relation with Teachers—Sense of Self-Efficacy. *9*(2). https://doi.org/10.12973/iji.2016.924a.

Muijs, D., Aubrey, C., Harris, A., & Briggs, M. (2004). How do they manage? A review of the research on leadership in early childhood. *Journal of Early Childhood Research*, *2*(2), 157–69. https://doi.org/10.1177/1476718X04042974.

PAUD. (2014). PERMENDIKBUD No. 137 Th. 2014. Jakarta: Kementerian Pendidikan dan Kebudayaan.

Ponte, P. (2005). How teachers become action researchers and how teacher educators become their facilitators. *Educational Action Research*, *10*(3), 399–422. https://doi.org/10.1080/09650790200200193.

Portin, B., DeArmond, M., Gundlach, L., & Schneider, P. (2003). Making sense of leading schools: A national study of the principalship. Retrieved from http://scholar.google.com/scholar?hl=en&btnG=Search&q=intitle:MAKING+SENSE+OF+leading+schools:+A+National+Study+of+the+Principalship#0%5Cnhttp://scholar.google.com/scholar?hl=en&btnG=Search&q=intitle:Making+sense+of+leading+schools:+A+national+study+of+th.

Raihani. (2008). An Indonesian model of successful school leadership. *Journal of Educational Administration*, *46*(4), 481–96. https://doi.org/10.1108/09578230810882018.

Slater, L., & Slater, L. (2009). Educational management administration and leadership pathways to building leadership capacity. https://doi.org/10.1177/1741143207084060.

Spillane, J. P., Halverson, R., & Diamond, J. B. (2004). *Towards a theory of leadership practice: a distributed perspective. Journal of Curriculum Studies, 36.* https://doi.org/10.1080/0022027032000106726.

Stanavage, A. (2015). Educational leader: An authentic role. *The bulletin of the National Association of Secondary School Principals*, *51*(322), 3-17.

Sumintono, B., Sheyoputri, E. Y. A., Jiang, N., Misbach, I. H., & Jumintono. (2015). Becoming a Principal in Indonesia: Possibility, Pitfalls and Potential. *Asia Pacific Journal of Education*, *35*(3), 342–52. https://doi.org/10.1080/02188791.2015.1056595.

Wahlstrom, K. (2011). Principals as cultural leaders. *EBSCOhost*. Retrieved from http://web .b.ebscohost.com.nl.idm.oclc.org/ehost/pdfviewer/pdfviewer?vid=51&sid=e953cd08-aa34–4e0a-97d6- e0c1180f0451%40sessionmgr120.

Woodrow, C., & Busch, G. (2008). Repositioning early childhood leadership as action and activism. *European Early Childhood Education Research Journal*, *16*(1), 83–93. https://doi.org/10.1080/13502930801897053.

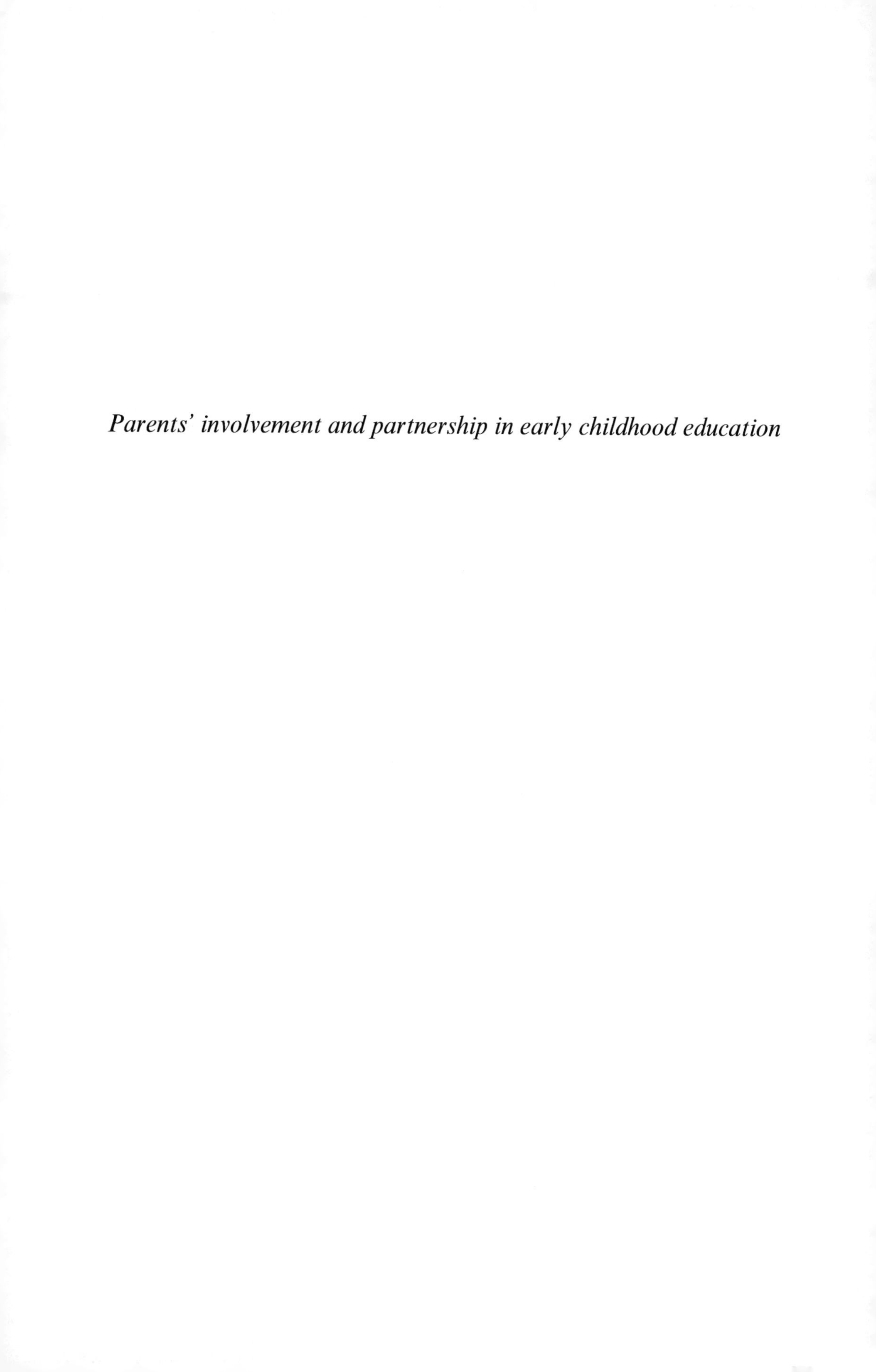

Parents' involvement and partnership in early childhood education

Early Childhood Education in the 21st Century – Yulindrasari et al. (eds)
© 2020 Taylor & Francis Group, London, ISBN 978-1-138-35203-2

Rethinking parental support system in daycare to improve cognitive development of early childhood

Alfiasari, S.S. Zainuddin & N.M. Surianti
Department of Family and Consumer Sciences, Faculty of Human Ecology, Bogor Agricultural University, Bogor, Indonesia

ABSTRACT: The growing number of working mothers in Indonesia has impacted on the increasing need for competent caregiver in order to optimize children's development. The aim of this chapter is to examine the relations among parental involvement, mother-caregiver communication, caregiver-children interaction and attachment, and children's cognitive development. This chapter is based on a study conducted at four daycare centers. The study involved 39 preschool children aged 2.5 to 6.4 years alongside their fathers and mothers as well as 16 caregivers who interact with the children daily. The results indicate that the mothers' involvement and mother-caregiver communication, in addition to the interaction and attachment between the caregivers and the children, has positive correlation to the cognitive development of the children. The findings also indicate that parental support system should become an important factor to be developed in daycare centers for optimal cognitive development. Therefore, the development of parental support system becomes a strategic issue in early childhood education particularly in daycare centers in Indonesia. This chapter also discussed the development of the strategy to create parental support system in daycare centers.

1 INTRODUCTION

There has been an increase in number of dual-earner families, which portrays a shift in family types. According to Rahilly and Johnston (2002), this phenomenon is due to the increase of the participation of women in the labor market. The rising number of dual-earner families adds to the risk of conflict between family and work. According to Puspitawati (2013), the risk of conflict between family and work might be greater in working mother with preschooler children. The increasing number of other caregivers replacing those working mothers in terms of mothering duties requires further research.

The increased participation of working mothers has led to the growth of institutions for the replacement of parental childcare known as daycares in Indonesia (Hastuti, 2015). One of the qualities of daycare services is shown by the role of caregivers in stimulating child development. From birth, child relationships with adults are very important in determining the social and emotional development of healthy children and function as mediators of language and cognitive development (Carl, 2007). As with parent-child relationships, one important component in daycare is interaction and attachment between child and the caregiver. Therefore, similar to the interaction of parents and children, the interaction between caregivers and children also plays an important role. Duncan (2003) stated that early childhood requires responsive and stimulating interactions with parents and other caregivers to improve children's cognitive development. Attachment between a child and their caregiver is an emotional bonding that could be seen from the behavior of the child who feels safe with their respective caregiver, which plays essential role on the child's development. Previous research indicates that children who have secure attachment to their caregivers shows better adjustment and adaptation (Commodari 2013), better receptive abilities in languages (Murray & Yingling, 2000), and better social and emotional development (Mooney, 2015).

The fact that caregivers are involved in children's development, not only mothers but also fathers or other caregivers, including teachers, plays an important role in caring (Commodari, 2013). As a family partner, daycare requires good collaboration with family in optimizing child development. Parental involvement and communication with caregivers are important as a form of parental support to their children development. Parental involvement is defined as active participation of parents (father and mother) in child's development, which in this study is divided between involvement at daycare and involvement at home. In addition, mother-caregiver communication will be able to connect the care received at home and at the daycare to the child's experience (Rentzou, 2012). Reedy and McGrath (2010) found that communication about children and their activities is very important in maintaining trust between parents and daycare.

This study expects to be able to explain the recent phenomena in dual-earner families as well as the role of daycare in early childhood cognitive development by elaborating the findings from both institutions. Hence, the aim of this study is to examine the role of parental involvement, mother-caregiver communication, caregiver-child interaction and attachment on early childhood cognitive development in daycare. It is expected that the findings would be able to provide empirical-based understanding to develop appropriate program in order to improve child development based on family and daycare partnership in Indonesia.

2 METHOD

2.1 *The design and the participants*

This study used a cross-sectional study design. The site selection was conducted purposively in Cibubur, Tangerang, Bekasi, and Bogor as the buffer zones of Jakarta. Participants of this study were 39 preschool children with their respective fathers and mothers, and 16 caregivers. The sampling frame was done purposively, with the requirements of the participants to be children aged 2.5–6.4 years in selected daycares with both father and mother working and willing to participate. The gender ratio was approximately equal with 51.3% boys and 48.7% girls at the average age of 3.9 years.

2.2 *The data collection procedures*

Data collection was conducted by observing both the interactions carried out by the caregiver to the child as well as the child's behavior in forming an attachment with the caregiver in the daycare. This was obtained through the observation process of one caregiver with -two to three children in one day starting at 8 am to 4 pm. Meanwhile, observation of child's cognitive development was done by measuring children's cognitive development outcomes.

The Family Involvement Question (FIQ) taken from Fantuzzo, Tighe, and Child (2000) are instruments used to measure parental involvement after certain modification by the researchers. FIQ consists of 34 statements. However, in this study the researchers only used 21 FIQ statements contained in two dimensions of involvement, involvement at daycare and involvement at home. Meanwhile, data of mother-caregiver communication was obtained using the Caregiver-Parent Partnership Scale instrument by Owen, Ware, and Barfoot (2000). This instrument was different from the original in which the researchers only take two dimensions of communication about children into account, namely five questions regarding shared information and three questions regarding searched information. In addition, caregiver-child interaction was measured using the Caregiver Interaction Scale (CIS) from Arnett (1989), while the caregiver-child attachment was measured using the Q-Sort instrument developed by Vaughn and Water (1990) which consists of 26 statements regarding children's secure attachment behavior. Finally, children's cognitive development was measured using the instrument of the Ministry of National Education for ages 2.5–3.4 years, 3.5–4.4 years, 4.5–5.4 years, and 5.5–6.4 years. Likert scale was used to measure the response (0 = cannot, 1 = can answer but not true, 2 = can answer correctly one or partially, 3 = can answer correctly two or whole). All of the instruments are reliable at Cronbach's alpha > 0.60.

Data analysis carried out was descriptive and inferential statistical analysis. Before conducting data analysis, the raw data was transformed into index (0–100) in order to standardize the variables. Descriptive analysis carried out is an analysis of minimum-maximum values, averages and standard deviations. Meanwhile, inferential analysis carried out is the correlation test to determine the relationship between the characteristics of children and family, parental involvement, mother-caregiver communication, caregiver-child interaction and attachment with early childhood cognitive development in daycare.

3 RESULT

3.1 *Parental involvement*

The average age of father in this study was 34.9 years while the mother's age was 33.0 years which showed that parents who entrusted their children in daycare were young working parents, therefore they had to entrust their early-aged children to a daycare. Parental involvement measured in this study is defined as parent participation in supporting children's development and education, i.e., father's involvement at daycare and at home as well as mother's involvement at daycare and at home. Father's involvement at home (61.28) resulted in a higher average index score compared to father's involvement at daycare (16.67). Similarly, mother's involvement at home (73.50) yielded a higher average index score compared to mother's involvement at daycare (38.14). Overall, the average index score of mother's involvement (60.03) was higher than father's involvement (44.28).

3.2 *Mother-caregiver communication*

The mother-caregiver communication, defined as the exchange of information between mothers and caregivers in the daycare in regards of sharing and searching information about children. The study found that the average index score of information seeking dimension (69.52) was higher than sharing information (60.85). Overall, mother-caregiver communication had an average index score of 64.10.

3.3 *Caregiver-child interaction*

Caregiver-child interaction is a reciprocal relationship between caregivers in the daycare in their approach to the child, such as asking them to talk or asking them to play. The average index of caregiver-child interaction was 76.16.

3.4 *Caregiver-child attachment*

Caregiver-child attachment at daycare refers to the secure attachment of the child with the caregiver, characterized by the child's desire to approach the figure they are attached to, especially in a state of fear or pressure in order to feel safe and comfortable. The result shows that the average index score of the caregiver-child attachment was 68.51.

3.5 *Child's cognitive development*

This study measured early childhood cognitive development including the ability of children to recognize colors, distinguish objects, pair objects, as well as language skills, that is, mentioning their own words, gender, age and name. Additionally, children should be able to concentrate on listening to stories. The results of the study found that the average index of the child's cognitive development was 77.55 with standard deviation value of 15.06.

Based on age groups, the highest average index score occured in the age group of 2.5–3.4 years at 84.03. In this age group, all children have been able to state their nickname. Furthermore, more than half of the children have been able to mention and match four colors as well as choosing objects based on size and pairing known objects (animal puzzles). Moreover, half of the children have been able to mention their own gender.

3.6 *The relationship between variables*

This research found that there is a correlation between parental involvement, mother-caregiver communication, caregiver-child interaction and attachment and cognitive development of the child. The variables with the most significant correlation to child's cognitive development was mother-caregiver communication ($r = 0.417$; $p < 0.01$). Based on the dimension, sharing information ($r = 0.398$; $p < 0.05$) and searching information ($r = 0.365$; $p < 0.05$) also has significant and positive correlation.

Moreover, the mother's involvement ($r = 0.352$; $p < 0.05$) was significantly and positively correlated to the child's cognitive development. The dimensions of mother's involvement at home ($r = 0.359$; $p < 0.05$) had a significant and positive correlations with cognitive development. The caregiver-child interaction and attachment ($r = 0.351$; $p < 0.05$) correlated significantly and positively as well.

4 DISCUSSION

This study found that there is still a lack of involvement of fathers or mothers. However, mother's involvement is notably better than father's. The results of this study confirm that the mother's involvement in supporting the development and education of children both at home and at daycare results in increased cognitive development of children. This is consistent with the research of Topor, Susan, Terri, and Susan (2010) which states that increasing parental involvement has a significant relationship with children's cognitive abilities.

Mothers who leave their children at daycares not only share information with caregivers, they also often search information from caregivers during the time the child is in daycare. The better the communication between mothers and caregivers in sharing and searching information about children both when children are at home and at daycare is, the better the improvement of children's cognitive development. According to Berns (2012), mothers who can build good and optimal communication with caregivers can support the cognitive development of children entrusted at daycare.

Moreover, this study also found that there is a significant positive correlation between the interaction and attachment of caregivers with cognitive development. In line with Duncan's (2003) study, there is a relationship between caregiver-child interactions with early childhood cognitive development. This is because children's cognitive outcomes involve interactions with their caregivers, characterized by inclination of talking, focusing on children, and teaching children to explore. In particular, children who have a secure attachment have better language and psychomotor abilities (attention and meta-cognition abilities) and cognitive abilities (e.g., reading, writing and mathematics) than children with unsecure attachments (Commodari, 2013).

Based on the findings, rethinking parental support system in Indonesian daycares to improve cognitive development is one of critical issues in early childhood intervention. The increasing amount of dual-earner families followed by the increase of daycares centres need appropriate approach. The parental support system should be realized as an important factor to child development, eventhough the children has already been entrusted to daycare centres. The daycare centres should make the parental support as the main prerequisite for children enrollment.

5 RECOMMENDATION

The development of parental support system in daycares should increase the contribution of parents, in order to foster good cooperation between parents and daycare. Daycare as a parenting partner is able to increase the bonding activities for children and parents such as fieldtrip or meetings between staff and parents. In addition, the interaction and attachment to caregivers is very important for a child's development. Caregiver's interactions with children can be enhanced by holding educational session for parents regarding to communicate to children well in particular. It views the development of co-parenting by daycares as strategic solution where daycares plays deeper role than only as parenting replacement institution but also as provider of parenting education for the parents.

REFERENCES

Arnett. (1989). *Smart start evaluation team. Caregiver interaction scales.* UNC-Chapel Hill: FPG Child Development Institute.

Berns, R. (2012). *Child, family, school, community: socialization and support.* 8th edn. Belmont, CA: Nelson Education.

Carl, B. (2007). Child caregiver interaction scale. Theses and Dissertations (All). 66. http://knowledge.library.iup.edu/etd/66.

Commodari, E. (2013). Preschool teacher attachment and attention skills. *SpringerPlus, 2*(1) (12), 1–12. doi:http://dx.doi.org/10.1186/2193-1801-2-673.

Duncan, G. J. (2003). Modeling the impacts of child care quality on children's preschool cognitive development. *Child Development, 74*(5), 1454–1475. doi:10.1111/1467-8624.00617.

Fantuzzo, J., Tighe, E., & Child, S. (2000). Family Involvement Questionnaire: A Multivariate Assessment of Family Participation in Early Childhood Education. *Journal of Educational Psychology, 92*(2), 367–376.

Hastuti, D. (2015). *Pengasuhan: teori, prinsip, dan aplikasinya di indonesia* (parenting: theoris, principals, and its application in indonesia). Bogor: IPB Press.

Mooney, R. (2015). The Preschool Playground: A Longing for a Mother to a Need for Friends. *Infant Observation, 18*(1), 36–51. doi:10.1080/13698036.2015.1010445.

Murray, A. D., & Yingling, J. L. (2000). Competence in Language at 24 Months: Relations with Attachment Security and Home Stimulation. *The Journal of Genetic Psychology, 161*(2) (06), 133–140.

Owen, M. T., Ware, A. M., & Barfoot, B. (2000). Caregiver-Mother Partnership Behavior and the Quality of Caregiver-Child and Mother-Child Interactions. *Early Childhood Research Quarterly, 15*(3), 413–428. https://doi.org/10.1016/S0885-2006(00)00073-9.

Puspitawati, H. (2013). *Pengantar ilmu keluarga (introduction to family studies).* Bogor (ID): Penerbit IPB Press.

Rahilly, S., & Johnston, E. (2002). Opportunity for Childcare: The Impact of Government Initiatives in England upon Childcare Provision. *Social Policy & Administration, 36*(5), 482–495.

Reedy, C. K., & McGrath, W. H. (2010). Can You Hear Me Now? Staff-Parent Communication in Child Care Centres. *Early Child Development & Care, 180* (3): 347–357. doi:10.1080/03004430801908418.

Rentzou, K. (2011). Parent–caregiver Relationship Dyad in Greek Day Care Centres." *International Journal of Early Years Education, 19*(2), 163–177. doi:10.1080/09669760.2011.609045.

Topor, D. R., Susan P. K., Terri, L. S., & Susan, D. C. (2010). Parent involvement and student academic performance: A multiple mediational analysis. *Journal of prevention & intervention in the community, 38*(3), 183–197.

Vaughn, B. E., & Everett, W. (1965). Attachment Behavior at Home and in the Laboratory: Q-Sort Observations and Strange Situation Classifications of One-Year-Olds. *Child Development, 61*(6), doi:10.2307/1130850.

Early Childhood Education in the 21st Century – Yulindrasari et al. (eds)
 ISBN 978-1-138-35203-2

Effective partnership between school, family, and society in early childhood education

L. Halimah, S.Y. Margaretha & R. Roni
Universitas Pendidikan Indonesia, Jawa Barat, Indonesia

ABSTRACT: Effective partnerships between school, family, and society has a very positive impact in the success of child's education. This chapter will try to explain the work of literature study, where effective partnerships between school, family, and society are applied. The result of this literature study is expected to provide inputs and advices to teachers who are always sought out to be the initiator in the building of such partnership.

1 INTRODUCTION

The collaborative cooperation between the school, family, and society are considered as a need for the school program to work (Allen, 2007; Fitzgerald, 2004; Sheridan, & Kim, 2015). One important subject for the development of education in children is the creation of partnership that is firm between the school, family, and society (Fitzgerald, 2004; Berns, 2013; Sheridan & Kim, 2015). A lot of theories and researches has shown that the effective collaboration between school, family, and society are the key to the success for the child's education in all ages (De Bruïne, Willemse, D'Haem, Griswold, Vloeberghs, & Van Eynde, 2014; Pugh, 2001; Dowling & Osborne, 2003; Rimm-Kaufman, Pianta, Cox, & Bradley, 2003; Sheridan & Kim, 2015; Hornby, 2011; Anning, Cottrell, Frost, Green, & Robinson, 2006).

There will be a lot of risks that the child will face if there are no consistency between the school, family, and society in educating the child in their early childhood. However, it is without doubt that the positive impact will be felt and seen through all the child's development potential if the school, family, and society builds a collaborative relationship, with the presence of ongoing partnerships through teamwork as well as the presence of that same hope to develop the child's success, may it be through competences such as academic, behavior, social, and emotional (Christenson & Reschly, 2010; Stevenson, 2006; Dearing, Kreider, & Weiss, 2008).

It is very important that the collaboration between the school, family, and society to happen in order to achieve the child's education success. This chapter is intended to explain the results of the literature study, regarding what, how, and why is there a need for partnership. The result of this literature study is expected to be insightful, especially for teachers, so that they can optimize their partnership with the family and society.

2 WHAT IS SCHOOL, FAMILY, AND SOCIETY PARTNERSHIP?

School, family, and society partnerships are the act of sharing and communicating. The relationship that involves firm and tight cooperation between school, family, and society, as well as having the same rights and responsibility. This partnership focuses on improving the experience and results of children's education, as well as aspects such as academic, social, emotional, and behavioral (Sheridan & Kratochwill, 2007). Joint planning is needed for partnership to be created, and in the execution, it needs the organizing of the other functionary (stakeholders) needs. Effective partnership will be put into realization if there are commitments to share information and aim including results that are expected (Fitzgerald, 2004; Deslandes, 2006).

Adams, Harris, and Jones (2016) defines partnership as the process of two or more parties working side by side to reach a common goal and target. Effective partnership is based on the effort of all party in reaching the same common goal. In the context of school, family, and society partnerships in Indonesia, known widely the "Tri Sentra Pendidikan" or as translated, these are the three centres of education. The partnership of the three centers of education is the effort of teamwork between the unit of education, family, and society that is based on the principles of teamwork, equality, trusts, respect, and the willingness to sacrifice in order to build the ecosystem of education that grows characters and the culture of student achievements (Kemendikbud, 2016; Kemendiknas; 2010; Kemendikbud, 2015).

Hence, it can be explained in the end that partnership between school, family, and society is the effort of teamwork, coordination, and collaboration to increase the chance and the success of the child's education, in which it is based with the acknowledging of the common roles and responsibility between school, family, and society. There is found to be the key concept of school, family, and society partnership, in which it includes: realizing that collaborative partnership should be out of a person's own will and not by force; the need to share of resources' being responsible in decision making; has a goal that aims to reach the defined common goal' acknowledging each other's roles; the ability to work together intuitively to plan formal process of programs; and trusting as well as respecting each other.

3 HOW DOES THE PARTNERSHIP BETWEEN SCHOOL AND SOCIETY BE IMPLEMENTED?

Effective partnership between school, family, and society will be able to facilitate the growth and development of the child optimally. For that to happen, effective partnership can only happen if they are planned, organized, and executed through either informal or formal activities, which at the end is supervised and evaluated (Kemendikbud, 2016; Epstein, & Salinas, 2004; Epstein, 2001).

In relation to the implementation of school, family, and society partnership, refer to the research findings of Epstein (2001); Fitzgerald (2004) where ideal partnership involves the family and society to be a productive partner in the child's and school success. For that, the partnership should cover the six types of school, family, and society involvement, namely: parenting, communicating, volunteering, learning at home, decision making, and collaborating with the community. The purpose of each type of partnership are described here.

Parenting: this can help families, giving support to the child's development, and the provision of the household conditions that supports the child to learn according to his/her own age and grade level. This helps the school in understanding the background, culture, and purpose of the family for the child (Epstein & Salinas, 2004; Epstein, 2001). In parenting level, the school part like the teachers will give information to the family and the school will also listen to the information from the family regarding the need of the child to study (Fitzgerald, 2004).

Communicating: this activity involves family and society regarding the school's program and the development of their children. Creating a two-way channel of communication between the school and the home. According to Epstein and Salinas (2004); Fitzgerald (2004), effective communication between the teachers and the parents are very important for them to develop a partnership. The importance of communication in terms of delivering of communicating the information can be disseminated to the parents to help them understand the objectives of the education, the awareness of how their child develop in his/her studies, and how to help the children at home. Same rules are applied vice versa for the parents to deliver information about their child to the teachers. It is highly supported that effective communications are established. To maximize the opportunity of teacher-parent partnership, communication should include: Two-way (with the possibility of both parties to deliver information accurately); conducted through informal and informal methods; giving feedbacks to parents regarding their child's development; seeking the parents' opinion about their child's situation; involving parents in the process of decision making (Kraft & Dougherty, 2013; Sarmento, & Freire, 2012).

Volunteering: this is the act of recruiting, training, and schedule making to involve families as volunteer and as the audience at the school or other places. This allows the teacher to work voluntarily where the child's success and the school's is supported (Epstein & Salinas, 2004; Epstein, 2001). An example of this activity is through either the school or the teachers in optimizing the family members as the supporter, and the community to offer themselves to increase the accomplishment of the curriculum (Fitzgerald, 2004).

Learning at home: the involvement of family with their children in academic learning at home, including helping the children to do household chores, goal setting, and other activity in relation with the curriculum. For this to happen, teachers have to design household chores that allows the children to share and discuss the tasks that are interested within the family (Epstein & Salinas, 2004; Epstein, 2001). Every parent can help their children's academic success, and the effective involvement of family can happen within every home (National Education Association, 2011; Fitzgerald, 2004).

Decision making: the partnership act in involving the family and the society as participants in decision making for the school programs. In the decision-making process, family members and society, through the school committee, have a task and responsibility as the resources that are facilitative and supportive to the education service with high quality (Epstein & Salinas, 2004; Epstein, 2001; Fitzgerald, 2004).

Collaborating with community: the coordination of resources and services for the family, children, and school with the society group including businesses, institutions, organisations, as well as academies or universities (Epstein & Salinas, 2004; Epstein, 2001; Kemendikbud, 2016). The purpose of this strategy is so that the school can create opportunities as well as trainings for parents or the community, or vice versa through the society for the teachers and parents. (National Education Association, 2011).

In hopes of realizing all types of partnership as mentioned, the school should act as the following as indicated by Kemendikbud (2016), (1) the initiator in the partnership, which is the party that starts the partnership building. As example, the first day of school in which the school is represented by the homeroom leaders to meet with the parent/guardians in discussing about PAUD Program (PAUD: Pendidikan Anak Usia Dini) or in other word, early childhood program, and parents/guardian meeting agenda; (2) partnership facilitator, which is the party that facilitates the manifestation of partnership between families and society, like providing places for organizing a class for parents/guardian; and (3) the partnership controller, which is the party that controls proactively so that all partnership will be better, like doing evaluation on behavioral changes of the parents/guardian in the involvement to support the child's educational progress at home.

4 WHY SHOULD PARTNERSHIP BETWEEN SCHOOL, FAMILY, AND SOCIETY BE BUILT?

To answer this question, there is a juridical basis that stated the need to build partnership between school family, and society, where, Firstly, The National Educational Law No 20 Year 2003, states very firmly that education is the responsibility between government, its citizens, and the family. Secondy, Direktorat Pembinaan Pendidikan Anak Usia Dini (2015) stated that the head of PAUD should collaborate with parents, the society, and all stakeholders to share skills, opinions, and their help in creating a curriculum that is based on a high standard for the child's learning. Thirdly, Government Regulation No 137 Year 2014, stated that challenges for PAUD institution especially to early childhood teachers, should be able to build partnership with parents and society in creating a development program for children in their early childhood education.

Apart from basing it only from a juridical basis point of view, there are a lot of literature study that's been published in which the partnership school, family, and society needs to be intertwined. As Marjoribanks (2002) and Fitzgerald (2004) state that in the globalization era, school has to be able to optimize a good and effective interactions between the parents and society. Furtherm Dowling and Osborne (2003) also argue that the school and families/parents should undergo partnership, as school and parents are two systems of education that has a very important impact in the child's development. In addition to these, Dimerman (2009) and Gurian (2001) stated that it is believed publicly that families have the sole responsibility in educating their children, whereas school and other instances that exists in the society are only giving support and the character building strengthening of the children as how the family already expected of them. (4) Gurian (2001); Adams, Harris, and Jones (2016) stated that especially for early childhood education, it is really important that the school undergo partnership with the families. Even it is needed to be ensured that the child's brain development at home is stimulated properly, so that when children enter the school, they are ready to learn. With that, in a lot of ways like parents acting as teachers for their kid, the tasks and responsibility of the parents are very much alike with teachers, as the term *'parents as teachers.'* (5) Borrell and Artal (2014); Hornby (2011) stated that along with the changing conditions of parents nowadays, complaints are often heard from teachers that they felt a hard time in working with families. Because of those conditions, the schools in this matter need the teachers to realize such changes that happens to the families in the globalization era, and have to give a strategic respond, and create proper relationships with parents in all circumstances no matter how hard or easy it is.

Referring to some statements made earlier, it is very clear that collaborative teamwork between school, family, and society is vital for the school programs created to work (Allen, 2007; Fitzgerald, 2004; Sheridan & Kim, 2015). Many theories are research finding shows that effective partnership between school, family, and society is a critical factor in all aspect of the child's development. Parents' and society's involvement in the progress development of children's education is the key to its success (De Bruïne et al., 2014; Pugh, 2001; Dowling & Osborne, 2003; Rimm, Pianta, & Bradley, 2003; London, Molotsi, & Palmer, 2015). One of the reason being parents as the first child's educator and the most lasting. With that being said, the success of school is really supported by the presence of cooperation between the educator, the education workforce, parents, and society. In accordance with said opinion, Sheridan and Kim (2015); Hornby (2011); Bruckman and Blanton (2003); Amatea, 2013, Mixon and McCarthy (2012); Goshen (2016); Helterbran and Fennimore (2004); Hirschland (2008); Rimm-Kaufman et al. (2003); Minke and Anderson (2005) stated that a lot of research findings shows the collaboration between the school and the parents as well as the society significantly have an impact to the success of the child's education. In school, the teachers have a role that is very strategic in educating children. However to optimize the strategic role, the teacher should ideally align with the parents and the society. The form of cooperation that could be done is to place parents and the society as a facilitator and source/spokesperson in all activities that supports the success of the school's program.

5 SUMMARY

The result of the literature study expresses that partnership between school, family, and society are the effort of teamwork and cooperation between all party in helping the success of the child's education.the effort of this teamwork and cooperation is made happened upon the common interests that is based on the agreement on sharing the responsibilities according to their own roles, giving meaningful contributions, and respecting each other with the sole focus on facilitating the success of the child's education in all their growth development aspects.

For that to happen, effective partnership can only happen should it be planned; organized; executed in numerous activity may it be informal or formal, which at the end is supervised and evaluated. During its implementation, the school in involving the families and society to be a productive partner in the child's success and the school, should do the following six types that involves the activity of: parenting, communicating, volunteering, learning at home, decision making, and collaborating with the community.

The reason that there is a need in building partnership with school, family, and society juridically is firmly emphasized that education itself is the collective responsibility between the government, family, and the society. It has been proven that the effective partnership between school, family, and society have a positive impact on the child's development.

REFERENCES

Adams, D., Harris, A., & Jones, M. S. (2018). Teacher-parent collaboration for an inclusive classroom: Success for every child. *MOJES: Malaysian Online Journal of Educational Sciences*, *4*(3), 58–72.

Allen, J. (2007). *Creating welcoming schools: A practical guide to homeschool partnerships with diverse families*. New York: Teachers College Press, Columbia University.

Amatea, E. S., Mixon, K., & McCarthy, S. (2013). Preparing future teachers to collaborate with families contributions of family systems counselors to a teacher preparation program. *The Family Journal, 21* (2), 136–145.

Anning, A., Cottrell, D., Frost, N., Green, J., & Robinson, M. (2006). *Developing multi- professional teamwork for integrated children's services: Research, policy and practice*. New York: Open University Press.

Berns, R. M. (2013). *Child, family, school, community: Socialization and support*. Australia: Wadsworth, Cengage Learning.

Borrell, S. R., & Artal, C. U. (2014). Formación docente y cultura participativa del centro educativo: claves para favorecer la participación familia-escuela. *Estudios sobre educación*, *27*, 153–168. DOI:10.15581/004.27.153–168.

Bruckman, M., & Blanton, P. W. (2003). Welfare-to-work single mothers' perspectives on parent involvement in head start: Implications for parent–teacher collaboration. *Early Childhood Education Journal*, *30*(3), 145–150.

Christenson, S. L., & Reschly, A. L. (2010). (Editor). *Handbook of school-family partnerships*. New York: Routledge Taylor & Francis Group.

Dearing, E., Kreider, H., & Weiss, H. B. (2008). Increased family involvement in school predicts improved child–teacher relationships and feelings about school for low-income children. *Journal Marriage & Family Review*, 43(3–4), 226–254.

De Bruïne, E. J., Willemse, T. M., D'Haem, J., Griswold, P., Vloeberghs, L., & Van Eynde, S. (2014). Preparing teacher candidates for family–school partnerships. *European Journal of Teacher Education*, *37*(4), 409–425.

Deslandes, R. (2006). Designing and Implementing School, Family, and Community Collaboration Programs in Quebec, Canada. *School Community Journal*, *16*(1), 81–106.

Dimerman, S. (2009). Character is the key: How to unlock the best in our children and ourselves. Canada: John Wiley & Sons Canada, Ltd.

Dowling, E., & Osborne, E. (2003). (Editor). *The family and the school: A Joint systems approach to problems with children*. London: KARNAC.

Epstein, J. L. (2001). *School, family, and community partnerships – Preparing educators and improving schools*. Boulder, CO: Westview.

Epstein, J. L., & Salinas, K. C. (2004). Partnering with families and communities. *Schools as Learning Communities*, 61(8),12–18.

Fitzgerald, D. (2004). *Parent partnership in the early years*. London: Continuum.

Goshen, O. (2016). Collaboration between parents and kindergarten teachers. *Early Childhood Research Quarterly*, *19*(3), 413–430.

Gurian, M. (2001). *Boys and girls learn differently: A Guide for Teachers and Parents*. United States: Jossey-Bass.

Helterbran, V. R., & Fennimore, B. S. (2004). Collaborative early childhood professional development: Building from a base of teacher investigation. *Early Childhood Education Journal*, *31*(4), 267–271.

Hirschland, D. (2008). *Collaborative intervention in early childhood: consulting with parents and teachers of 3-to 7-year-olds*. Oxford: Oxford University Press.

Hornby, G. (2011). *Parental involvement in childhood education: Building effective school-family partnerships*. London: Springer Science & Business Media.

Molotsi, P. H., & Palmer, A. (2015). Collaboration of family, community, and school in a reconstructive approach to teaching and learning. *The Journal of Negro Education*, *53*(4), 455–463.

Kemendiknas. (2010). Kerangka acuan pendidikan karakter [Reference frame for character education]. Jakarta: Kemendiknas; Direktorat Ketenagaan; Direktorat Jenderal Pendidikan Tinggi; Kemetrian Pendidikan Nasional.

Kemendikbud. (2015). Pedoman penanaman sikap pendidikan anak usia dini [Guidelines for planting attitudes to early childhood education]. Jakarta: Kementerian Pendidikan dan Kebudayaan Direktorat Jenderal Pendidikan Anak Usia Dini dan Pendidikan Masyarakat Direktorat Pembinaan Pendidikan Anak Usia Dini.

Kemendikbud. (2016). Petunjuk teknis kemitraan satuan pendidikan anak usia dini (PAUD) dengan keluarga dan masyarakat [Technical guidance on partnership for early childhood education units with

families and communities]. Jakarta: Kementerian Pendidikan dan Kebudayaan; Direktorat Jenderal PAUD dan Pendidikan Masyarakat Direktorat Pembinaan Pendidikan Keluarga.
Kraft, M. A., & Dougherty, S. M. (2013). The effect of teacher–family communication on student engagement: Evidence from a randomized field experiment. *Journal of Research on Educational Effectiveness*, *6*(3), 199–222.
Marjoribanks, K. (2002). *Family and school capital: Towards a context theory of students' school outcomes*. Australia: Springer Science & Business Media.
Minke, K. M., & Anderson, K. J. (2005). Family—School Collaboration and Positive Behavior Support. *Journal of Positive Behavior Interventions*, *7*(3), 181–185.
National Education Association. (2011). Family-school-community partnerships 2.0: Collaborative strategies to advance student learning. *Washington, DC: Author. Retrieved February, 13*, 2012.
Pugh, G. (2001). (Editor). *Contemporary issues in the early years: Working collaboratively for children*. London: Paul Chapman Publishing in association with Coram Family.
Rimm-Kaufman, S. E., Pianta, R. C., Cox, M. J., & Bradley, R. H. (2003). Teacher-rated family involvement and children's social and academic outcomes in kindergarten. *Early Education and Development*, *14*(2), 179–198.
Sarmento, T., & Freire, I. (2012). Making school happen: Children-parent-teacher collaboration as a practice of citizenship. *Education Sciences*, 2(4),105–120 doi:10.3390/educsci2020105.
Sheridan, S. M., & Kratochwill, T. R. (2007). *Conjoint behavioral consultation: Promoting family-school connections and interventions*. Springer Science & Business Media.
Sheridan, S. M., & Kim, E. M. (2015). (Editors). *Processes and pathways of family-school partnerships across development*. New York: Springer International Publishing.
Stevenson, N. (2006). *Young Person's Character Education Handbook*. United States: JIST Publishing, Inc.

Early Childhood Education in the 21st Century – Yulindrasari et al. (eds)
© 2020 Taylor & Francis Group, London, ISBN 978-1-138-35203-2

Parents' perspectives toward home literacy environment and relationship to the impact of early literacy and interest in early childhood

N. Karimah & B. Zaman
Universitas Pendidikan Indonesia, Bandung, Indonesia

ABSTRACT: The importance or focus on literacy is in line with helping children face the challenges of a more complex future. Therefore parents need to teach literacy early at home. Home is the first place for children to get an education. The purpose of this study is to describe parents' perspectives on home literacy environment. The method used was a meta-analysis of previous research related to literacy in the home environment. The analysis shows two main themes from ten articles analyzed, namely: 1) parental perceptions of literacy skills at home, 2) the relationship of parents' perspectives on children's initial literacy and interests. It is also expected that teachers will help and guide parents in creating good home literacy environments.

1 INTRODUCTION

In order to face the intense competition of globalization and the Asean Economic Community (MEA), as well as to make provisions for the golden generation in 2045, RI 4.0 and SDGs, it is necessary to build up Character Education Strengthening (CES). There are 5 (five) values that are the focus of CES, namely (1) nationalism, (2) integrity, (3) independence, (4) mutual cooperation, and (5) religion. This is inseparable from the Nawa Cita program which became the vision of President Joko Widodo. The Jokowi-Kalla duo has declared *Nawa Cita* which is a Sanskrit term for Nine National Priority Agenda (*Nawa Cita*) to improve the quality of life and prosperity of Indonesian people. To make realize it, it is necessary to plant these values early on. One of the ways to teach children about these values is through literacy.

According to statistical data from UNESCO, from a total of 61 countries, Indonesia is ranked 60 with a low literacy level. Rank 59 is filled by Thailand and the last rank is filled by Botswana whereas Finland was ranked first with a high literacy rate, nearly reaching 100%. This data clearly shows that reading interest in Indonesia is still far behind Singapore and Malaysia. Literacy skills and good reading interest are needed in order to face future developments. Poor literacy will cause a person to be left behind.

On the contrary, related to the benefits of literacy, the results of the 2006 PIRLS study in 45 countries studied showed that children who came from families who had stimulated children's literacy abilities early on, had higher literacy skills (Mullis, Martin, Kennedy, & Foy, 2007). Good preschool literacy skills help children to learn to read more easily and increase the level of success of children in school (Senechal & LeFreve, 2002). The results of the meta-analysis conducted by the National Early Literacy Panel (NELP) in 2008 showed that literacy skills of preschool children predict subsequent literacy skills at the moderate to high levels.

Research conducted in Indonesia on 84 children aged 3-6 years in Surakarta showed that the majority of them only read less than 15 minutes a day, reading book facilities owned by children were still less than 10, and parents did not yet have the habit of reading children's story books (Ruhaena, 2015). This shows the low awareness of parents in introducing literacy to children from an early age.

Ernawulan and Agustin (2008: 2) explain that early childhood (0-6 years) also called the golden age and after this development passes, no matter how much intelligence the child achieves, it will not experience any improvement. In this golden age, proper and appropriate stimulation is needed. Learning to read, write and conventional numeracy is not recommended. On the other hand, it is strongly encouraged to develop literacy and numeracy that are appropriate to the child's development stage. The most important part of creating a good home literacy environment is the creation of a space that is organized, stimulating, comfortable and attractive to motivate children to follow it. Based on this argument, it is necessary to conduct further research on parents' perspectives toward home literacy environment which influences the initial assessment and interest of early childhood. Discussions also pay attention to socio-cultural elements in Indonesia that are different from those in other countries.

2 LITERATURE REVIEW

The paper uses the perspective of developmentalism with a constructivism approach. Vygotsky emphasized the importance of one's active role in constructing knowledge. Asri (2005) explains that a person's cognitive development is also actively determined by the individual himself, and by the active social environment.

Child literacy activities with parents can be an important stimulation in preschool literacy skills as explained by Vygotsky's socio-cultural theory. The process of achieving language skills and literacy regarding preschool children is seen as a result of social interaction. Children learn better in the context of parent-child relations (Mullis et al., 2004). Concerning development of reading and writing, it is more recommended to involve parents (Reese, 2010). Some intervention programs geared towards improving parents' knowledge and skills in stimulating the development of children's literacy have been proven effective (Saint-Laurant & Giasson, 2005; Rasinski, 2005; Reutzel et al., 2006). This means that this ability is achieved because there is social interaction with the presence of someone who has greater ability. The home literacy environment includes variables such as literacy artifacts, functional use of literacy, verbal references to literacy, library use, parental encouragement and reading values, teaching of parents 'skills, children's interests, modeling of parents' literacy behavior, parental education, and people's own attitudes towards education, Lay, Winston & Charis (2014).

Weigel et al. (2006) found that children were more interested in reading when they were involved in literacy and language activities with their parents. According to Nutbrown, Hannon, & Morgan (2005), child literacy develops when parents do/give 4 things: opportunity, recognition, interaction, and example. This framework is used to examine or plan literacy activities at home or school. Parents need to provide opportunities for literacy to become a practice and develop it into abilities. The purpose of this study is to broaden parents' perspectives toward home literacy environments and relationship to the impact of early literacy and interest in early childhood.

3 METHODOLOGY

This paper uses the meta-analysis methodology (Elyasir, 2015). This study analyzed 10 studies on literacy in the home environment in early childhood. The research itself is part of research that broadens the literacy of the home environment about parents' perspectives and their role in children's interests

4 FINDINGS AND DISCUSSION

The analysis shows two main themes from ten articles analyzed, namely: 1) parental perceptions of literacy skills at home, 2) the relationship of parents' perspectives on children's initial literacy and interests.

4.1 *Parent perception of literacy skills at home*

The analysis of the study by Garcia (2011) found that parents' perceptions of the initial literacy skills of learning bilingual languages played an important role in the initial literacy process. In addition, parents also show interest in helping children read and play an active role in the initial literacy experience through playing. The conclusion is that the language used everyday will help children succeed academically. Teachers must bridge the gap between parents and teachers so that schools can use the home environment as a resource.

The results of Lay's research, Winston & Charis (2014), show that parental involvement in reading at home include interactions where parents directly teach their children how to read words, how to write their own names, write simple words, and how to read story books or information material. This is an active component in literacy of the home environment that facilitates initial literacy. Two beliefs that arise from parents when influencing children's reading skills: a sense of success in parents preparing their children for school and the influence of parents when they read with their children. Studies have found a positive relationship between parental reading, trust and parental involvement in literacy and language activities (DeBaryshe, 1995; Weigel et al., 2006).

Another perspective arises from socio-cultural differences about literacy. Parents are often unaware of instructional techniques, but the practice of raising children affected by cultural values is usually in accordance with how children will live in their adult years (Bjorklund & Causey, 2017). This sociocultural perspective emphasizes that development is influenced by the interaction of adults and children in a cultural context that determines how, where and when this interaction takes place. Ruhaena (2015) The socio-cultural review explains the acquisition of preschool language literacy and abilities in the context of everyday life which is meaningful through active involvement in real activities in the microsystem environment, namely the family. Research by Ruhaena (2015) resulted in conclusions that facilities in the process of developing interest and early literacy skills at home preschoolers have used media primarily in the form of storybooks and multimedia technologies such as television and computers. Mothers have also been involved in preschool literacy activities where housewives are more routine in conducting literacy activities than working mothers. The main factor that contributes to children's reading motivation is modeling parents' reading as a fun activity.

4.2 *Relationship of parents perspectives on early literacy and children's interest*

Recent research shows that literacy environments in homes vary, in that different environmental components can affect the results of different mental and educational developments. For example, Griffin and Morrison (1997) broadly measured literacy environments at home such as magazine and newspaper subscriptions, library use, watching television and reading books, and found that it was positively related to receptive kindergarten vocational skills, reading skills, and the introduction of math skills. The results of further research conducted by Christian et al. (1998) showed that the conceptualization of the home literacy environment was positively related to the achievement of verbal reading of kindergarten children.

The results of Lay's research, Winston & Charis (2014) showed that literacy in the home environment, which was conceptualized as family oriented literacy activities and parents' beliefs about reading, had a positive and significant relationship with reading competencies arising from preschoolers and their reading motivation. The involvement of parents has great strategic value in making an important contribution to the success of children in learning to read and write (Paratore in Reutzel et al., 2006). A home environment that is responsive to children improves language skills (receptive and expressive) and literacy (Roberts et al., 2005).

To influence reading and writing activities and children's interests, as revealed by Ruhaena (2015), children enjoy reading and writing activities and are not easily bored orlazy, ie mothers must be creative in creating play situations and improving the quality of interactions in literacy activities .

Other studies show that reading a shared book, parents' assessment of literacy, the quality of the home environment and overall support for the home environment are positively related to preschool literacy skills (Burgess, 1997; Senechal et al., 1998). Reading motivation is important because research has shown that children's interactions in literacy predict early literacy skills.

5 CONCLUSION

According to Lay, Winston & Charis (2014), literacy in the home environment is very important during the preschool years in encouraging the development of reading in children. Regardless of national or cultural boundaries, in any country where children are located, literacy development in the home environment will certainly have an initial and potentially lasting influence on language, reading and development. Therefore it is important for parents to be encouraged and empowered to facilitate literacy in an active home environment earlier in order to support the development of their children's literacy.

REFERENCES

Bjorklund, D. F., & Causey, K. B. (2017). *Children's thinking: Cognitive development and individual differences.* Sage Publications.

Asri, B. (2005). *Belajar dan pembelajaran.* Jakarta: PT. Rineka Cipta.

Burgess, S.R. (2002). The influence of speech perception, oral language ability, the home literacy environment, and pre-reading knowledge on the growth of phonological sensitivity: a one-year longitudinal investigation. *Reading And Writing: An Interdisciplinary Journal* 15: 709–737

Garcia, N. C. (2011). Early literacy experiences in the home: a parent's perspective. *Journal of Border Educational Research*, 9.

Mullis, I., Martin, M., Kennedy, A., & Foy, P. (2007). *Progress in international reading literacy study.* Pirls 2006 report. In: Lynch School of Education, Boston College, Chestnut Hill, MA: TIMMS & PIRLS International Study Centre. International Association for the Evaluation of Educational Achie-vement (IEA).

Mullis, R. L., Mullis, A. K., Cornille, T. A., Ritchson, A. D. & Sullender, M. S. (2004). *Early literacy outcomes and parent involvement.* Tallahassee, Fl: Florida State University

Rasinski, T., & Stevenson, B. (2005). The effects of fast start reading: A fluency-based home involvement reading program, on the reading achievement of beginning readers. *Reading Psy-chology*, *26*, 109-125. doi: http://dx.doi.org/10.1080/02702710590930483

Rasinski, T., & Stevenson, B. (2005). The effects of fast start reading: A fluency-based home involvement reading program, on the reading achievement of beginning readers. *Reading Psychology*, *26*, 109-125. http://dx.doi.org/10.1080/02702710590930483

Reese, E., Sparks, A., & Leyva, D. (2010). A review of parent intervention for preschool children's language and emergent literacy. *Journal of Early Childhood Literacy*, *10*(1), 97-117. doi: http://dx.doi.org/10.1177/1468798409356987

Reutzel, D. R., Fawson, P. C., & Smith, J. A. (2006). Words to go!: Evaluating the first-grade parent involment program for 'making' words at home. *Reading Reasearch and Instructions*, *45*, 119-159.

Roberts, J., Julia, J., Margaret, B., & Fran, P. G. (2005). The role of home literacy practices in preschool children's language and emergent literacy skills. *Journal of Speech, Language, and Hearing Research*, *48*, 345–359. 1092-4388/05/4802-0345.

Ruhaena, L. (2015). Model multisensori: solusi stimulasi literasi anak prasekolah. *Jurnal Psikologi*, *42*(1), 47–60.

Ruhaena, L., & Juni, A. (2015). Pengembangan minat dan kemampuan literasi awal anak prasekolah di rumah. *The 2nd University Research Coloquium*, 2407-9189.

Saint-Saint-Laurant, L., & Giasson, J. (2005). Effects of family literacy adapting parental intervention to first graders' evolution of reading and writing abilities. *Journal of Early Childhood Literacy*, *5*(3), 253-278.

Senechal, M., & LeFreve, J. (2002). Parental Involvment in the development of children's reading skill: a five-year longitudinal study. *Child Development*, *73*, 445-460.

Weigel, J.D., Sally, S.M., & Kymberley, K.B. (2006). Contributions of the home literacy environment to preschool-aged children's emerging literacy and language skills. *Early Child Development and Care*, *176*(3-4), 357-378, doi: 10.1080/03004430500063747.

Early Childhood Education in the 21st Century – Yulindrasari et al. (eds)
© 2020 Taylor & Francis Group, London, ISBN 978-1-138-35203-2

The relationship between parental communication and children's interpersonal intelligence towards their conflict resolution in the kindergaten of Aisiyah Butsanul Atfhal Magelang city

N. Rahmah & P.Y. Fauziyah
Universitas Negeri Yogyakarta, Yogyakarta, Indonesia

ABSTRACT: This study aims at determining the relationship of parental communication with conflict resolution, parental communication relations partially related to interpersonal intelligence and the relationship of interpersonal intelligence with conflict resolution as well as to determine the relationship of parental communication and interpersonal intelligence with conflict resolution in the kindergaten of *Aisiyah Butsanul Atfhal*, Magelang City. This research was done with quantitative approach and correlational types. It is to reveal the influence of parental communication skills towards children interpersonal intelligence and conflict resolution in kindergarten. This research can be categorized as ex-post facto which reveals the reality or symptoms of occurred events and has implications for various subsequent actions that are estimated as the object being studied. The researcher did not give any treatment on the variables. The results of this study showed that all null hypotheses (Ho) are rejected and all alternative hypotheses (Ha) are accepted, with the details as follows. Hypothesis 1) relationship (R) was 0.356 and obtained the coefficient of determination (R2) of 0.127, which implied that the relationship of communication and intrapersonal intelligence was 12.7%. Hypothesis 2) the relationship value (R) was 0.534 and the determination coefficient was obtained (R2) of 0.285, which implied that the relationship of parental communication with conflict resolution ability was 28.5%. Hypothesis 3) relationship (R) was 0.520 and obtained the coefficient of determination (R2) of 0.271, which implied that the relationship of parental communication with conflict resolution ability was 27.1%. Hypothesis 4) relationship (R) was 0.640 and obtained the coefficient of determination (R2) of 0.410, which implied that the relationship of parental communication with conflict resolution ability was 41%.

1 INTRODUCTION

Education is a basic capital to prepare qualified human beings. National education is to develop the capacity and to shape the character and state civilization which is useful in the framework of educating the life of the people (Departemen Pendidikan Nasional, 2003). To realize this goal, the role of early childhood education is crucial to shape students' character. However, early childhood education is not olny emphasized on early childhood institution but also families. Family is the first place for children to learn social life that has big influence on children's growth and development.

A child who has interpersonal intelligence can socialize and solve problems properly. On the other hand, the socialization process may generate conflict in various aspects of social life including in early adult phase. As a mutual problem solving action, conflict resolution is needed. It is one of social abilities that must be considered in the future because, in this children age conflict, resolution is one of the important steps in developing other social skills.

However, based on observations in *Aisyiyah Bustanul Athfal*, Magelang City, the role of teachers, principals and parents in the development of interpersonal intelligence related to conflict resolution has not been optimal yet. Therefore, this study tries to reveal the relationship of parental communication with children's interpersonal intelligence and conflict resolution in *Aisyiyah Bustanul Athfal*, Magelang City.

2 RESEARCH METHOD

This research was done with quantitative approach and correlational types to determine the influence of parental communication skills on children's interpersonal intelligence and conflict resolution in kindergartens. This research can be categorized as ex-post facto. This research was carried out for two months, February 2018 - April 2018 in *Aisyiyah Bustanul Athfal* in Magelang city which belongs to Muhammadiyah Foundation.

The research population was a total of 253 parents and children of Aisyiyah Bustanul Athfal, Magelang City. Meanwhile the research sample was using the sampling formula by Slovin formula (Husain, 2004) with margin of error of 5%, so the samples was 144 respondents which randomly selected from 8 schools. The sample determination was based on purposive sampling technique where each school was taken by respondents at least 18 parents and children aged 5-6 years.

Data collection techniques using questionnaire and observation. Data generated from questionnaire with a range of 1-4 with alternative answers of multilevel scores. The instrument validity was using construct validity and the calculation of correlation coefficient was done computerically with the assistance of SPSS 22.0 package. The reliability testing was carried out by Cronbach Alpha Formula technique. Meanwhile, quantitative analysis was to analyze data based on respondents' answers to questionnaires.

3 FINDINGS AND DISCUSSION

Based on descriptive analysis with the help of Excel program, for parental communication variables (X_1) was 1.7346, the highest value of 158, the lowest score was 65, the mean was of 120, mode was 130, 22 for st. deviation. The following is the frequency distribution table of parental communication category.

Table 1. Distribution of parental communication categories.

No	Catagory	Interval	Number of respondent	Percentage (%)
1	Very Good	$X \geq 148$	37	25.42
2	Good	$111 > X \geq 147$	46	32.20
3	Moderate	$74 > X \geq 110$	56	38.98
4	Bad	$X < 74$	5	3.40
Total			144	100.00

The data indicate that parental communication of in Kindergarten of *Aisyiyah Bustanul Athfal* was not good. It can be seen that the score of "very good" category was 37 respondents (25.42%), for" good" category was 46 respondents (32.20%), for "moderate" category was 56 respondents (38.98%) and for "bad" category was 5 respondents (3.40%).

In case of interpersonal intelligence variables (X) it can be seen the number of 6,192, highest score was 58, lowest was 23, mean score was 141.98, mode was 44, standard deviation was 8. The frequency distribution table interpersonal intelligence students as follows.

Table 2. Distribution of students' interpersonal intelligence.

No	Category	Interval	Number of Respondent	Percentage (%)
1	Very Good	X ≥ 45	37	25.42
2	Good	35 > X ≥ 40	54	37.30
3	Moderate	28 > X ≥ 35	27	18.64
4	Bad	X < 28	27	18.64
Total			144	100.00

These data showed that interpersonal intelligence was good. It showed that the score for "very good" category was 37 respondents (25.42%), for "good" category was 54 respondents (37.30%), for moderate category was 27 respondents (18.64%) and for "bad" category was 27 respondents (18.64%).

In case of students' conflict resolution variable (Y) for the number of 15,429 showed the highest score of 148, the lowest of 58, the mean score of 107, mode of 109, st. deviation of 19.

Table 3. Distribusi Kategori Resolusi konflik siswa.

No	Category	Interval	Number of Respondent	Percentage (%)
1	Very Good	X ≥ 138	71	49.15
2	Good	138 > X ≥ 78	24	16.59
3	Moderate	78 > X ≥ 63	34	23.73
4	Bad	X < 63	15	10.17
Total			144	100.00

The data showed that the students' conflict resolution was very good. It indicated that the score for "very good" category was 71 respondents (49.15%), for the "good" category was 24 respondents (16.59%), for the "moderate" category was 34 respondents (23.73) and for "bad" category was 15 respondents (10.17%).

Then, the prerequisite of data analysis testing was carried out before the data analysis. It consisted of normality test, linearity test. normality test. The results of the normality test on the research variables showed that all the research variables had a significance score that bigger than 0.05 in (P>0.05), so it can be concluded that all research variables were normally distributed. The results for the multicollinearity test indicated that the two variables had tolerance of 1,000 and VIF of 1,000. Due to the tolerance score of the two independent variables was more than 0.1 and VIF was less than 10, there was no multicollinearity problem in the regression model. Meanwhile, heteroscedasticity test results was using SPSS for windows 23.00 as shown in Figure 1.

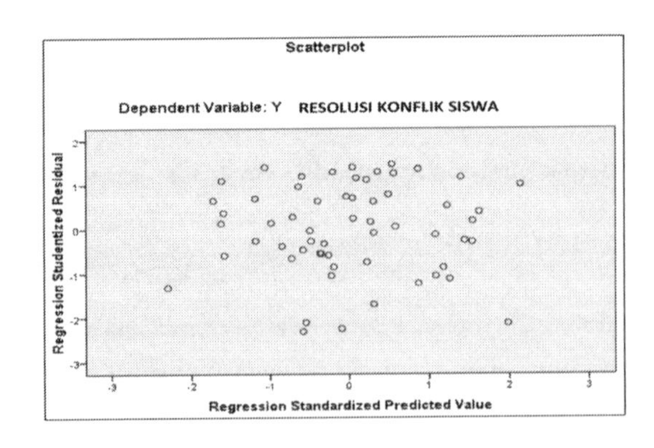

Figure 1. Heteroscedasticity test results.

The scatterplots above showed that the points spread with unclear patterns and below 0 on the Y axis, the coefficients table showed the value of Sig. bigger than 0.05, it meaned that there were no heteroscedasticity problems in the regression model. In addition, in case of Autocorrelation test with Durbin-Watson with significance of 0.05 and the amount of data (n) 144, and k = 2 (the number of independent variables) obtained dL values of 1.7400 and dU of 1.6974. Because its value was located between dU and 4-dU (4-1.6974 = 2.3503), the null hypothesis is accepted or there is no autocorrelation.

4 HYPOTHESIS TESTING

The first hypothesis states that there is a positive and significant relationship between parental communication and interpersonal intelligence. The basis of decision making by using t_{count} between variables of parental communication ability and interpersonal intelligence. The results of the analysis showed the value of the relationship (R) as 0.356 and the determination coefficient was (R2) of 0.127, with the understanding that the communication relationship and intrapersonal intelligence was 12.7%, while the rest wre influenced by other variables. Anova analysis showed that F_{count} = 20,581> F_{table} 3,06 with a significance level of 0,000. The significance score of the analysis was smaller than the 0.05 significance value, or there has been a significant correlation between the two variables. Meanwhile, the score in column B in constanta (a) was 27,278 and the value of parental communication was 0.131 which was positive because its value was positive, there has been a positive correlation between communication between parents and interpersonal intelligence.

The first hypothesis testing was using n = 144 and k = 2, the score of t_{table} with a significance level of 5% was 1,977. It was obtained that parental communication t_{count} of 4,537> t_{table} 1,977 which was bigger than t_{table}, so there has been a correlation between parental communication and interpersonal intelligence. It can be concluded that there was a positive and significant relationship between parental communication and interpersonal intelligence, so the first hypothesis is proven to be true.

The second hypothesis stated that there was a positive and significant relationship between parental communication with students' conflict resolution abilities. The analysis results explain the value of the relationship (R) was 0.534 and the obtained coefficient of determination (R2) for 0.285, meaning that parental communication relationship with conflict resolution ability was 28.5%, while the rest were influenced by other variables. It explained whether there was a real (significant) correlation of creativity variables with intrapersonal intelligence. Table score showed F_{count} = 56,719> F_{table} 3.06 with a significance level of 0.000. The significance score of the analysis was smaller than the significance score of 0.05 or there has been a significant relationship between the two variables.

It has been a positive relationship between parental communication and conflict resolution abilities. The third hypothesis testing was using n = 144 and k = 2, where the value of t_{table} with a significance level of 5% was 1,977. The parental communication score of 7.531> t_{table} 1.977 which was bigger than t_{table}, it can be concluded that there was positive and significant relationship between parental communication and interpersonal intelligence, so the first hypothesis is proven true.

The third hypothesis states that there is a positive and significant relationship between interpersonal intelligence and students' conflict resolution abilities. It was 0.520 and the obtained coefficient of determination (R2) of 0.271, which implies that the communication relationship of parental with conflict resolution ability was 27.1% while the rest were influenced by other variables.

The analysis results explain whether there was a significant (significant) relationship of creativity variables with intrapersonal intelligence. F_{count} = 52.662> F_{table} 3.06 with a significance level of 0.000. The significance score of the analysis was smaller than the 0.05 significance score or there has been a significant relationship between the two variables. It means there has been a positive relationship between parental communication and conflict resolution abilities.

The first hypothesis testing was using n = 144 and k = 2, the score of t_{table} with a significance level of 5% was 1,977. t_{count} of 7,257> t_{table} 1,977 which is bigger than t_{table}, then based on the analysis and explanation of the third hypothesis testing, it indicated a positive and significant relationship between parental communication with interpersonal intelligence, so the first hypothesis is proven to be true.

The multiple regression test analysis results showed, the relationship score (R) of 0.640 and the obtained coefficient of determination (R2) of 0.410, that the communication relationship of parental with conflict resolution ability was 41%, while the rest were influenced by other variables. The significance score of the analysis was smaller than the significance value of 0.05 or there has been a significant relationship between the two variables. It indicates the relationship is positive or there has been a positive relationship between parental communication and conflict resolution abilities.

The first hypothesis testing was using n = 144 and k = 2, so that the score of t_{table} with a significance level of 5% was 1,977. It was obtained that the parental communication was 5.777, the count of interpersonal intelligence was 5.460> t_{table} 1.977, there is a correlation between parental communication and interpersonal intelligence with conflict resolution ability. It can be concluded that there is a positive and significant relationship between parental communication and interpersonal intelligence, so the first hypothesis is proven true.

4.1 *The relationship between parental communication and children's conflict resolution abilities*

Parental communication has a positive relationship or good parental communication will increase the ability of children conflict resolution. Parents workplace influences the pattern of communication with children and parents' education experience influences the relationship between parents and their children (Na'imah, 2012). The parents' ability to maintain communication in any situation needs to be continuously trained so that it positively affects the children behavior. It is line with idea (Andayani, 2004) that the bad relationship between husband and wife with many conflicts can reduce their parenting control.

4.2 *Relationship between parents' communication skills and interpersonal intelligence*

Stereotyping is a method that is commonly found to assess others to certain categories. Individuals or groups in this case are parents can respond to experience and the environment by treating family members differently or tend to oriented on gender, intellegence, diligenty or laziness (Nuraedah, Fahmi, Nurwindiyastuti, & Wisnu, 2014). Certo (2012) explains that:

> To be a successful interpersonal communicator, a manager must understand the following: how interpersonal communication works, the relationship between feedback and interpersonal communication, the importance of verbal versus nonverbal interpersonal communication.

It can also be related to a family where father and mother are an important part of the family. Parents must be able to create communication atmosphere that can make interpesonal intelligence work or providing time for their child with heart to heart talk.

4.3 *Relationship between interpersonal intelligence and children's conflict resolution abilities*

A child who has good interpersonal skills will first observe friends in preschool classrooms and interact with other during free playing activities, such as blocks building or imaginative games as community workers (Macsata, 2015). The child's ability in establishing a good social relationship can motivate them to find new things and those who have good interpersonal intelligence can solve problems that occur in their social relations with a win-win solution approach, and prevent the emergence of problems in their social relations. So that, the children ability to build relationships can relate on how children get ego and conflict management that occur and correlate with learning and students' motivation (Powel & Powel, 2004)

4.4 *The relationship between parental communication abilities and children's conflict resolution abilities*

Good communication with children can make child and parent relationships more comfortable and warm. It can help parents to understand their children development. Parental communication is one of the important things as stated by Bruneau et al. (1991) that:

> Parent communications need to be a continuous part of a holistic literacy curriculum. In the case described here, it took a follow-up letter in the spring to complete communication with the parents.

Parental communication needs to be a sustainable part of the curriculum literacy. Problem solving or cases need to be followed-up in the form of report to complete communication with parents.

5 CONCLUSION

The results of this study shows that all null hypotheses (Ho) are rejected and all alternative hypotheses (Ha) are accepted, with the details as follows. Hypothesis 1) relationship (R) was 0.356 and obtained the coefficient of determination (R2) of 0.127, which implied that the relationship of communication and intrapersonal intelligence was 12.7%. Hypothesis 2) the relationship value (R) was 0.534 and the determination coefficient was obtained (R2) of 0.285, which implied that the relationship of parental communication with conflict resolution ability was 28.5%. Hypothesis 3) relationship (R) was 0.520 and obtained the coefficient of determination (R2) of 0.271, which implied that the relationship of parental communication with conflict resolution ability was 27.1%. Hypothesis 4) relationship (R) was 0.640 and obtained the coefficient of determination (R2) of 0.410, which implied that the relationship of parental communication with conflict resolution ability was 41%.

REFERENCES

Departemen Pendidikan Nasional. (2003). Undang-Undang Nomor 20 Tahun 2003 tentang sistem pendidikan nasional pasal 3, Jakarta: Depdiknas

Husain, U. (2004). *Metode penelitian untuk skripsi dan tesis bisnis.* Cetakan 6, Jakarta: PT Raja Grafindo Persada.

Na'imah, T. (2012). Pendidikan karakter (kajian dari teori ekologi perkembangan). *Prosiding Seminar Nasional Psikologi Islami*, 159–166.

Andayani, B. (2004). Tinjauan Pendekatan Ekologi Tentang Perilaku Pengasuhan orangtua. *Buletin Psikologi*, *12*(1), 44–60. doi: https://doi.org/10.22146/bpsi.7468

Nuraedah, A., Fahmi, A., Nurwindiyastuti, D., & Wisnu, K. (2014). *Hubungan interpersonal. makalah: hubungan interpersonal (Pengertian, Teori, Tahap, Jenis, dan Faktor yang Mempengaruhi Hubungan Interpersonal) Disusun.*

Certo, S. C. (2012). *Concepts and skills (modern management) twelfth edition samuel.* Prentice hall: Pearson.

Macsata, K. M. (2015). Working it out together: teaching the steps of conflict resolution to preschoolers working it out together.

Early Childhood Education in the 21st Century– Yulindrasari et al. (eds)
© 2020 Taylor & Francis Group, London, ISBN 978-1-138-35203-2

Organization does better: Family resilience strategy through women's organizational social movement

S. Komariah, Wilodati & F.N. Asyahidda
Universitas Pendidikan Indonesia, Indonesia

ABSTRACT: Family resilience will affect the resilience of a nation. Indeed, the understanding of maintaining the integrity of family life is always taught to every generation. However, in practice, conflicts often occur which lead to divorce. Therefore, the social movement in building capable family resilience in the midst of greater family challenges needs to be done. At the moment, it is certainly not the time to just stick to the resolution of problems or repressive patterns, but a preventive pattern is needed as a prevention of the growing problems in the family as a result of the lack of understanding of family conflict resolution. In practice, the Islamic Wife Organization has carried out the social movement, the fund contributes greatly to creating a "creative space" for women's social movements to implement a more tangible role in educating children, families, and communities so that they can realize family resilience. This study uses a qualitative approach with descriptive method successfully describes how the patterns of social movements carried out by the Islamic Wife Organization in providing teaching and understanding about how to maintain family life can be implied not only for families in organization but also family life in general.

1 INTRODUCTION

Achieving goals in a family is certainly not an easy matter. Strong commitment is needed to bring harmony in the midst of ups and downs to challenge family problems (Bongaarts et al., 2015), because when a family is when individuals as couples have a sense of oneness of life to devote themselves to family interests. Simply put, when the roles and functions of family members cannot be met to the fullest, it is not impossible if the family will be easily shaken up by problems that make its stability threatened (Brown et al., 2015). When family conditions are unstable, divorce is an option in solving problems (Roth, Harkins, & Eng, 2014). It is undeniable that divorce is one result of problems that no longer find a way out in a family manner. The meaning of the marital relationship that has been faded along with the instability of family conditions. Not only that, the instability of family conditions is also influenced by the development and change in the meaning of marriage in a wider social environment, so that divorce is no longer a taboo thing to do (Blake, 2015).

Problems in the family can simply arise due to difficulties or disruptions to the inability to carry out social functions properly, this has a negative impact on meeting the needs of life, both physically, spiritually and socially (Voena, 2015). Socio-economic and demographic factors such as low levels of education and marriage at a young age which are relatively unstable make divorce a solution for solving family problems (Roth et al., 2014). Some factors leading to divorce include husbands not being responsible for family needs, incidences of husbands' infidelity, better economic independence of women, lack of knowledge of religious knowledge, and prolonged family economic turmoil (Amato, 2016). Rotz (2016) shows that there are at least four main reasons couples decide to divorce, which is a relationship that is not harmonious, lack of partner responsibility, the presence of third parties, and family economic problems.

Women are individuals who fight for the same rights as men (Shah & Shah, 2012). Perhaps the helplessness of women has made her comfortable in second place in the family, but in reality, the high divorce rate figures illustrate that at present social change has brought women to struggle in various aspects (Tong, 2007). Women nowadays think more critically in demanding neglected rights, do not care about the single parent paradigm which is not impossible to have a big burden that must be borne in the family (Bowman, 2016).

Getting married at a young age is also a major cause of divorce, as an impact of lack of physical and psychological maturity (Rouquette, Lepetre-Mouelhi, & Couvreur, 2019). Psychological families become vulnerable because often quarrels occur as a result of mental, emotional, social, economic, and cultural immaturity in dealing with problems (Riany, Meredith, & Cuskelly, 2016). Whereas family life is believed to be going well and prosperously when it is able to fulfill basic family life needs such as clothing, food, shelter and work. It is increasingly complicated when there are differences in backgrounds that make couples difficult to make adjustments, it is difficult to create a situation of family harmony, and cause chaos or conflict in the family. Getting married at a young age leads to a higher vulnerability to family problems.

Family resilience, large or small, will affect the resilience of a nation, because the family is a pillar of society. The decline in family resilience in the face of conflict leads to high divorce rates in Indonesia. There is a strong suspicion that the husband and wife are unable to manage and resolve conflicts in a qualified manner in the family, so the decision to divorce is chosen as a solution to the family problem being faced. This shows that something must be improved from the concept of marriage in Indonesia, both in terms of knowledge about marriage, to various ways that can be done as an effort to overcome family conflicts.

Socially, divorce is no longer taboo. The stigma about marriage has shifted into a declaration of marital relations, or even the neglect of possible efforts in solving family problems. The culture of the community has been thick with the notion of not easy to form a qualified family, a family that can handle conflict well, so that this shifts the perception of learning in the direction of the inability to solve family problems. This does not mean denying some problems that can be the cause of divorce but, from the data obtained divorce is dominated by individual negligence in marriage commitments.

The development of the roles and functions or activities of women and men also contributed to the weight of the morning decision-making for divorce (Beer & Duong, 2016). Not a few men consider that women are no longer able to carry out their duties as "women", as well as men who are considered to be unable to fulfill their obligations as "men" in the traditional construction of society towards both sexes (Gustafsson & Björklund, 2008). It does not seem excessive to say that on some sides it seems that the times is not able to encourage the development of the roles and functions of women and men in the domestic, public, and social sectors (Branisa, Klasen, Ziegler, Drechsler, & Jütting, 2014). Social movements in building family resilience that are qualified in the midst of family challenges that are increasingly needed to be done. At this time, of course, it is not the time to just stick to problem solving or repressive patterns, but a preventive pattern is needed to prevent the increase in family problems as a result of the lack of understanding of family conflict resolution (Kalil, 2003). Social movements are the most important types of collective behavior. Some sociologists also revealed that social movements are seen as a collective action compared to collective behavior, and a social movement always contains actions aimed at increasing the status, power, or influence of the entire group, not just focusing to someone or just a few people (Nejaime, 2016; Gill Anitha, 2009).

2 THEORETICAL FRAMEWORK

2.1 *The social problem of family life*

In reality and its implementation, the family does not escape from problems until eventually it will experience a family crisis. The family crisis is a family condition that is very unstable where two-way communication in democratic conditions is gone so the family becomes chaotic, irregular and directed, parents lose the authority to control the lives of their children, especially teenagers, children who fight parents, and occur continuous quarrel between mother and father, especially regarding the matter of educating children. The culmination of a prolonged family crisis is divorce, which often harming the children (Stack & Scourfield, 2014).

Furthermore, causes of family crisis include: (1) lacking or breaking up of communication between family members, especially father and mother, this is due to the busyness of the father and mother (husband and wife) who work together so that it takes time for each other communicate with within the family; (2) egocentric attitude; (3) economic problems; (4) busyness problems; (5) education problems; (6) infidelity problems; and (7) movement away from religion (Rouquette et al., 2019; Eekelaar & Maclean, 2004) Social problems that are usually experienced and occur in the family include broken homes and divorce. The broken home family is described as a family which is parents are hostile to each other, some time they are aggressive to their family (Malihah & Nurbayani, 2016; Wulandari, Hufad & K, 2016). This is due to personal factors such as selfishness, intolerance, and lack of trust in each other. Furthermore, the special situation factors in the family such as working wives, the interference of parents, often leaving home due to busy-ness and others (Cohen, 2015).

2.2 *Family resilience*

Family resilience can be measured using a system approach that includes the input component (this component includes physical and non-physical resources), process (this component includes family management, family problems and how the family problem coping mechanism), output (this component shows the fulfillment of physical needs) and psychosocial), and outcomes (this component includes the impact that results from the three previous components) (Kalil, 2003). Inputs in family resilience include that family members fear God Almighty and adhere to prevailing social values/norms, family members have forward insight and gender insight, family members have knowledge of the sciences in building and running families, have the spirit optimistic, and every family member is able to access information resources (Matarrita-Cascante, Trejos, Qin, Joo, & Debner, 2017).

This input or input is supported by a process, namely that each family carries out family functions which include reproductive functions, affection, protection, socialization and education, religion, social culture and others (Card & Barnett, 2015). In the family applies resource management and household economic management, starting from time management, job management, financial management, stress management, and number of children planning between husband and wife (Renati, Bonfiglio, & Pfeiffer, 2016). Between family members apply an equal gender partnership, democratic, mutual respect and interdependence with each other, fellow family members have strong ties and communicate and interact well, each family member is jointly committed to realizing shared family goals (Walsh, 2016b).

From the input that is owned and the process that is carried out it will produce output/welfare output which is physically prosperous, socially prosperous, economically prosperous, psychologically/mentally prosperous, and spiritually prosperous (Masten & Monn, 2015). The impact that results from the components of input, process and output is to produce individuals with good character, a happy and satisfied family with everything that is owned and produced by family members as individuals or groups, living life in harmony and living harmoniously in family and society, economically and socially independent, living in equality and justice in the family and society, family members contribute and are useful to families, communities and nations (Henry, Morris, & Harrist, 2015).

3 METHOD

This research was designed to increase the resilience of the family before and after the divorce claim. So far the researcher has tried to give what is needed by the family from the initial formation through premarital seminars, family processes through counseling, and when there are problems in the family through patterns of conflict resolution designed based on needs and in-depth family analysis as the subject of research (Yoshihama, Blazevski, & Bybee, 2014). In an effort to provide the best for the family, this research was carried out using the action research method. The method was formed through a series of research activities in the form of describing problems in the field, interpretation of data or information obtained, to obtain a social situation as a basis for conducting research and realizing social movements that are right for the family (Andrews & Abawi, 2017). Various changes or interventions may occur, but this cannot be separated from the basic analysis of the objectives and capabilities of the research subject, and as an ingredient of improvement and active participation of researchers in planned programs (Starks, 2007).

Research carried out through social movements as an optimization of family resilience in divorce decision making was conducted with several families in the city of Bandung through collaboration with Islamic Wife Organization in Bandung city, West Java–Indonesia. The election of the Islamic Wife Organization is based on the view that Islamic organizations have their own way of instilling family values and resolving conflicts within the family. Because of that, the informants in this study consisted of husbands, wives, administrators of the Islamic Association of the Wife, and community leaders. In the research process, this was used as the basis for planting the values of family resilience through social movements planned together as a result of in-depth analysis of the conditions of the latest family resilience in the city of Bandung. The amount of news about the increase in divorce rates in the city of Bandung, especially in the case of divorce laws is used as a first step in preparing the program and material to be provided.

4 FINDING AND DISCUSSION

This study emergency from high divorce rates in Bandung city phenomenon. The study revealed that divorce cases in Bandung became the highest in West Java. Based on data obtained from January to September 2017, a total of 4,725 reports were received. The city of Bandung is also a place for many divorced female civil servants, namely 40 divorce cases throughout 2017, and 70% of them are in the age 40–58 years old. The causes vary from economic issues, practices, to social media (especially Facebook), which are 10 out of 25 cases.

The Bandung Religious High Court noted that there were 5,415 divorce claims throughout 2017. 4,113 were divorce rates, which meant that women filed for divorce, while only 1,302 lawsuits were filed by men. This figure increases by 5% every year, because 90% of complaints filed are in the form of divorce applications, while others are in the form of polygamy and Islamic marriage permits. Interestingly, this number dropped dramatically when entering the holy month of Ramadan. Only 190 divorce claims were filed during the holy month of Ramadan. This figure is far compared to ordinary days which penetrate 451 lawsuits or 15 cases per day. This shows that religion has a significant influence on the decision to divorce.

The next stage of Coalescence is cooperation with Islamic organizations, namely the Islamic Wife Organization, to describe the social movements of Islamic Wife Organization in realizing family resilience. Then the Bureaucratization stage (bureaucratization stage is the Formation of members of the Social Movement committee as an Effort to Increase Family Resilience, and Decline (stage of decline/decline) Internalization of family resilience and conflict resolution in solving family problems. members, counseling training, and Family Consultation Institutions (Marri & Walker, 2008; Xiang, Du, Ma, & Fan, 2017).

The social movement through activities in the field of family resilience is an effort to realize good quality of marriage, namely: (1) having a maintained commitment; (2) honesty, loyalty and trust; (3) a sense of responsibility; (4) willingness to adjust; (5) flexibility and tolerance in every aspect of marriage including sexual life; (6) Consider the wishes of the couple; (7) open communication, with full empathy and mutual respect for partners; (8) establish relationships between partners with affectionate affection; (9) comfortable friendship between partners; (10) ability to overcome crises in every situation in togetherness, and maintain spiritual values between marriages and their offspring (Sadarjoen, 2010; Schoors, Caes, Verhofstadt, Goubert, & Alderfer, 2015; Walsh, 2016a).

Moreover, pre-marital activities are important things to do because in pre-marital activities there is a time for deeper marriage counseling. It is intended that couples have more time to formulate appropriate problem solving through ways of mutual respect, tolerance, and communication so as to achieve a happy family life motivation (Gill & Anitha, 2009; Hefner, 2016). The effectiveness of activities can be maximized by involving a number of skills possessed by the couple.

Social movements through premarital activities are expected to be able to provide positive supplies and stimuli for the continuity of the family. Knowledge of a qualified marriage and family is expected to be able to give more consideration to solving family problems so as not to lead to divorce. At least, this activity is able to give an overview in optimizing the roles and functions of families from various sides, physical, mental, social, and cultural economic conditions (Masten & Monn, 2015). This is expected to be able to provide activities that are in line with the current conditions in the field, which include dealing with violence, trauma and crisis, welfare, social justice, and individual identity in the family as well as wider society (Card & Barnett, 2015).

5 CONCLUSION

Factors that encourage women to sue for divorce show the weakness of family resilience, which is an unstable. economic condition which results in economic difficulties in the family which subsequently become a factor that weakens the physical security and economic security of the family. The negative attitudes shown by one another in family conditions in conflict or problem are factors that weaken socio-psychological resilience as well as socio-culture. Factors need to be built that will strengthen family resilience including the implementation of family roles and functions, gender partnerships are a factor that strengthens family integrity. Interpersonal communication built by family's shapes relationships between families and social relations well, so as to strengthen the socio-cultural resilience of the family. Parental support and guidance as well as positively embedded family values, maintaining the confidentiality of the household, as well as the culture of a religious society form social-psychological resilience and socio-cultural resilience. The efforts to maintain or strengthen family resilience. The Islamic Wife Organization as an autonomous part of Islamic organizations exactly carried out social movements in realizing family resilience which was the implementation of its *jihad* programs in the form of *halaqoh*, member education, counseling training and Family Consultation Institutions. This is done in stages starting from the leadership of the congregation, branches, branches, regions, regions, to the centre.

REFERENCES

Amato, P. R. (2016). The consequences of divorce for adults and children: An update trends in the divorce rate. https://doi.org/10.5559/di.23.1.01.

Andrews, D., & Abawi, L. (2017). Three-dimensional pedagogy: A new professionalism in educational contexts. *Improving Schools*, *20*(1), 76–94. https://doi.org/10.1177/1365480216652025.

Gill, A & Anitha, S. (2009). Coercion, consent and the forced marriage debate in the UK. In *Marital Right* (133–152). Routledge, https://doi.org/10.1007/s10691–009–9119–4.

Beer, Z. W.De., & Duong, T. A. (2016). The divorce of sporothrix and ophiostoma: solution to a problematic relationship, (1907), 165–191. https://doi.org/10.1016/j.simyco.2016.07.001.

Blake, W. (2015). *The Marriage of Heaven and Hell Table of Contents*. Charles River Editors via PublishDrive.

Bongaarts, J., Mauldin, W. P., Phillips, J. F., Mauldin, W. P., Phillips, J. F., & Bongaarts, J. (2015). The demographic impact of family programs planning. *Studies in Family Planning*, *21*(6), 299–310.

Bowman, C. G. (2016). Recovering socialism for feminist legal theory in the 21st century. *Connecticut Law Review*, *49*(1), 117–170.

Branisa, B., Klasen, S., Ziegler, M., Drechsler, D., & Jütting, J. (2014). The institutional basis of gender inequality: the social institutions and gender index (SIGI). *Feminist Economics*, *20*(2), 29–64. https://doi.org/10.1080/13545701.2013.850523.

Brown, W., Ahmed, S., Roche, N., Sonneveldt, E., Darmstadt, G. L., & Foundation, M. G. (2015). Impact of family planning programs in reducing high-risk births due to younger and older maternal age, short birth intervals, and high parity. *Seminars in Perinatology*, *39*(5), 338–344. https://doi.org/10.1053/j.semperi.2015.06.006.

Card, N. A., & Barnett, M. A. (2015). Methodological considerations in studying individual and family resilience. *Family Relation*, *64*(1), 120–133. https://doi.org/10.1111/fare.12102.

Cohen, A. B. (2015). Religion's profound influences on psychology: Morality, intergroup relations, self-construal, and enculturation. *Current Directions in Psychological Science*, *24*(1), 77–82. https://doi.org/10.1177/0963721414553265.

Maclean, M., & Eekelaar, J. (2004). Marriage and the moral bases of personal relationships. In *Law and Families* (105–133). Routledge.

Gustafsson, U., & Björklund, F. (2008). Women self-stereotype with feminine stereotypical traits under stereotype threat. *Current Research in Social Psychology*, *13*(18), 219–231.

Hefner, C.-M. (2016). Models of achievement: Muslim girls and religious authority in a modernist islamic boarding school in indonesia. *Asian Studies Review*, *40*(4), 1–19. https://doi.org/10.1080/10357823.2016.1229266.

Henry, C. S., Morris, A. S., & Harrist, A. W. (2015). Family resilience : moving into the third wave. *Family Relation*, *64*(1), 22–43. https://doi.org/10.1111/fare.12106.

Kalil, A. (2003). *Family resilience and good child outcomes a review of the literature*. New zealand: center for social research and evaluation, ministry of social development.

Malihah, E., & Nurbayani, S. (2016, April). Pedophilia and the Lack of Social Control (A Case Study of Sudajaya). In *1st UPI International Conference on Sociology Education*. Atlantis Press.70–72.

Marri, A. R., & Walker, Æ. E. N. (2008). 'Our leaders are us': youth activism in social movements project, 5–20. https://doi.org/10.1007/s11256–007–0077–3.

Masten, A. S., & Monn, A. R. (2015). Child and family resilience: A call for integrated science, practice, and professional training, 55455(February), 5–21. https://doi.org/10.1111/fare.12103.

Matarrita-Cascante, D., Trejos, B., Qin, H., Joo, D., & Debner, S. (2017). Conceptualizing community resilience: revisiting conceptual distinctions. *Community Development*, *48*(1), 105–123. https://doi.org/10.1080/15575330.2016.1248458

Renati, R., Bonfiglio, N. S., & Pfeiffer, S. (2016). Challenges raising a gifted child : Stress and resilience factors within the family. https://doi.org/10.1177/0261429416650948.

Riany, Y. E., Meredith, P., & Cuskelly, M. (2016). Understanding the influence of traditional cultural values on Indonesian parenting. *Marriage & Family Review*, 4929(March), 1–20. https://doi.org/10.1080/01494929.2016.1157561.

Roth, K. E., Harkins, D. A., & Eng, L. A. (2014). Parental conflict during divorce as an indicator of adjustment and future relationships: a retrospective sibling study. *Journal of Divorce and Remarriage*. https://doi.org/10.1080/10502556.2013.871951.

Rotz, D. (2016). Why have divorce rates fallen?: the role of women's age at marriage. *Journal of Human Resources*, *51*(4), 961–1002.

Rouquette, M., Lepetre-mouelhi, S., & Couvreur, P. (2019). Adenosine and lipids: a forced marriage or a love match? Author. *Advanced Drug Delivery Reviews*. https://doi.org/10.1016/j.addr.2019.02.005.

Schoors, M.Van Caes, L., Verhofstadt, L. L., Goubert L.,, & Alderfer, M. A. (2015). Systematic review : family resilience after pediatric cancer diagnosis, 40(June), 856–868. https://doi.org/10.1093/jpepsy/jsv055.

Shah, S., & Shah, U. (2012). Women, educational leadership and societal culture. *Education Sciences*. https://doi.org/10.3390/educ2010033.

Stack, S., & Scourfield, J. (2014). Recency of divorce, depression, and suicide risk, (July 2013). https://doi.org/10.1177/0192513X13494824.

Starks, H., & Brown Trinidad, S. (2007). Choose your method: A comparison of phenomenology, discourse analysis, and grounded theory. *Qualitative health research*, *17*(10), 1372–1380.

Tong, R. (2007). Feminist thought in transition: Never a dull moment. *Social Science Journal*, 44(1), 23–39. https://doi.org/10.1016/j.soscij.2006.12.003.

Voena, A. (2015). Yours, mine, and ours: Do divorce laws affect the intertemporal behavior of married couples?. *American Economic Review*, *105*(8), 2295–2332.

Walsh, F. (2016a). Applying a family resilience framework in training, practice, and research: mastering the art of the possible, *55*(4), 616–632. https://doi.org/10.1111/famp.12260.

Walsh, F. (2016b). Family resilience: a developmental systems framework. *European Journal of Developmental Psychology*, *13*(3), 313–324.
Wulandari, P., Hufad, A., & K., S. N. (2016). The status and role of women in the community of suku dayak hindu budha bumi segandhu indramayu. UPI ICSE 2015, 155–158.
Xiang, Z., Du, Q., Ma, Y., & Fan, W. (2017). A comparative analysis of major online review platforms: Implications for social media analytics in hospitality and tourism. *Tourism Management*, *58*, 51–65. https://doi.org/10.1016/j.tourman.2016.10.001.
Yoshihama, M., Blazevski, J., & Bybee, D. (2014). Enculturation and attitudes toward intimate partner violence and gender roles in an asian indian population: Implications for community-based prevention. *American Journal of Community Psychology*, *53(3–4)*, 249–260. https://doi.org/10.1007/s10464-014-9627-5.

Early Childhood Education in the 21st Century– Yulindrasari et al. (eds)
© 2020 Taylor & Francis Group, London, ISBN 978-1-138-35203-2

Parenting: How to break the risks of the digital era on children

E. Malihah & F.N. Asyahidda
Universitas Pendidikan Indonesia, Indonesia

ABSTRACT: Parenting is an important factor in fostering children's character. All parents are desperate for their children to have good qualities that are in accordance with social values and norms in society. However, at this time the existing of any kind of parenting style has experienced in line with the times. The digital age is an age where all forms of parenting that have been ideal are forced to experience extreme changes, thus giving rise to a gap between previous parenting and the development of the times. By using a qualitative approach and descriptive method of analysis and a cross-sectional survey strategy on young parents from 25 to 33 years old that have children 5 to 10 years old. The results show that parents with diverse technological skills have formed parenting according to the times. It is expected that the findings of this study can have implications for the development of child care in the digital age for further study to be carried out so as to produce appropriate parenting in dealing with children in the digital age.

1 INTRODUCTION

Parents play an important role in child development. As we know, that role is not only important for physical but also for psychological development. In practice, the role of parents is supported by various types of parenting that are used to educate children to be able to face all forms of age challenges (Riany, Meredith, & Cuskelly, 2016). However, technological development, digitalization, and the rapid flow of information are major threats to children's personality development (Nucci, Narvaes, & Krettenauer, 2014). Then, this is made worse by their limitation of technology information, even though at this time technology is a daily friend of their children. By that, parents are required to think far ahead about all the impacts that will occur from the various relationships of their children (Racz et al., 2015). In the global context the role of parents is faced debates of the 21st-century parenting role (Frey, 2018), the debate lies in the types of parenting that are currently able to shape the character of children who are able to face all the possibilities that will occur in millennial era. Therefore, this raises questions about proper parenting in the face of children in the digital age.

The millennial generation, is an example of generation that grew and developed in digital era (Mastrolia & Willits, 2016), several studies showed that the millennial generation experienced moral degradation that was very extraordinary because of the onslaught information technology is so fast (McGrath & Walker, 2016). This shows that there is a need for parenting mediators who are able to provide understanding to children about the impact of the digital era, where parents can give an idea of what will happen if a child takes certain actions, so the child will think about what they will or are do (Odenweller et al., 2014).

The digital era is a big challenge for parents (Amaral, Fonseca, Tiago, & Tiago, 2016). Indeed, digitalization as one of the factors in the emergence of the digital era is intended to facilitate all forms of human life where at first it runs analogously transforms into digital and aims to improve work efficiency. In the global context, digitalization plays an important role in the dissemination of information, and its existence is used as a major commodity in advancing a nation (Gallardo-Echenique, Marqués-Molías, Bullen, & Strijbos, 2015). The faster the reception and dissemination of information, the more advanced the nation will be. However, in practice, the digital era that is not addressed wisely actually brings new social disasters, including moral degradation, deviation, increased crime, loss of respect for parents, and understanding of the values of premature life (Agboola, 2011). Some previous research shows that the morale of the younger generation is currently decreasing from year to year due to the high use of unwise digital devices, especially smartphones (Mastrolia & Willits, 2016).

The moral decline of the younger generation in the digital era is basically caused by several factors, one of which is the lack of parents' understanding of the digital age itself, such as the omission of cellphone use and omission in watching television because what is displayed on cellphones and television often highlights things that not in accordance with the child's growth and maturity in their personality (Sikkens, San, Sieckelinck, & Winter, 2017). In addition, sociologically, children who develop together with the digital era tend to be passive and less sensitive to the circumstances of the surrounding community (Calandra, Mauro, Cutugno, & Martino, 2016). They are more closed and tend not to want to get along with peers in the environment where they live, and prioritize interactions in cyberspace (Przybylski et all, 2014). Social media is the main means by which they interact and not a few of them use social media for things that are negative and have an impact on moral decline.

Investigate what is happening now, especially in parenting that has an impact on children's development. It is very important for parents to be able to understand the impact of the digital era. They even have to understand more than their children and go beyond what they have gotten from their parents first (Racz et al., 2015). This chapter seeks to examine subsequent parenting based on the transcendence of behavior which explains that parents must have a holistic understanding of the impact of the digital age can be used as an alternative in dealing with these challenges (Wilson, 2010). The parenting practices in digital era parents must reflect on the experiences they get from their parents (Inwardly) (Mascheroni, Ponte, & Jorge, 2018). Then, what is the form of interaction with the community in the place where they live based on the results of their parents' parenting and compare with the current life of the child (Outwardly) (Patrikakou, 2016). Furthermore, using these experiences to develop the best actions to prevent various types of risks that can occur in the digital age (Odenweller et al., 2014). Related to this, this article tries to give an idea of how parenting works in the daily lives of parents who provide care for their children in the digital age, and the extent to which they know the impact of the digital era and try to provide an understanding of the importance of developing parenting digital.

2 THEORETICAL FRAMEWORK

Today, parents throughout the world are faced with the use of technology in the care of their children. Most of them rely on technology, especially in the form of smartphones or tablets to calm their children (Higgins, Xiao, & Katsipataki, 2012). However, there are many adverse effects when there are omissions and lack of supervision in the use of these gadgets. The number of moral values are based on these facts (Ferris, Hershberg, Su, Wang, & Lerner, 2015). In fact, today's young generation has experienced extraordinary moral degradation. This fact is based on the impact of industry globalization and the acceleration of information that can be accessed wherever and whenever (Luo & Zhong, 2015). Worse, the insecurity of parents is lacking in what is happening now as if to let that happen. The lack of understanding of how to educate children in this digital era actually brings new disasters, namely the disintegration of social values and cultural values possessed by previous generations (Francis et al., 2018). This happens because what is displayed from the gadgets that children use shows Western cultures or other packaged cultures more attractive than their original culture.

From this point of view, the role of parents is very important in monitoring what their children see in social media (Malihah & Nurbayani, 2016). However, debate occurs when in practice children feel too supervised so that they often do things beyond reason without their parents (Narangajavana Kaosiri, Callarisa Fiol, Moliner Tena, Rodríguez Artola, & Sánchez García, 2017). In addition, a new phenomenon arises where parents are more concerned with their existence through social media (Simpson, 2013). There are several cases where parents become apathetic towards the development of their children (Sobo, Huhn, Sannwald, & Thurman, 2016). Here we can see some of the negative effects of the digital age which can affect children and parents. Therefore, it is very important to find new patterns in responding to and addressing the impact of the digital era.

2.1 *Parenting in Digital Age*

Children's access to online world faced a new challenge for parents to reduce the adverse effects. In cyberspace, children can assess information without being mediated by their parents anywhere and anytime without leaving the room (Livingstone & Bober, 2004). In addition, most parents don't know about social networking technology as a tool in the digital era. In fact, the fact that some of their children's behavior is often based on what they see in online media (Xiang & Gretzel, 2010). A study in Canada shows that the online platforms used by their children have a very significant effect on changing their children's behavior (Seijts, Crossan, & Carleton, 2017). On the other hand, parents do not understand what really happened to their children; they tend to think of it as a natural change towards the next process of the growth of their children in the digital age (Izabela, Joanna, & Alberto, 2015).

The evidence shows that parents tend to underestimate the amount of time their children spend in cyberspace, and their ignorance of the negative effects of Internet use results in high rates of delinquency in adolescents (Lerner, 2018). For example, research conducted in Canada and Brazil shows that the level of social deviations committed by adolescents has experienced a significant increase, and the practice is motivated by Internet use not monitored by their parents (Izabela et al., 2015) (Alperin, 2015) In addition, research conducted in Pennsylvania shows that cyberbullying carried out by their students begins to be out of control because supervision is only done by teachers in schooling, while when at home children tend to be freed by their parents to access what they want (Bleakley, Ellithorpe, & Romer, 2016). Some researchers explained that the teenager high access to online world with the lack of parents capability to control them, it has changed adolescents behavior (Wilson, 2010). Teenagers at the moment seem to be free in choosing relationships that are considered to be in accordance with their passion. Parents only become audiences of social change in their children (Maria, Hughes, Mckee, & Young, 2016).

However, apart from the challenges of parents in dealing with children in the digital era. Parents can make this situation an opportunity to learn what and how best parenting in this digital era is done (Lim, 2016). Some assumptions on parenting in the digital era show three basic indicators, namely: (1) Parental warmth, (2) Parental monitoring, (3) Parenting style and (4) subgroup differences where in this article the indicator becomes a reference in the focus of research conducted (Racz et al., 2015).

3 METHOD

Studying parents' experiences in caring for children in the digital era in fact has implications for new discoveries where some existing parenting styles are considered to require renewal that is in line with the acceleration of the information industry (Ciszek, 2016). From this point of view, this chapter uses a qualitative approach with descriptive analytical methods to be able to clearly get a picture of how they are taking care. The focus of this research is young parents (27–33 years) in educating children in the digital age.

The trend that occurs according to the results of this study is that young parents have more knowledge in using digital devices not only to communicate but also to improve their existence (Patrikakou, 2016). This shows the basic speculation that parents with high technological skills can shape good personality in children even though their children are also very addictive to technology (Racz et al., 2015). On the other hand, parents who have teenagers tend to be passive towards technology while their children are very active in using social media (Alper,

2013). Speculation arises when parents with low-tech understanding are very likely to form a deviant child's personality because their children always see what's in cyberspace (Przybylski et all, 2014).

The strategy used in this study was a cross-sectional survey to five young parents who have children aged 5–10 years. Questions that were raised were related to their involvement in parenting and forming children's character in the digital age (Price, Allen, Ukoumunne, Hayes, & Ford, 2017). Previous studies with similar methods showed that the many irregularities in parenting resulted in the children of the resource being apathetic towards the environment and tending to rely on pseudo-experiences in their friendship (Odenweller et al., 2014).

4 FINDING AND DISCUSSION

The high intensity of children in interacting in cyberspace raises several impacts that may not have existed before. For Baby Boomers and Generation X parents, the parent authorities are in the extended family, it mean not only fathers and mothers provide supervision but also their closest family (Wright, 2015). Parents in that generation want their children to be in the house, because family autonomy in nurturing and educating can be done optimally (Daniels & Zurbriggen, 2016). At that time, the level of delinquency was very low because family autonomy tended to form obedience and fear of parents (Agboola, 2011). This is then inversely proportional to the current situation where the X generation who has now become parents is considered to be irrelevant in applying the parenting they received from their parents (Bejtkovský, 2016).

The Y generation or the millennial generation is very different from its predecessors. These generations tend to have more freedom to interact and interact, plus technological sophistication that is in line with their growth has implications for their daily behavior (Darmo, 2015). This is where the problem arises, because the generation gap occurs where parents with old parenting still apply this pattern to their millennial generation. The results showed that five young parents who had children aged 5–10 years thought that their knowledge of parenting from their parents had changed a bit because they were not in accordance with the conditions of the times. On the other hand, the young parents combine the good values of their parents' upbringing with the information they get on the internet to care for their children. As a result, three out of five young parents think that information on Internet assistance is very helpful for them to provide understanding to children about the effects of good and bad using the Internet. Then, this understanding is combined with the parenting they get from their parents, such as prohibiting using the gadget too long and prioritizing learning.

While two out of five young parents think that they always seek information from the internet in supervising and educating children. They tend to use digital devices that are integrated with gadgets owned by children. The young parents assume that, by doing so, it will make supervision much more efficient and effective. However, differences in the form of parenting lead to different results in the behavior and character of the child. Parents with combination parenting have an impact on children's attitudes that are more humanistic, have high social awareness, and are communicative in bringing opinions (Agboola, 2011) (Maria et al., 2016). This is different with parents where digital devices are the main tool in providing care. As a result, their children tend to be passive, difficult to get along with, and difficult to express opinions.

This can be analyzed through a functional structural approach where a community system is formed as well as human organs where an organ does not function it will change a system (Sivo, Karl, Fox, Taub, & Robinson, 2017). Here the role of parents who prioritize their gadgets is to change the structure and function itself, where parents as the main structure in parenting are replaced by tools. The result is pseudo-behavior that children catch because the essence of parenting is a sense of belonging between parents and children (Chu & Gruhn, 2017). The tendency of child traits found in the research is passive and attractive at one time. Passive in the sense of their lives in the real world as if they don't care about the environment around where they live both with peers and their own families. However, they are also active in a virtual environment that they are very fond of. Research on adolescents with high intensity of use of technology shows that they in cyberspace have a very friendly and very friendly attitude when on the internet but vice versa when in the real world (Sivo et al., 2017).

The results of these studies indicate that the digital era can shape the character and attitudes of children through what they see on the Internet. Ambiguous oversight will lead to feelings of confusion in children and tend to be apathetic towards their real environment (Price et al., 2017). The results of this study found a solution to the problem of parenting in the digital era, namely parents must prioritize children's experiences in determining actions in parenting. In addition, reward and punishment is very important to do because the high level of knowledge that is generated from the use of the internet often makes children become addicts in using these digital devices. As well as restrictions on internet use also should not be done too tightly because it will lead to high curiosity in children so that children tend to be dishonest with what they see on the Internet.

5 CONCLUSION

Parenting in the digital era is very prioritizing the role of parents as a supporting system for children. Parents are no longer a central role in determining what their children will do because technological skills in children have formed a critical attitude towards whatever their parents choose. Digital generation children always compare the effects of good and bad according to what they understand and dare to reject what their parents think is right. Therefore, parenting in the digital era does not allow parents to be authoritarian in making decisions. Intense two-way discussion is needed in the formation of children's character. In addition, it is not recommended for parents who only rely on false communication in educating children, because pseudo-interaction does not create a sense of belonging between children and parents. The impact is worse, it will form an apathy, self-determination, and tend to be anti-social. This research is expected to be able to provide more knowledge and add more understanding to science about parenting in the digital age.

REFERENCES

Agboola, A. A. (2011). Managing deviant behavior and resistance to change. *International Journal of Business and Management*, 6(1), 235–243. doi: https://doi.org/10.1111/1467–8551.00040.

Alper, M. (2013). The parent app: Understanding families in the digital age, (September), 37–41. doi: https://doi.org/10.1080/17482798.2013.830382.

Alperin, J. P. (2015). Geographic variation in social media metrics: An analysis of Latin American journal articles. doi: https://doi.org/10.1108/AJIM-12–2014–0176.

Amaral, F., Fonseca, J., Tiago, M., & Tiago, F. (2016). Digital natives 3.0: social network initiation. *World Journal of Business and Management*, 2(2), 19. doi: https://doi.org/10.5296/wjbm.v2i2.9876.

Bejtkovský, J. (2016). The current generations: The baby boomers, x, y and z in the context of human capital management of the 21st century in selected corporations in the czech republic. *Littera Scripta*, 9(2), 25–45.

Bleakley, A., Ellithorpe, M., & Romer, D. (2016). The role of parents in problematic Internet use among us adolescents, 4, 24–34.

Calandra, D. M., Mauro, D.Di, Cutugno, F., & Martino, S. Di. (2016). Navigating wall-sized displays with the gaze: A proposal for cultural heritage. *CEUR Workshop Proceedings*, 1621(June 2000), 36–43. https://doi.org/10.1023/A.

Chu, Q., & Gruhn, D. (2017). Moral judgments and social stereotypes. *Social Psychological and Personality Science*. doi: https://doi.org/10.1177/1948550617711226.

Ciszek, E. L. (2016). Digital activism: How social media and dissensus inform theory and practice. *Public Relations Review*, 42(2), 314–321. doi: https://doi.org/10.1016/j.pubrev.2016.02.002.

Daniels, E. A., & Zurbriggen, E. L. (2016). It's not the right way to do stuff on Facebook: An investigation of adolescent girls' and young women's attitudes toward sexualized photos on social media. *Sexuality and Culture*, 20(4), 936–964. doi: https://doi.org/10.1007/s12119–016–9367–9.

Darmo, I. S. (2015). Millennials green culture: The opportunity and challenge (a case study of higher education student). In *The 3rd International Multidisciplinary Conference on Social Sciences*, 21–28. doi: https://doi.org/ISSN 2460–0598

Ferris, K. A., Hershberg, R. M., Su, S., Wang, J., & Lerner, R. M. (2015). Character development among youth of color from low-SES backgrounds: An examination of Boy Scouts of America's ScoutReach program. *Journal of Youth Development*, 10(3), 14–30.

Francis, L. J., Pike, M. A., Lickona, T., Lankshear, D. W., Francis, L. J., Pike, M. A., ... Lankshear, D. W. (2018). Evaluating the pilot Narnian virtues character education English curriculum project: A study among 11- to 13-year-old students. *Journal of Beliefs & Values*, 7672, 1–17. doi: https://doi.org/10.1080/13617672.2018.1434604.

Frey, W. H. (2018). The millennial generation: A demographic bridge to America's diverse future. Available at: https://digitalscholarship.unlv.edu/brookings_lectures_events/126

Gallardo-Echenique, E. E., Marqués-Molías, L., Bullen, M., & Strijbos, J. W. (2015). Let's talk about digital learners in the digital era. *International Review of Research in Open and Distance Learning*. doi: https://doi.org/10.19173/irrodl.v16i3.2196.

Higgins, S., Xiao, Z., & Katsipataki, M. (2012). The impact of digital technology on learning: a summary for the education endowment foundation full report. Durham,UK: Education Endowment Foundation and Durham University.

Izabela, M. B., Joanna, T., & Alberto, M. (2015). Supportive communication with parents moderates the negative effects of electronic media use on life satisfaction during adolescence, 189–198. doi: https://doi.org/10.1007/s00038–014–0636–9.

Lerner, R. M. (2018). Character development among youth: Linking lives in time and place. International *Journal of Behavioral Development*, 42(2), 267–277. doi: https://doi.org/10.1177/0165025417711057.

Lim, S. S. (2016). Through the tablet glass: Transcendent parenting in an era of mobile media and cloud computing, 2798(February), 20–29. doi: https://doi.org/10.1080/17482798.2015.1121896.

Livingstone, S., & Bober, M. (2004). Taking Up Online Opportunities? Children's Uses of the Internet for Education, Communication and participation. *E-Learning and Digital Media*, 1(3), 395–419.

Luo, Q., & Zhong, D. (2015). Using social network analysis to explain communication characteristics of travel-related electronic word-of-mouth on social networking sites. *Tourism Management*, 46, 274–282. doi: https://doi.org/10.1016/j.tourman.2014.07.007

Malihah, E., & Nurbayani, S. (2016, April). Pedophilia and the Lack of Social Control (A Case Study of Sudajaya). In *1st UPI International Conference on Sociology Education*. Atlantis Press. 70–72.

Maria, E. L.-R., Hughes, H. E., Mckee, L., & Young, H. (2016). Early adolescents as publics: A national survey of teens with social media accounts, their media use preferences, parental ... *Public Relations Review* (March). doi: https://doi.org/10.1016/j.pubrev.2015.10.003.

Mascheroni, G., Ponte, C., & Jorge, A. (2018). Digital Parenting: The Challenges for Families in the Digital Age. Nordicom University of Gothenburg: The International Clearinghouse on Children, Youth and Media

Mastrolia, S. A., & Willits, S. D. (2016). Of moral panics & millennials. in m. sigmond, n., myers, c., bellolit, j. d., gallagher (ed.), *Northeastern Association of Business*, Pennsylvania. *Economics and Technology*, 190–207.

McGrath, R. E., & Walker, D. I. (2016). Factor structure of character strengths in youth: Consistency across ages and measures. *Journal of Moral Education*, 45(4), 400–418. doi: https://doi.org/10.1080/03057240.2016.1213709

Narangajavana Kaosiri, Y., Callarisa Fiol, L. J., Moliner Tena, M. Á., Rodríguez Artola, R. M., & Sánchez García, J. (2017). User-Generated Content Sources in Social Media: A New Approach to Explore Tourist Satisfaction. *Journal of Travel Research*, 004728751774601. doi: https://doi.org/10.1177/0047287517746014.

Nucci, L., Narvaes, D., & Krettenauer, T. (Eds).(2014). *Handbook of moral and character education*. New York: Routledge.

Odenweller, K. G., Booth-butterfield, M., Weber, K., Odenweller, K. G., Booth-Butterfield, M., & Weber, K. (2014). Investigating helicopter parenting, family environments, and relational outcomes for millennials investigating helicopter parenting, family environments, and relational outcomes for millennials, (October), 37–41. doi: https://doi.org/10.1080/10510974.2013.811434.

Patrikakou, E. N. (2016). Parent Involvement, Technology, and Media: Now What? *School Community Journal*, 26(2), 9–24.

Price, A., Allen, K., Ukoumunne, O. C., Hayes, R., & Ford, T. (2017). Examining the psychological and social impact of relative age in primary school children: a cross-sectional survey. *Child: care, health and development*. *43*(6), 891–898. doi: https://doi.org/10.1111/cch.12479.

Przybylski, A. K., Mishkin, A., Shotbolt, V., & Linington, S. (2014). A Shared Responsibility: Building Children's Online Resilience (research paper). Virgin Media, University of Oxford and The Parent Zone.

Racz, S. J., Johnson, S. L., Bradshaw, C. P., Cheng, L., Jensen, S., Johnson, S. L., & Bradshaw, C. P. (2015). Parenting in the digital age: Urban black youth' s perceptions about technology-based communication with parents, 9400 (April 2016). doi: https://doi.org/10.1080/13229400.2015.1108858.

Riany, Y. E., Meredith, P., & Cuskelly, M. (2016). Understanding the influence of traditional cultural values on Indonesian parenting. *Marriage & Family Review*, 4929(March), 1–20. doi: https://doi.org/10.1080/01494929.2016.1157561.

Seijts, G., Crossan, M., & Carleton, E. (2017). Embedding leader character into HR practices to achieve sustained excellence. *Organizational Dynamics*, 46(1), 30–39. doi: https://doi.org/10.1016/j.orgdyn.2017.02.001.

Sikkens, E., San, M. Van, Sieckelinck, S., & Winter, M. De. (2017). Parental Influence on Radicalization and De-radicalization according to the Lived Experiences of Former Extremists and. *Journal for Deradicalization*, (12), 192–226.

Simpson, J. E. (2013). A divergence of opinion: How those involved in child and family social work are responding to the challenges of the Internet and social media, 94–102. doi: https://doi.org/10.1111/cfs.12114.

Sivo, S., Karl, S., Fox, J., Taub, G., & Robinson, E. (2017). Structural Analysis of Character Education: A Cross- Cultural Investigation.

Sobo, E. J., Huhn, A., Sannwald, A., & Thurman, L. (2016). Information Curation among Vaccine Cautious Parents : Web 2.0, Pinterest Thinking, and Pediatric Vaccination Choice (January). doi: https://doi.org/10.1080/01459740.2016.1145219.

Wilson, A. (2010). Grown up digital: how the net generation is changing your world. *International Journal of Market Research, 52*(1), 139–140. doi: https://doi.org/10.2501/S1470785310201119.

Wright, T. A. (2015). Distinguished scholar invited essay: Reflections on the role of character in business education and student leadership development. *Journal of Leadership and Organizational Studies*, 22(3), 253–264. doi: https://doi.org/10.1177/1548051815578950.

Xiang, Z., & Gretzel, U. (2010). Role of social media in online travel information search. *Tourism Management*, 31(2), 179–188. doi: https://doi.org/10.1016/j.tourman.2009.02.016

Early Childhood Education in the 21st Century – Yulindrasari et al. (eds)
© 2020 Taylor & Francis Group, London, ISBN 978-1-138-35203-2

Author Index